FEATURES OF LIQUID CRYSTAL DISPLAY MATERIALS AND PROCESSES

Edited by **Natalia V. Kamanina**

Features of Liquid Crystal Display Materials and Processes
Edited by Natalia V. Kamanina

Published by InTech

Janeza Trdine 9, 51000 Rijeka, Croatia

Publishing Process Manager Ivona Lovric

Technical Editor Teodora Smiljanic

Cover Designer InTech Design Team

Additional hard copies can be obtained from orders@intechopen.com

Features of Liquid Crystal Display Materials and Processes, Edited by Natalia V. Kamanina
p. cm.
ISBN 978-953-307-899-1

Meet the editor

Dr. Sci. PhD. Natalia Vladimirovna Kamanina was born in Kaliningrad, Russian Federation, 1957. She graduated with an Honor Diploma from Leningrad Polytechnical Institute (1981), St. Petersburg, Russia, and received a PhD (Physics & Mathematics) at Vavilov State Optical Institute, St.-Petersburg, Russia (1995), as well as a Dr. Sci. (Physics & Mathematics) at the same institution (2001). She is currently a Head of the Lab for "Photophysics of media with nanoobjects" at Vavilov State Optical Institute St.-Petersburg, Russia and has been involved in collaboration research with many researchers and scientists all over the world since 1995, publishing about 170 technical papers in the area of Laser-Matter Interaction. Parallel to her scientific activity, she has also been lecturing from 2001, as a Professor of Physical Electronics and Opto-Electron Devices. Dr. Kamanina is the Chair of the Electronics Department (St. Petersburg Electrotechnical University "LETI"), and a Professor of the Optical Physics and Modern Natural Science and the Chair of the Photonics and Optoinformation Department (St. Petersburg Technical University "IFMO").

Contents

Preface

Since the First International Congress on Liquid Crystals (Lcs), held at Kent State University, OH, USA, in 1965, the implications of these systems associated with various aspects of telecommunications, laser, display, automobile, aerospace technologies, thermo-optics, medicine and biology have been the subject of considerable debate among researchers, scientists and engineers. Indeed, LCs, being a unique mesomorphic phase of matter, combine properties of both solids (long-range orientation order, manifestations of Bragg diffraction, etc.) and liquids (fluidity, viscosity, etc.). Important features of LCs are weak dispersion forces between organic molecules and strong orienting fields. An intrinsic characteristic of the organic liquid crystal state is unidirectional (nematic structure) or bidirectional (smectic structure) ordering, albeit not in three dimensions as in a real inorganic crystals. In other words, this state is more structured than the liquid one, but less than the solid phase. Moreover, the orienting power of LCs is used in the development of composite materials. LC aligns suspended particles, acting as matrices easily controllable by elastic forces and by thermal, magnetic, light and electric fields. The order parameter of an LCs is the degree of its regularity characterized by the deviation of the direction of the long axis of a molecule from that of the LC director. Peculiarities of electrical schemes to control LC systems and features of LC molecules orientation along, perpendicular or at some pretilt angle on the substrates, coated with conducting and alignment layers, predict the operation of LC devises and generally display technology (TN, IPS, MVA, etc.) with good advantage.

By the way, an electric field applied to a liquid crystal or an electric current passing through a medium produces effects that do not occur in other electro-optical media, and are responsible for most LC devices technical characteristics, such as: resolution, contrast, speed, sensitivity, grey level, etc. These parameters can be improved using new studies and searching for the new theoretical methods and practical approach. This book includes advanced and revised contributions, covering theoretical modeling for optoelectronics and nonlinear optics, along with including experimental methods, new schemes, new approach and explanations which extend the display technology for laser, semiconductor device technology, medicine equipment, biotechnology, etc.

The advanced idea, approach and information described here will be fruitful for the readers to find a sustainable solution in a fundamental study and in the industry approach. The book can be useful to students, post-graduate students, engineers, researchers and technical officers of optoelectronic universities and companies.

Acknowledgements

The editor would like to thank all chapter authors, reviewers and to all who have helped to prepare this book. The editor would also like to acknowledge Ms. Ivona Lovric, Process Manager at InTech – Open Access Publisher, Croatia for her good and continued cooperation.

Natalia V. Kamanina, Dr.Sci., PhD, Head of the Lab for
"Photophysics of media with nanoobjects",
Vavilov State Optical Institute,
Saint-Petersburg,
Russia

Part 1

Materials and Interfaces

1

Polyimides Bearing Long-Chain Alkyl Groups and Their Application for Liquid Crystal Alignment Layer and Printed Electronics

Yusuke Tsuda
Kurume National College of Technology
Japan

1. Introduction

Polyimides exhibit excellent thermal and mechanical properties, and have extensive engineering and microelectronics applications. Aromatic polyimides such as polyimides based on pyromellitic dianhydride are prepared from aromatic diamines and aromatic tetracarboxylic dianhydrides *via* poly(amic acid)s. Since conventional aromatic polyimides are insoluble, these polymers are usually processed as the corresponding soluble poly(amic acid) precursors, and then either thermally or chemically imidized. However, owing to the instability of poly(amic acid)s and the liberation of water in the imidization process, problems can arise (Fig. 1). Extensive research has been carried out to improve the solubility of polyimides and successful recent examples involve the incorporation of fluorine moieties, isomeric moieties, methylene units, triaryl imidazole pendant groups, spiro linkage groups, and sulfonated structure. Soluble polyimides bearing long-chain alkyl groups have also been reported, and their applications mainly involve their use as alignment layers for liquid crystal displays (LCDs).

Our research group has systematically investigated the synthesis and characterization of soluble polyimides based on aromatic diamines bearing long-chain alkyl groups such as alkyldiaminobenzophenone (ADBP-X, X = carbon numbers of alkyl chain) (Tsuda et al., 2000a) alkoxydiaminobenzene (AODB-X) (Tsuda et al., 2000b), diaminobenzoic acid alkylester (DBAE-X) (Tsuda et al., 2006), and alkyldiaminobenzamide (ADBA-X) (Tsuda et al., 2008), and the results from these research are described in the original papers and the review paper (Tsuda, 2009). Our recent paper has described soluble polyimides having dendritic moieties on their side chain, and it was found that these polyimides having dendritic side chains were applicable for the vertically aligned nematic liquid crystal displays (VAN-LCDs) (Tsuda et al., 2009). These dendronized polyimides were synthesized using the novel diamine monomer having a first-generation monodendron, 3,4,5-tris(n-dodecyloxy)benzoate and the monomer having a second-generation monodendron, 3,4,5-tris[-3′,4′,5′-tri(n-dodecyloxy)benzyloxy]benzoate.

Some soluble polyimides were synthesized from the diamine monomer having three long-chain alkyl groups; aliphatic tetracarboxylic dianhydride; 5-(2,5-dioxotetrahydrofuryl)-3-methyl-3-cyclohexene-1,2-dicarboxylic anhydride (Cyclohexene-DA) or aromatic tetracarboxylic dianhydride; 3,3′,4,4′-diphenylsulfone tetracarboxylic dianhydride (DSDA)

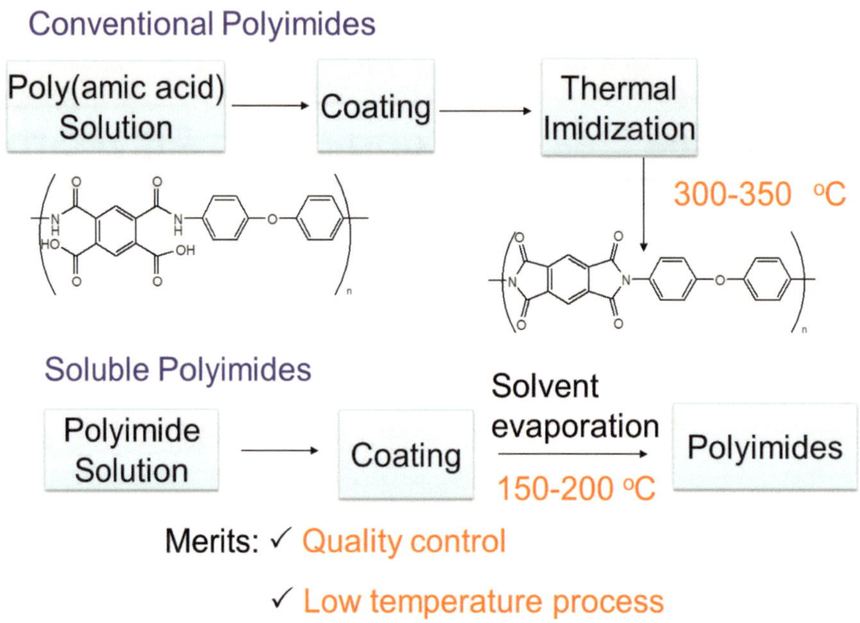

Fig. 1. Conventional polyimides and soluble polyimides

or 3,4′-oxydiphthalic anhydride (3,4′-ODPA) as a dianhydride, and 4,4′-diaminodiphenylether (DDE) as a diamine co-monomer. Thin films of the obtained polyimides were irradiated by UV light (λmax; 254 nm) , and the contact angles for the water decreased from near 100° (hydrophobicity) to near 20° (hydrophilicity) in proportion to the irradiated UV light energy. Thus, the surface wettability of polyimides bearing long-chain alkyl groups can be controlled by UV light irradiation, such methods are expected to be applied in the field of organic, flexible and printed electronics (Tsuda et al., 2010, 2011a, 2011b).

In this chapter, the author reviews the synthesis and basic properties of soluble polyimides bearing long-chain alkyl groups, and their application for liquid crystal alignment layer and printed electronics.

2. Results and discussion

In this section, the synthesis of aromatic diamine monomers having long-chain alkyl groups and corresponding soluble polyimides bearing long-chain alkyl group (Fig. 2), their basic polymer properties, and the application for VAN-LCDs and printed electronics are described.

2.1 Synthesis of aromatic diamine monomers containing long-chain alkyl groups

The synthesis routes for aromatic diamines bearing single long-chain alkyl groups are illustrated in Fig. 3. Alkyldiaminobenzophenones (ADBP-9~14) were prepared *via* two steps using 3,5-dinitrobenzoyl chloride as the starting material. The Friedel-Crafts reaction of 3,5-dinitrobenzoyl chloride with alkylbenzene catalyzed by aluminum chloride in nitrobenzene gave 3,5-dinitro-4′-alkylbenzophenones in good yields. The reduction of 3,5-dinitro-4′-

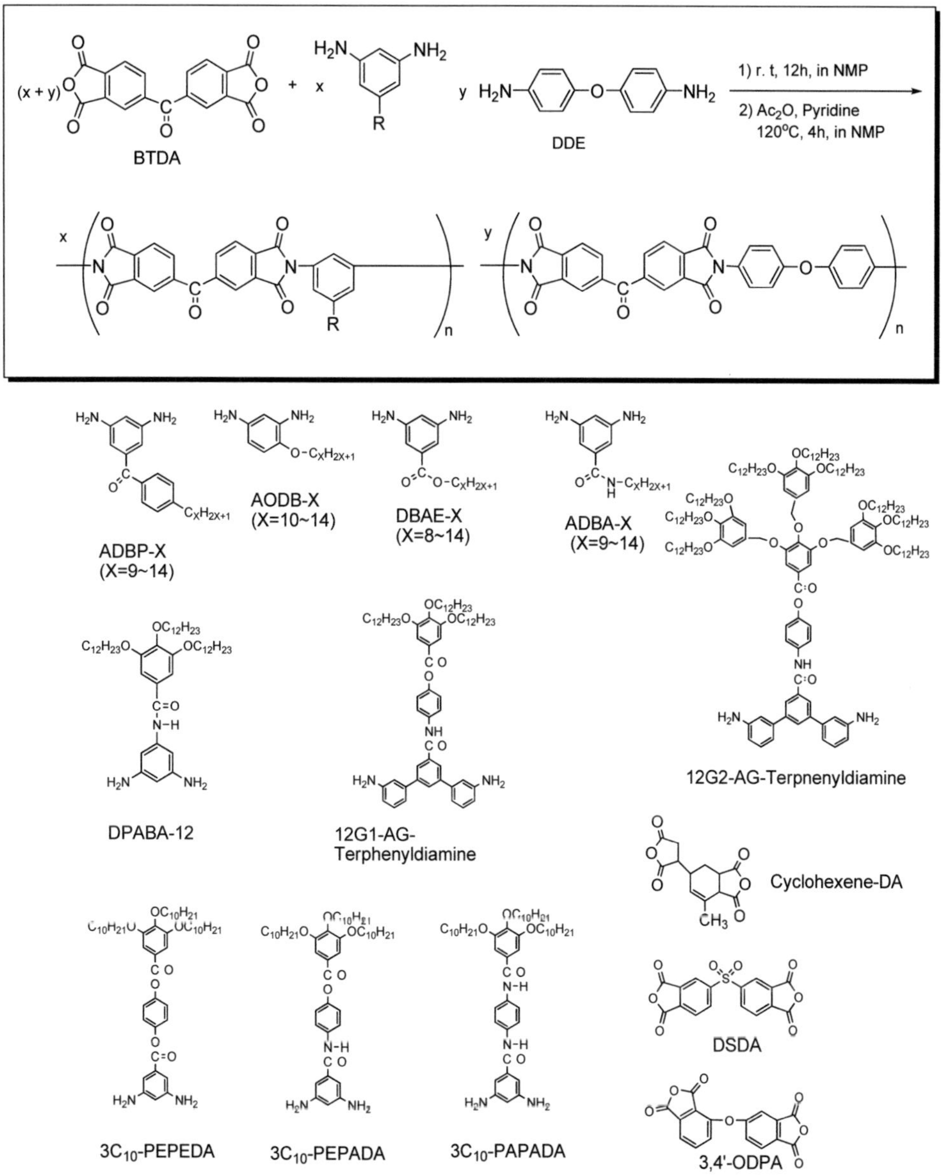

Fig. 2. Soluble polyimides bearing long-chain alkyl groups

alkylbenzophenone was performed by catalytic hydrogenation using palladium on carbon and hydrogen gas introduced by 3-5 L gas-bag. Although hydrazine hydrate/ethanol system is sometimes used for the reduction of nitro compounds, this system is not preferred because the carbonyl group in 3,5-dinitro-4′-alkylbenzophenones reacts with hydrazine.

Alkyloxydiaminobenzenes (AODB-10~14) were prepared in two steps using 2,4-dinitrophenol as the starting material. The Williamson reaction using 2,4-dinitrophenol and 1-bromoalkanes catalyzed by potassium carbonate in DMAc gave 1-alkyloxy-2,4-dinitrobenzenes in satisfactory yields. The reduction of 1-alkyloxy-2,4-dinitrobenzenes was performed by catalytic hydrogenation using Pd/C and hydrogen gas at 0.2-0.3 Mpa.

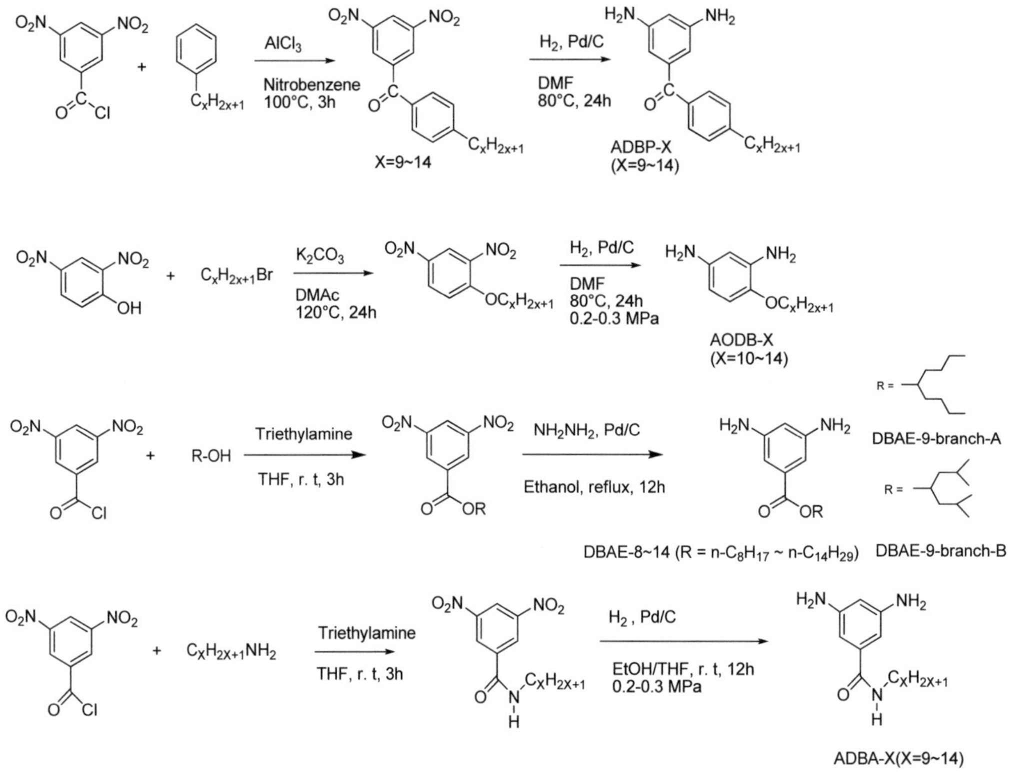

Fig. 3. Synthesis of aromatic diamines having single long-chain alkyl groups

Although the hydrazine hydrate/ethanol system can be used for the reduction of nitro-compounds, the medium pressure system is preferable due to better yields and purity of the products.

Diaminobenzoic acid alkylesters (DBAE-8~14) were prepared in two steps using 3,5-dinitrobenzoyl chloride as the starting material. The esterification reaction using 3,5-dinitrobenzoyl chloride and aliphatic alcohols having long-chain alkyl groups catalyzed by triethylamine in THF gave alkyl 3,5-dinitrobenzoate in satisfactory yield. The reduction of alkyl 3,5-dinitrobenzoate was performed by catalytic hydrogenation using Pd/C as a catalyst and hydrazine hydrate/ethanol as a hydrogen generator. The relatively mild hydrogenation using hydrazine hydrate/ethanol system seemed to be preferable in the case of alkyl 3,5-dinitrobenzoate, because the scissions of ester linkages were sometimes recognized besides the hydrogenation of nitro-groups in the use of medium pressure hydrogenerator.

Alkyldiaminobenzamides (ADBA-9~14) were prepared in two steps using 3,5-dinitrobenzoyl chloride as the starting material. The condensation reaction using 3,5-dinitrobenzoyl chloride and aliphatic amines having long-chain alkyl groups catalyzed by triethylamine in THF gave N-alkyl-3,5-diaminobenzamides in satisfactory yields. The reduction of N-alkyl-3,5-diaminobenzamide was performed by catalytic hydrogenation using Pd/C and hydrogen gas at 0.2-0.3 MPa in a medium pressure hydrogenerator in satisfactory yield (60-80%).

The aromatic diamines containing first-generation dendritic moieties, N-(3,5-diaminophenyl)-3,4,5-tris(alkoxy)benzamide (DPABA-X, X=6,12), were synthesized

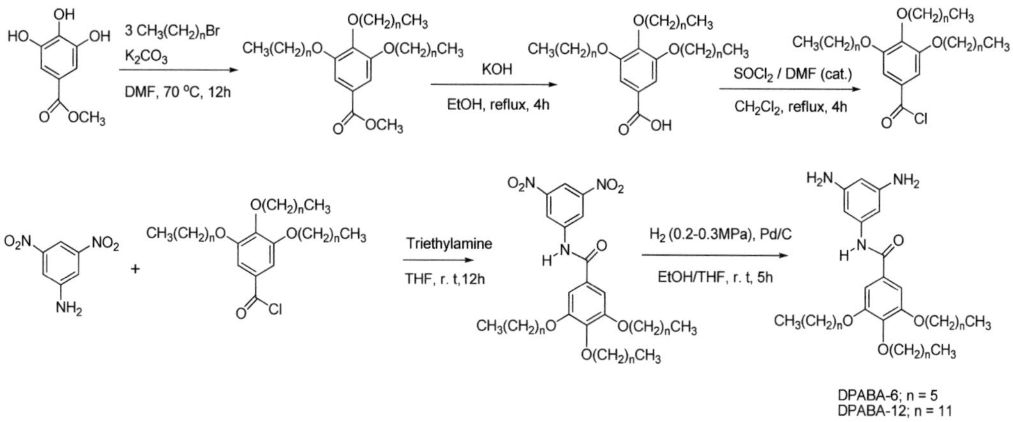

Fig. 4. Synthesis of aromatic diamines having triple long-chain alkyl groups

following the method shown in Fig. 4. 3,4,5-Trialkyloxybenzoyl chloride, known as the building block for Percec-type dendrons, was synthesized from 3,4,5-trihydroxybenzoic acid methyl ester (gallic acid methyl ester) followed by Williamson-etherification using alkylbromide catalyzed by potassium carbonate, hydrolysis of ester groups by potassium hydroxide, then acid chlorination using thionyl chloride. The condensation reaction using the above acid chloride and 3,5-dinitroaniline catalyzed by triethylamine gave the dinitro-precursor of DPABA, and this was finally hydrogenated to DPABA.

4-[3,5-Bis(3-aminophenyl)phenyl]carbonylamino]phenyl 3,4,5-tris (n-dodecyloxy)benzyloxy benzoate (12G1-AG-Terphenyldiamine) and 4-[3,5-Bis (3-aminophenyl) phenyl] carbonylamino] phenyl 3,4,5-tris[3′,4′,5′-tris(n-dodecyloxy) benzyloxy] benzoate (12G2-AG-Terphenyl diamine) were synthesized by the method shown in Fig. 5 using the first- and second- generation Percec-type monodendrons. These synthesis routes include the condensation reactions with 3,5-dibromo benzoic acid and 3′,4′,5′-tris (n-dodecyloxy)benzyloxy chloride with 4-aminophenol, followed by Suzuki coupling reaction with 3-aminophenyl boronic acid. It is considered that these synthetic methods of aromatic diamine monomers using Suzuki coupling are the versatile method as the synthesis of aromatic diamines without the severe reduction that sometime causes the side reaction.

Novel diamine monomers, such as $3C_{10}$-PEPEDA, $3C_{10}$-PEPADA and $3C_{10}$-PAPADA having three long-chain alkyl groups connected by phenylester and/or phenylamide linkages were recently synthesized *via* several step reactions from Gallic acid methyl ester using protect group synthetic technique. The detail description of these monomer syntheses will be reported elsewhere.

2.2 Synthesis of soluble polyimides bearing long-chain alkyl groups

The synthesis route for the polyimides and copolyimides based on BTDA (Cyclohexene-DA, DSDA, 3,4′-ODPA), DDE and aromatic diamines bearing long-chain alkyl groups is illustrated in Fig. 2. Two-step polymerization systems consisting of poly(amic acid)s synthesis and chemical imidization were performed. The poly(amic acid)s were obtained by reacting the mixture of diamines with an equimolar amount of BTDA at room temperature for 12 h under an argon atmosphere. Polyimides were obtained by chemical imidization at 120°C in the presence of pyridine as base catalyst and acetic anhydride as dehydrating agent. These are the optimized synthesis conditions previously developed for the synthesis

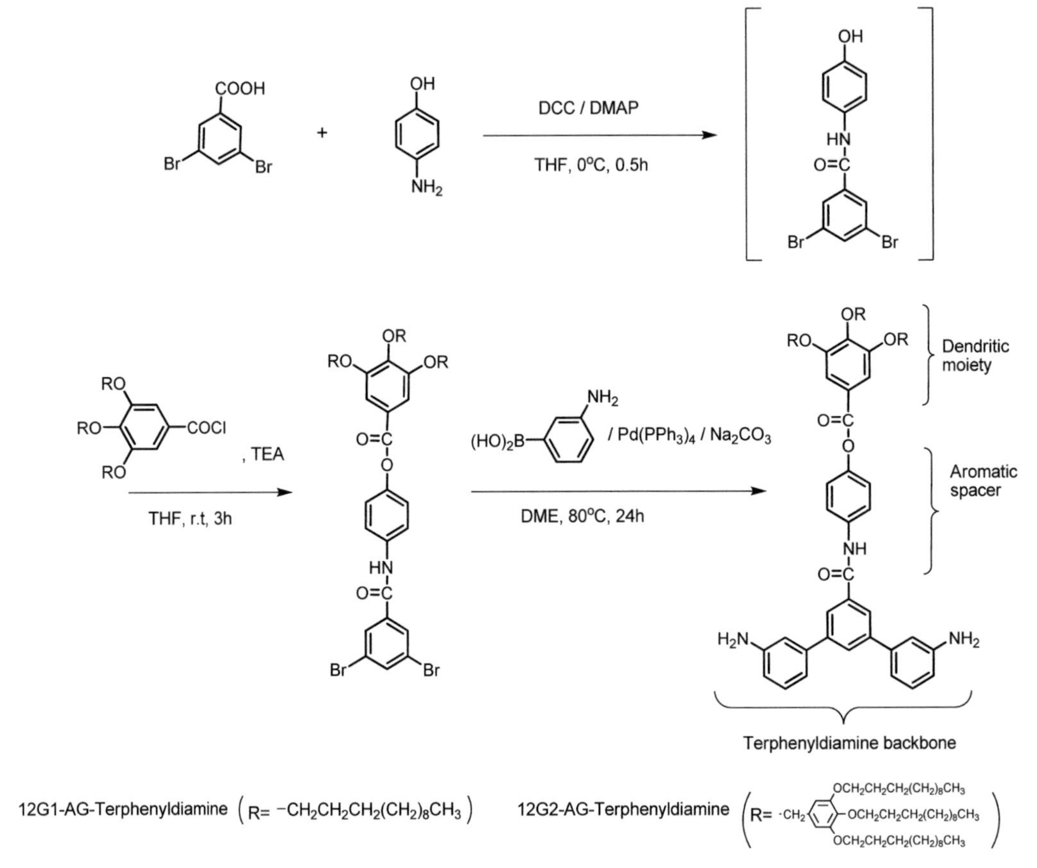

Fig. 5. Synthesis of aromatic diamines having multiple long-chain alkyl groups (dendritic terphenyl diamines)

of soluble polyimides in our laboratory. BTDA, DSDA and 3,4'-ODPA, these are highly reactive and common aromatic tetracarboxylic dianhydrides were mainly used as a dianhydrides monomer, and DDE that is highly reactive and a common aromatic diamine was used as a diamine co-monomer. In the case of soluble polyimides, clear polyimide solutions were eventually obtained. In other cases, clear poly(amic acid) solutions were obtained, however, gelation or precipitation occurred in the course of imidization process. The polymerizations based on the dendritic diamine monomers, 12G1-AG-terphenyldiamine and 12G2-AG-terphenyldiamine were firstly investigated using NMP as a solvent. Although viscous poly(amic acid)s solution were obtained, precipitation sometime occurred during the imidization process. It was speculated that the hydrocarbon and phenyl moiety of dendritic diamine monomers reduces the solubility of polyimides in NMP; therefore, a polar aromatic solvent, *m*-cresol or pyridine were sometime used to improve the solubility of dendritic moieties.

2.3 Properties of soluble polyimides bearing long-chain alkyl groups

From the continuous investigation in our laboratory, various precious data was obtained. The representative results are shown in this section.

2.3.1 Solubility

As far as the solublity of polyimides based on long-chain alkyl groups is concerned, the following interesting results have been obtained. Experimental results of homopolymerization and copolymerization based on BTDA/ADBP-12, AODB-12, DBAE-12, ADBA-12, DPABA-12/DDE are summarized in Table 1. Although all polyamic(acid)s were soluble in NMP which is a solvent used for polymerization, however, the solubility of homopolyimides and copolyimides depended on polymer structures. BTDA/ADBP-12 homopolyimides and BTDA/ADBP-12/DDE copolyimides containing 40 mol% of ADBP or more were soluble in NMP. Thus, the effect of long-chain alkyl group in ADBP for the enhancement of solubility was confirmed. BTDA/AODB-12 homopolyimides and BTDA/AODB-12/DDE copolyimides containing 25 mol% or more of AODB-12 units were also soluble in NMP. Judging from the results of copolymerization based on BTDA/ADBP-9~14/DDE and BTDA/AODB-10~14/DDE, it is recognized that AODB bearing alkyl groups *via* an ether linkage were more effective for the enhancement of solubility in comparison to ADBP.

On the other hand, all homopolyimides and copolyimide based on BTDA/DBAE-8~14/DDE were insoluble in NMP probably due to the rigid ester linkage groups. The experimental results of copolymerization based on BTDA/ADBA-12/DDE are quite unique. Although BTDA/ADBA-12 homopolyimide was insoluble, the copolymers, BTDA/ADBA-12/DDE (100/75/25) and BTDA/ADBA-12/DDE (100/50/50) were soluble in NMP. The solubility of these copolyimides may be improved by the randomizing effect based on copolymerization as well as the entropy effect of long chain linear alkyl groups. Based on the fact that all copolyimides BTDA/DBAE-8~14/DDE were insoluble in NMP, ADBA is more effective for the enhancement of solubility in comparison to DBAE. Fig. 6 summarizes the effect of functional diamines, AODB-X, ADBP-X, ADBA-X and DBAE-X bearing long-chain alkyl groups for the enhancement of solubility investigated in our laboratory, and it is concluded that the effect of functional diamines are increased as AODB (ether linkage) > ADBP (benzoyl linkage) > ADBA (amide linkage) > DBAE (ester linkage) (Fig. 6). The polyimides and copolyimides based on BTDA, DPABA-6 or DPABA-12, and DDE containing 50 mol % or more DPABA were soluble, showing that the effect of DPABA for the enhancement of solubility was larger than ADBA. It is speculated that the three long-chain alkyl groups in DPABA enhance the solubility of polyimides.

Furthermore, several important results concerning on the structure-solubility relationships of the polyimides bearing long-chain alkyl groups are obtained and concluded as follows: (1) ADBP with an even number of carbon atoms were effective in enhancing the solubility, while polymers based on ADBP with an odd number of carbon atoms remained insoluble. It can be assumed that the conformation around C-C bonds of the long-chain alkyl groups and alignment of benzene ring attached with these alkyl groups and carbonyl group affect this odd-even effect. (2) Copolymerization using the conventional aromatic diamine, DDE resulted in the improvement of both the molecular weight and the thermal stability. (3) The copolymerization study based on AODB-10~14 and DDE demonstrated that AODB-12 having 12 methylene units was the most effective in enhancing the solubility. (5) DBAE having branched alkyl chains such as nonan-5-yl 3,5-diaminobenzoate (DBAE-9-branch-A) and 2,6-dimethylheptane-4-yl 3,5-diaminobenzoate (DBAE-9-branch-B) were introduced in these polyimides, and the homopolyimides based on BTDA/ DBAE-9-branch-A and BTDA/ DBAE-9-branch-B, and copolyimides containing more than 50% of DBAE-9-branch-A or DBAE-9-branch-B were soluble in NMP. Thus, it was found that the introduction of branched alkyl chains enhances solubility.

Diamine[a]		Poly(amic acid)	Polyimide							
Long-chain-DA	DDE	η_{inh}[b]	Solubility	η_{inh}[b]	Tg[c]	10% Weight loss temperature[d]		Molecular Weight[e]		
mol%		dLg^{-1}	in NMP	dLg^{-1}	°C	in Air	in N_2	*Mn*	*Mw*	*Mw/Mn*
						°C	°C			
ADBP-12										
0	100	1.15	insoluble							
25	75	0.44	insoluble							
50	50	0.49	soluble	0.37	264	467	500			
75	25	0.49	soluble	0.46	261	469	481			
100	0	0.34	soluble	0.37	254	468	464			
AODB-12										
0	100	1.15	insoluble							
25	75	0.39	soluble	0.29	262	460	456			
50	50	0.21	soluble	0.23	264	456	457			
75	25	0.14	soluble	0.19	284	447	452			
100	0	0.14	soluble	0.16	277	436	441			
DBAE-12										
0	100	1.15	insoluble							
25	75	0.48	insoluble							
50	50	0.45	insoluble							
75	25	0.40	insoluble							
100	0	0.31	insoluble							
ADBA-12										
0	100	1.15	insoluble							
25	75	0.95	insoluble							
50	50	0.66	soluble	0.57	247[f]	474	468	43700	97000	2.2
75	25	0.59	soluble	0.36	260[f]	452	435	27900	54200	1.9
100	0	0.45	insoluble							
DPABA-12										
0	100	1.15	insoluble							
25	75	0.96	insoluble							
50	50	0.83	soluble	0.65	253, 241[f]	453	446	45300	119100	2.6
75	25	0.60	soluble	0.39	325[f]	400	441	31500	77200	2.5
100	0	0.53	soluble	0.37	247[f]	352	429	25600	55300	2.2

[a]Equimolar amount of BTDA (3.3',4,4'-Benzophenonetetracarboxylic dianhydride) was used to the total molar amount of diamine. Reaction condition; r.t., 12 h poly(amic acid), Pyridine (5 molar) / Ac_2O (4 molar), 120 °C. [b]Measured at 0.5 g dL^{-1} in NMP at 30 °C. [c]Measured by DSC at a heating rate of 20 °C/min in N_2 on second heating. [d]Measured by TGA at a heating rate of 10° C/min. [e]Determined by SEC in NMP containning 10 mM LiBr using a series of polystyrenes standards having narrow polydispersities. [f]Softening temperature, measured by TMA at a heating rate of 10 °C/min

Table 1. Polyimides and copolyimides bearing long-chain alkyl groups

2.3.2 Molecular weight

As an index of molecular weight, the measurement of inherent viscosities (η_{inh}) and SEC measurement have been carried out in our laboratory. The inherent viscosities of all polymers were measured using Cannon Fenske viscometers at a concentration of 0.5 g/dL in NMP at 30 ˚C. Size exclusion chromatography (SEC) measurements were performed in NMP containing 10mM LiBr at 40°C with a TOSOH HLC-8020 equipped with a TSK-GEL ALPHA-M. Number average molecular weight (*Mn*), weight average molecular weight (*Mw*) and polydispersity (*Mw/Mn*) were determined by TOSOH Multi Station GPC-8020 calibrated with a series of polystyrenes as a standard. For examples, η_{inh} values for the

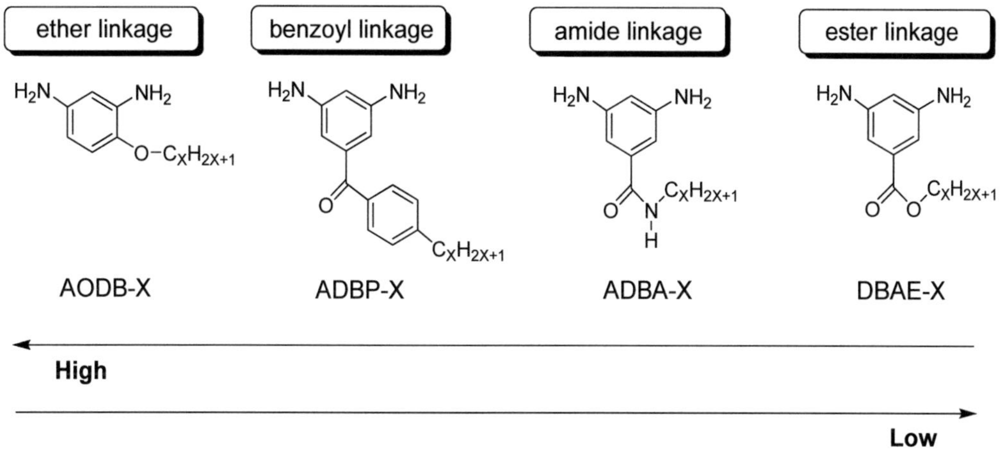

Fig. 6. Effect of aromatic diamines bearing long-chain alkyl groups on polyimide solubility

soluble polyimides in Table 1 are in the range of 0.16~0.65 dLg^{-1}. The weight average molecular weights of the polyimides based on ADBA-12 and DPABA-12 determined by SEC measurements are in the range of 54200 to 119100. These values indicated that the molecular weights of these polyimides were considered to be medium or rather lower values for polyimides, however, all polyimides show good film formation ability. In almost all cases, the molecular weights increased with the percentage of DDE, i. e. highly reactive diamine. The representative SEC traces are shown in Fig. 7, indicating that copolyimides based on BTDA/ADBA-11/DDE have typical monomodal molecular weight distribution, and their polydispersity is in the range of 2.2-2.4, which are typical values for polycondensation polymers.

2.3.3 Spectral analysis

^1H NMR spectra were measured on a JEOL JNM-AL400 FT NMR instrument in $CDCl_3$ or dimethylsulfoxide-d_6 with tetramethylsilane (TMS) as an internal standard. IR spectra were recorded on a JASCO FT/IR-470 plus spectrophotometer. ATR Pro 450-S attaching Ge prism was used for the ATR measurements of polyimide films.

The polyimide film samples for the measurement of ATR and thermomechanical analysis (TMA) mentioned in the next section were prepared by the following casting method. About five wt % polyimide solution in appropriate solvents such as NMP, chloroform, *m*-cresol on aluminum cup or glass substrate and the solution were slowly evaporated by heating on a hotplate at appropriate temperature (*ca.* 50 °C for chloroform, *ca.* 150 °C for NMP and *m*-cresol) until the films were dried, then the films were dried in a vacuum oven at 100 °C for 12 h. In case the molecular weights of polyimides were lower, the polyimide films tended to be brittle.

In the case of soluble polyimides, NMR measurements are convenient because solution samples can be prepared, and provide more quantitative data. For example, Fig. 8 shows the ^1H NMR spectrum of the copolyimide based on ADBA-12/DDE (50/50) that is soluble in DMSO-d_6 and the peaks support this polymer structure. The intensity ratio of CH_3 protons

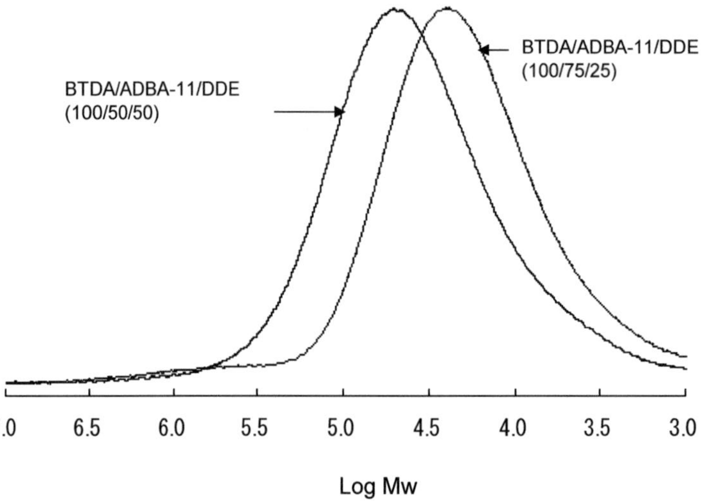

Fig. 7. Representative SEC traces of soluble polyimides based on aromatic diamines bearing long-chain alkyl groups. BTDA/ADBA-11/DDE (100/50/50): *Mn*, 49500; *Mw*, 118800; *Mw/Mn*, 2.4. BTDA/ADBA-11/DDE (100/75/25): *Mn*, 30700; *Mw*, 67900; *Mw/Mn*, 2.2

of long-chain alkyl groups and the aromatic proton H_A or H_B is approximately 3/4, meaning that copolymer composition corresponds to the monomers initial ratio. Imidization ratios of polyimides are generally determined by FT-IR measurements, comparing absorption intensities of amic acid carbonyl groups with those of imide carbonyl groups. However, FT-IR measurements give relatively less quantitative data in comparison with NMR measurements. In the case of these soluble polyimides, generally, a broad signal due to the NH protons of poly(amic acid) appears around 12 ppm in DMSO-d_6, while this signal disappears in the corresponding polyimide. The imidization ratios of these polyimides can be calculated from the reduction in intensity ratio of the NH proton signals in poly(amic acid)s and these values for the polyimides prepared in our laboratory are sufficiently high, near to 100 %.

ATR measurement is the useful method to measure IR spectrum of polymer films. Representative ATR spectrum of dendronized polyimides based on 12G1-AG-Terphenyldiamine and 12G2-AG-Terphenyldiamine were shown in Fig. 9 and these spectrum show the strong absorptions based on C-H bonds of long-chain alky groups and the strong absorptions of C-O bonds of alkyloxy groups, and these absorption intensities become stronger with the increase of long-chain alkyl ether segments in the polyimides.

2.3.4 Thermal properties

Differential scanning calorimetery (DSC) traces were obtained on a Shimadzu DSC-60 under nitrogen (flow rate 30 mL/min) at a heating rate of 20° C/min and the glass transition temperatures (Tg) were read at the midpoint of the heat capacity jump from the second heating scan after cooling from 250 °C. Thermomechanical analysis (TMA) was performed on a Shimadzu TMA-50 under nitrogen (30 mL/min) at a heating rate of 10 °C/min with a

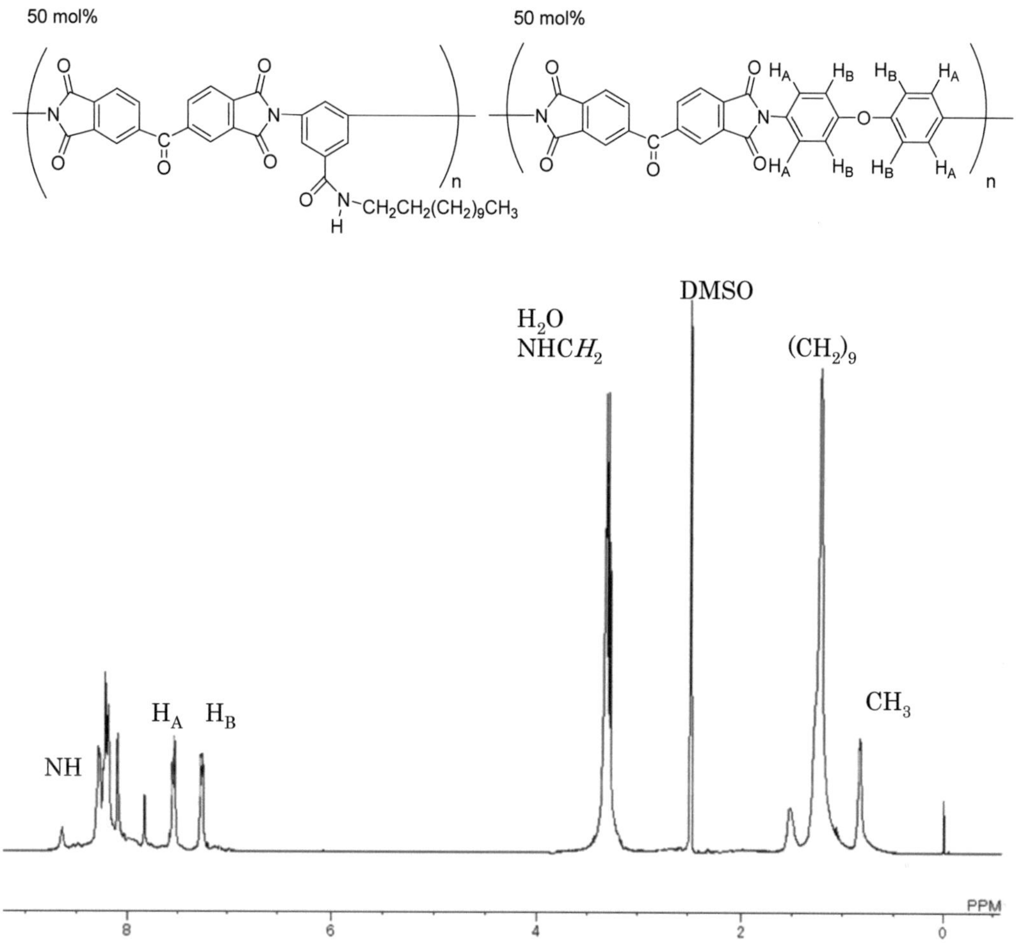

Fig. 8. ^{1}H NMR spectrum of a copolyimide based on BTDA/ADBA-12/DDE (100/50/50)

10 g load in the penetration mode using the film samples approximately 300 μm in thickness. Softening temperatures (Ts) were taken as the onset temperature of the probe displacement on the second TMA scan after cooling from 220 °C. Thermogravimetric analysis (TGA) was performed on a Shimadzu TGA-50 in air or under nitrogen (50 mL/min) at a heating rate of 10 °C/min using 5 mg of a dry powder sample, and 0 (onset), 5, 10% weight loss temperatures (Td_0, Td_5, Td_{10}) were calculated from the second heating scan after cooling from 250 °C.

The Tg's of these polyimides sometimes were not recognized by DSC measurements, probably due to the rigid imide linkages. In these cases, TMA measurements were performed to determine the Tg. Many publications have described that the softening temperature (Ts) obtained from TMA measurements corresponds to the apparent Tg of polymers. As can be seen from Tables 1, the Tg values of these polyimides are in the range from 241-325 °C, showing similar values observed in soluble polyimides obtained from our laboratory (*ca.* around 250 °C) and are 100-150 °C lower than those of the conventional fully aromatic polyimides, however, are 100-150 °C higher than the commodity thermoplastics.

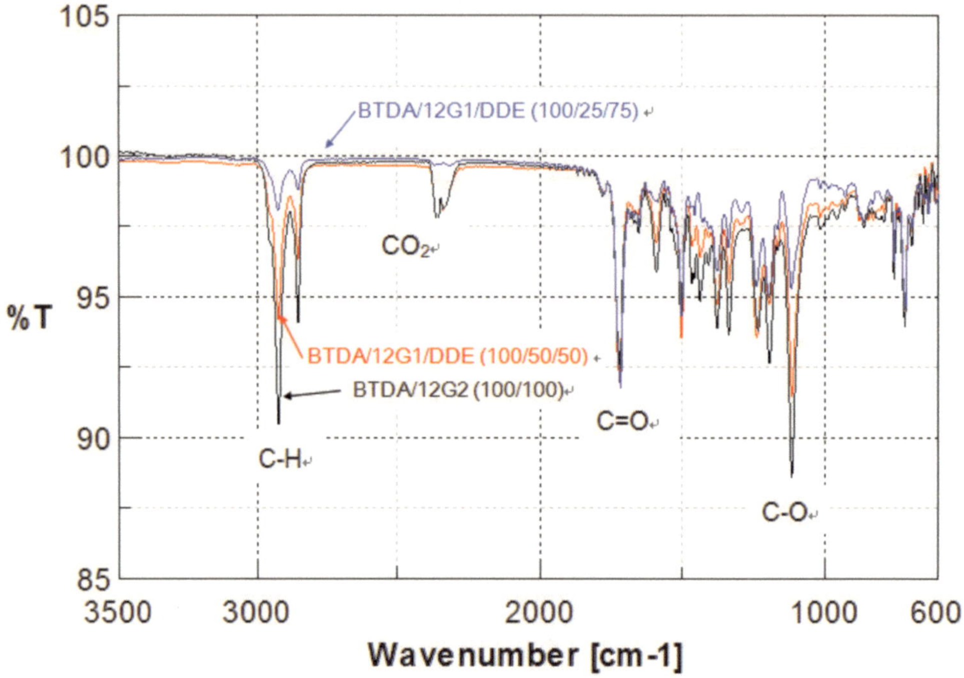

Fig. 9. Representative ATR spectrum of dendronized polyimides

Consequently, the physical heat resistance of these soluble polyimides bearing long-chain alkyl groups can be ranked as heat resistant polymers.

The Td_{10} values of these polyimides bearing long-chain alkyl groups in Table 1 are in the range 352~474 ^{o}C in air and 429~500 ^{o}C under nitrogen, showing similar values observed in soluble polyimides obtained from our laboratory (*ca.* 400~500 ^{o}C). In most cases, Td values in air were lower than Td values under nitrogen following the general fact that oxidative degradation proceed rapidly in air. As the incorporation of DDE resulted in a reduction of aliphatic components of the polyimides, the Td_{10} of these polyimides tends to increase with the increment of the DDE component (Table 1). These Td_{10} values of soluble polyimides obtained in our laboratory are 100~200 ^{o}C lower than those of wholly aromatic polyimides; however, the chemical heat resistance of these polyimides still can be ranked as heat resistant polymers. Fig. 10 shows the TGA traces of dendronized polyimides based on BTDA/12G1-AG-Terphenyldiamine (100/50/50). These TGA traces showed steep weight loss at the intial stage of degradation, and these weight loss percent almost correspond the calculated value of the weight percent of alkyl groups in polymer segments. Therefore, it is considered that the degradation of long-chain alkyl groups occurred at the initial stage of thermal degradation. Furthermore, these TGA traces also show the evidence that the long-chain alkyl groups exist in the polyimides and the cleavage of alkyl groups did not occurred during the polymerization.

2.4 Application for VAN-LCDs

The alignment layer application for VAN-LCDs using polyimides having dendritic side chains was performed at Cheil Ind. Inc., Korea. LCDs test cell properties were measured as

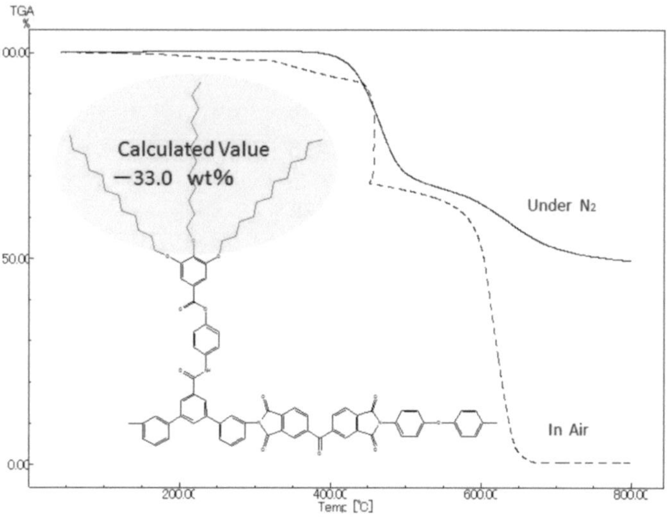

Fig. 10. Representative TGA traces of dendronized polyimides based on 12G1-AG-Terphenyldiamine {(BTDA/12G1-AG-Terphenyldiamine/DDE (100/50/50)}

follows: the polyimide solutions were spin-coated onto ITO glass substrates to a thickness of 0.1 μm, and cured at 210 °C for 10 minutes to produce liquid crystal alignment films. After the liquid crystal alignment films were subjected to a rubbing process, the alignment properties and the pretilt angles of the liquid crystal were measured. The surface of the alignment films were rubbed by means of a rubbing machine, two substrates were arranged anti-parallel to each other in such a manner that the rubbing direction of the each substrates were reverse, and the two substrates were sealed while maintaining cell gaps of 50 μm to fabricate liquid crystal cells. The liquid crystal cells were filled with the liquid crystalline compounds (Merk licristal). The alignment properties of the liquid crystal were observed under an orthogonally polarlized optical microscope. The pretilt angles of the liquid crystal were measured by a crystal rotation method. In order to examine the electrical properties, the test cells were prepared by the same manner as above except the cell gap, 5 μm. The voltage holding ratios were measured with VHRM 105 (Autronic Melchers). To evaluate the VHR, the applied frequency and voltage was 60 Hz, 1V with pulse of 64 μsec. The voltage versus transmittance and optical response properties, such like contrast ratio, response time, image sticking, etc., were measured using computer-controlled system in conjunction with an tungsten-halogen lamp, a function/arbitrary waveform generator, photomultiplier. The residual DCs were measured by C-V method using impedance analyzer.

The polyimide alignment layers containing 8 mol % of 12G1-AG-Terphenyldiamine were utilized for the vertical alignment mode (VA-mode). The synthesis of polyimide alignment layers containing 8 mol % of 12G1-AG-Terphenyldiamine was carried out in NMP as a solvent by conventional two step polymerization method regularly used for the synthesis of polyimide alignment layers for TN-LCDs , and 12G1-AG-Terphenyldiamine monomer was used as one of the diamine components. LCDs test cell properties are summarized in Table 2. PIA-DEN represents the test cell using the polyimide alighnment layers containing 8 mol % of 12G1-AG-Terphenyldiamine, and TN represents the test cell using the regular polyimide alignment layers. The pretilt angles of LC molecules were over 89° in PIA-DEN test cells, which are the suitable values for VAN-LCDs. It is speculated that an extremely

ITEM		PIA-DEN	TN mode
Pretilt angle (°)		>89	4~6
Surface energy (dyn/cm^2)[a]		39	48
VHR (%)	25°C	>99	>99
	60°C	>98	>95
Response time (ms)		<25	<30
Contrast ratio		580	250
Residual DC (mv)		<200	<200
Image sticking		<1	<1

[a] Surface energy of polyimide alignment films measured by a contact angle metod

Table 2. LCDs test cell properties using the alignment films containing dendronized polyimides

bulky and hydrophobic dendritic moieties affects the generation of pretilt angles between the surface of polyimide and liquid crystalline molecules as illustrated in Fig. 11. The considerably lower surface energy value of the PIA-DEN alignment film in comparison with the one of TN mode also indicate that the surface of polyimides containing dendritic moieties is more hydrophobic.

The various important properties of PIA-DEN test cells such as voltage holding ratio (VHR), response time, contrast ratio, residual DC, and image sticking are equivalent or advantageous in comparison with those of regular TN test cell. Fig. 12 shows a V-T (voltage-transmittance) curve of these test cells, and shows a dramatic change of T. Consequently, it is convinced that the dendritic monomers, and dendritic polyimides developed by our research can be applied for the alignment films for VAN-LCDs.

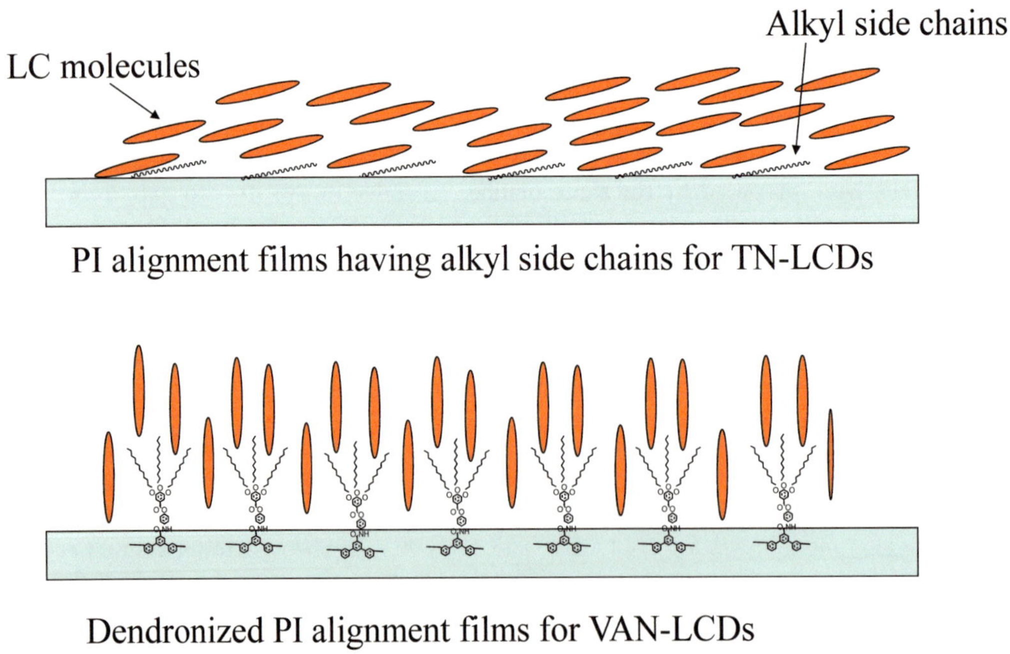

Fig. 11. Vertical alignment of LC molecules using dendronized polyimides as alignment layers

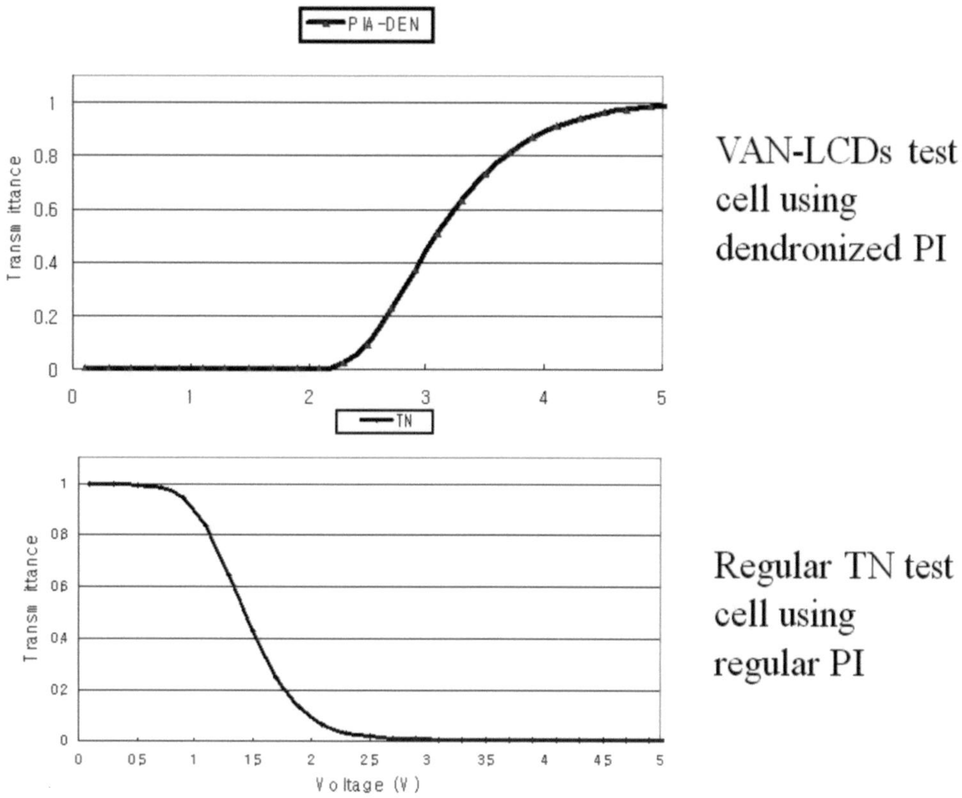

Fig. 12. Voltage-transmittance curves of LCD test cells using dendronized and conventional polyimides

2.5 Application for printed electronics

Recently, various printing methods such as an ink-jet print method have been investigated for manufacturing polymeric thin-films, and the surface wettability and their control methods have become important. Thus, the author has investigated the surface wettability control of these polyimides by UV light irradiation that is a conventional method for microlithography (Fig. 13). The soluble polyimides bearing long-chain alkyl groups used for this study were synthesized from 12G1-AG-Terphenyldiamine, $3C_{10}$-PEPEDA, $3C_{10}$-PEPADA or $3C_{10}$-PAPADA that have three long-chain alkyl groups, aliphatic tetracarboxylic dianhydride; Cyclohexene-DA or aromatic tetracarboxylic dianhydride; DSDA or 3,4'-ODPA, and DDE as a diamine co-monomer. Polyimide thin-films were obtained as follows: 0.5-2.0 wt % polyimide solution in NMP were cast on glass substrates and the solution were slowly evaporated by heating at approximately 100-120 ^{o}C until the films were dried, then the films were dried in a vacuum oven at 100 ^{o}C for 12 h. Water contact angles were measured by SImage mini (Excimer. Inc., Japan) and UV light irradiation were performed using UV lamp unit E50-254-270U-1 (254 nm, 6.0 mW/cm^2, Excimer. Inc., Japan) and a cool plate NCP-2215 (NISSIN Laboratory equipment, Japan) adjusted at 20°C that was used to neglect the effect of thermal degradation of polyimide films during UV irradiation process.

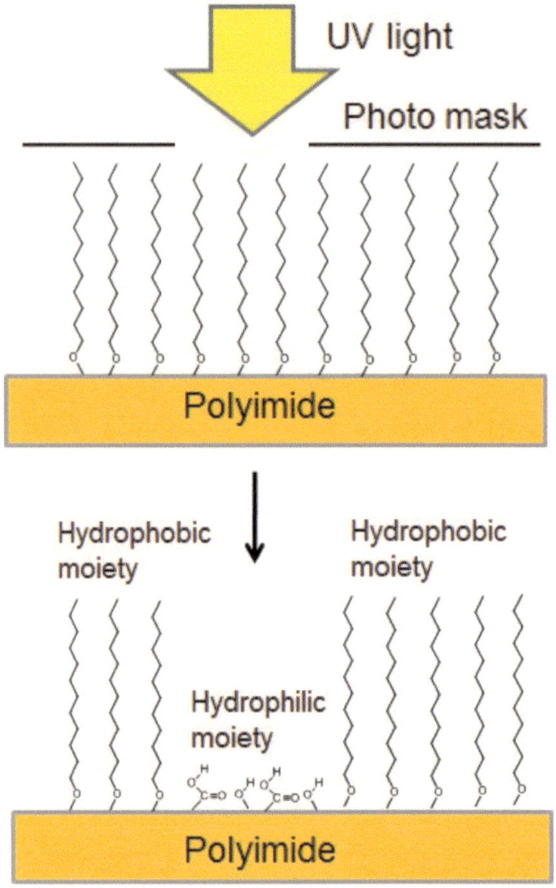

Fig. 13. Conceptual scheme of wettability control of the polyimide surface by UV irradiation

Thus, the polyimide thin films were irradiated by UV light, and the contact angles for the water decreased from near 100° (hydrophobicity) to the minimum value, 20° (hydrophilicity) in proportion to irradiated UV light energy. The thin film specimens after UV light irradiation were rinsed by isopropyl alcohol. The representative results using the polyimides based on $3C_{10}$-PEPADA are summarized in Table 3 and Fig. 14.

Although the water contact angles decreased after UV light irradiation, the degrees of changes depended on the polyimide structures. For examples, the contact angles of the polymides based on 3,4'-ODPA or DSDA/ $3C_{10}$-PEPADA /DDE remarkably decreased from around 100° to around 20-30° after UV light irradiation (254nm, 2-8J). These changes were less in the case of the polyimides based on Cyclohexene-DA/$3C_{10}$-PEPADA /DDE, and the changes were much less in the case of the polyimides based on Cyclohexene-DA/ DDE without long-chain alkyl groups.

It is considered that these changes of wettability of polyimides are mainly based on the photo-degradation or scission of long-chain alkyl groups, and that the generation of the hydrophilic functional groups such as COOH and OH groups occurred. ATR measurements

Monomer			Polyimide				
Tetracarboxylic dianhydride[a]	Diamine		Water contact angle after UV irradiation [b], () [c]				
	mol%		0J	2J	4J	6J	8J
Cyclohexene-DA	$3C_{10}$-PEPADA	DDE					
	100	0	104 (101)	96 (92)	95 (83)	86 (67)	81 (50)
	50	50	97 (96)	95 (81)	87 (64)	68 (59)	57 (38)
	0	100	80 (80)	73 (75)	67 (60)	58 (50)	38 (24)
DSDA	$3C_{10}$-PEPADA	DDE					
	100	0	104 (104)	88 (79)	76 (64)	60 (45)	44 (33)
	50	50	99 (95)	87 (76)	81 (72)	62 (60)	45 (54)
	0	100					
3,4'-ODPA	$3C_{10}$-PEPADA	DDE					
	100	0	100 (99)	80 (75)	57 (57)	36 (30)	24 (23)
	50	50	96 (94)	80 (73)	52 (57)	31 (32)	31 (30)
	0	100	78 (78)	77 (75)	44 (70)	42 (63)	36 (52)

[a] Equimolar amount of tetracarboxylic dianhydride was used to the total amount of diamines. [b] Water contact angles (deg) using contact angle meter (Excimer inc.,SImage mini)at 25℃. [c] Water contact angles (deg) after rinsing by isopropyl alcohol.

Table 3. Water contact angles of the polyimide surface after irradiation of UV light

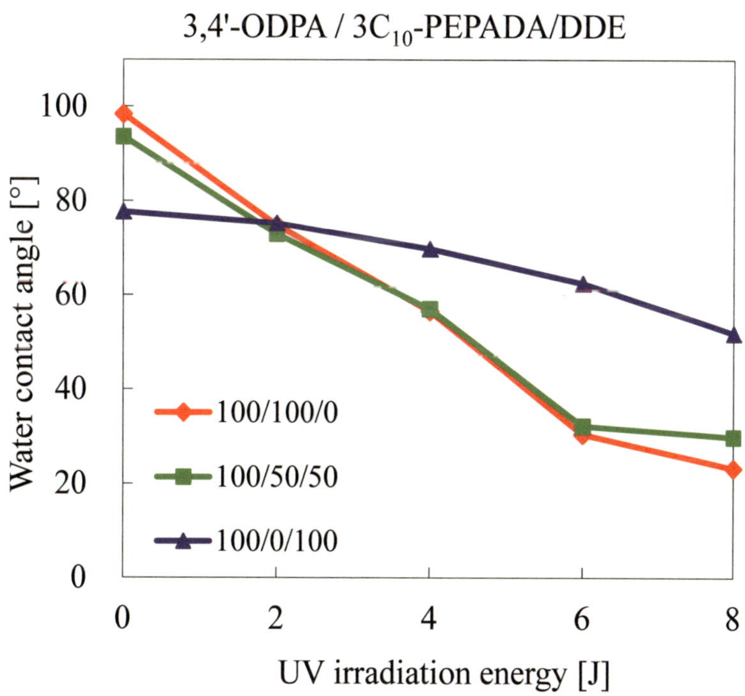

Fig. 14. UV irradiation energy dependence of water contact angles of polyimide films

of the polyimide surfaces after UV light irradiation support this assumption, and the absorption of OH groups around 3300 cm^{-1} increase, the absorption of alkyl groups around 2900 cm^{-1} decrease, and the absorption of ether groups around 1200 cm^{-1} decrease with the increase in the photo-irradiation energy (Fig. 15). The intensive surface analysys was examined using XPS and SFM. XPS measurements were carried out on an XPS -APEX (Physical Electronics Co. Ltd.) with an Al Kα X-ray source (150 W). Chamber pressure; 10^{-9} - 10^{-10} Pa; take off angles; 45° and SFM (SII Nanotechnology Inc., SPA 400) was operated in a dynamic force microscopic (DFM) mode. The generation of hydrophilic moieties was analyzed in detail by XPS narrow scan, and chemical shifts due to C-O and C=O bonds clearly increase after UV light irradiation (Fig. 16). The surface nm size roughness probably based on long-chain alky groups was observed by SFM analysis (Fig. 17), however, these micro roughness seemed not to change after UV light irradiation. Thus, the change of surface wettability of polyimides is occurred mainly by the changes of chemical structures of polyimide surface. It is speculated that the complicated photo-induced reactions such as auto-oxidation, cleavage of ester groups, Fries rearrangement, etc. occur on the surface of polyimides on the course of UV light irradiation (Fig. 18).

In conclusion, the surface wettability of polyimides bearing long-chain alkyl groups can be controlled by UV light irradiation, and these methods are expected to be applied in the field of printed electronics.

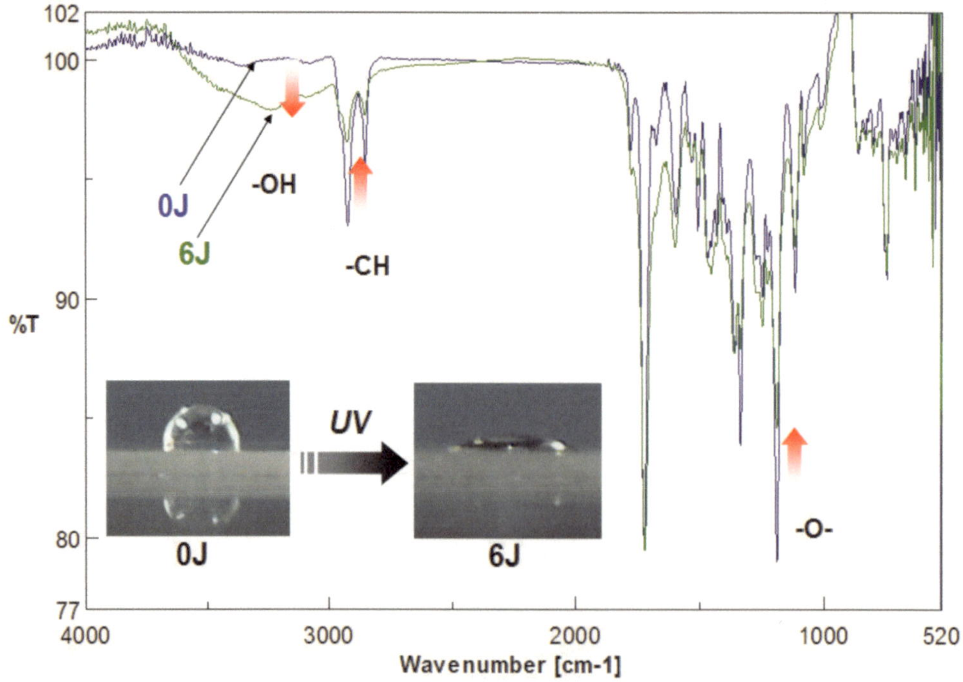

Fig. 15. Representative ATR spectrum of polyimides bearing long-chain alkyl groups before and after UV irradiation

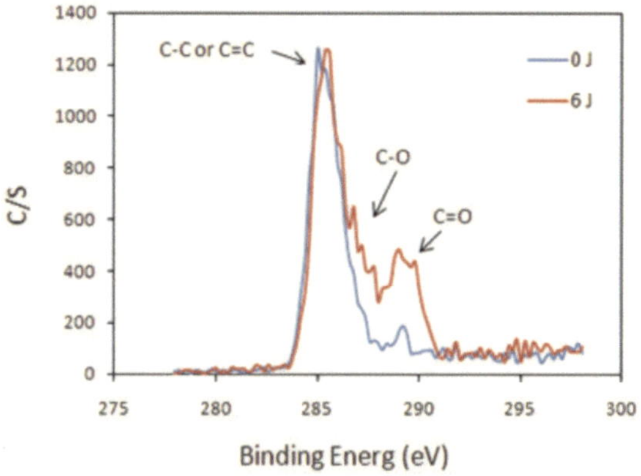

Fig. 16. XPS narrow scan of 3,4'-ODPA / $3C_{10}$-PEPADA

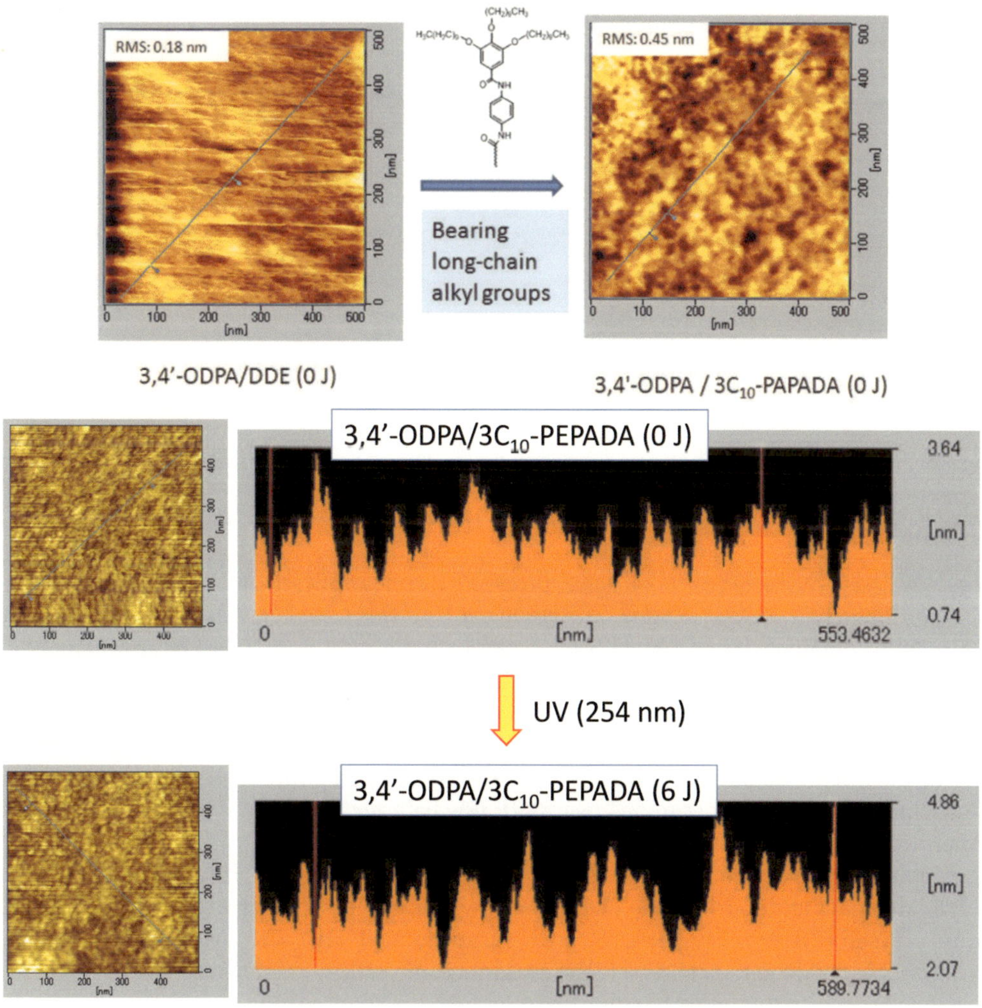

Fig. 17. SFM images of 3,4'-ODPA / DDE and 3,4'-ODPA /$3C_{10}$-PAPADA

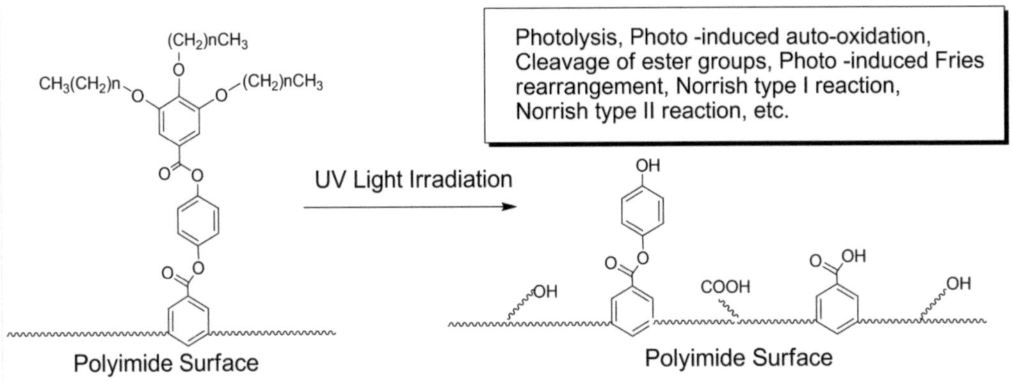

Fig. 18. Anticipated photochemical reactions on the surface of polyimides

3. Conclusion

The synthesis, characterizations, basic properties and applications of soluble polyimide bearing long-chain alkyl groups are reviewed in this chapter. These polyimides are successfully obtained based on the novel aromatic diamine monomers having long-chain alkyl groups. As these polyimides are soluble in various organic solvents, the spectral analyses such as NMR are possible, and the polymer structures are well characterized. The basic properties of these polyimides such as the solubility and the thermal stability are investigated in detail and the structure-properties relationships are well considered. Thus, it is concluded that these polyimides bearing long-chain alkyl groups are suitable polymeric materials for microelectronics applications.

The application as alignment layers for LCDs was investigated, and it was found that these polyimides having dendritic side chains were applicable for the vertically aligned nematic liquid crystal displays (VAN-LCDs). It is speculated that an extremely bulky and hydrophobic dendron moiety affects the generation of vertical alignment.

The thin films of polyimides bearing three long-chain alkyl groups were irradiated by UV light , and the contact angles for the water decreased from near 100° (hydrophobicity) to near 20° (hydrophilicity) in proportion to irradiated UV light energy. From the result of surface analyses, it is recognized that the hydrophobic long-chain alkyl groups on the polyimide surface decrease and the hydrophilic groups such as a hydroxyl group generate on their surface. Thus, the surface wettability of polyimides bearing long-chain alkyl groups can be controlled by UV light irradiation, and these methods are expected to be applied in the field of printed electronics.

4. Acknowledgment

The author thanks Dr. Atsushi Takahara of Kyushu University, Drs. Takaaki Matsuda and Tsutomu Ishi-I of Kurume National College of Technology for various advices. The author also thanks many students of Kurume National College of Technology for their help with the experiments. Financial supports from Cheil Industries Inc., Kyushu Industrial Technology Center, DYDEN Corporation, and Toyohashi University of Technology are gratefully acknowledged.

5. References

Tsuda, Y., Kawauchi, T., Hiyoshi, N. & Mataka, S. (2000a). Soluble Polyimides Based on Alkyldiaminobenzophenone. *Polymer Journal.* Vol. 32, No. 7, (June 2000), pp. 594-601, ISSN 0032-3896

Tsuda, Y., Kanegae, K. & Yasukouchi, S. (2000b). Soluble Polyimides Based on Alkyloxydiaminobenzene. *Polymer Journal.* Vol. 32, No. 11, (November 2000), pp. 941-947, ISSN 0032-3896

Tsuda, Y., Kojima, M. & OH, J.-M. (2006). Soluble Polyimides Based on Diaminobenzoic Acid Alkylester. *Polymer Journal.* Vol. 38, No. 10, (October 2000), pp. 1043-1054, ISSN 0032-3896

Tsuda, Y., Kojima, M., Matsuda, T. & OH, J.-M. (2008). Soluble Polyimides Based on Long-chain Alkyl Groups *via* Amide Linkages. *Polymer Journal.* Vol. 40, No. 4 (April 2008), pp. 354-366, ISSN 0032-3896

Tsuda, Y. (2009). Soluble Polyimides Based on Aromatic Diamines Bearing Long-chain Alkyl Groups, In: *Polyimides and Other High Temperature Polymers. Vol. 5,* Mittal, K. L. (Ed.), pp. 17-42,VSP/Brill, ISBN 978-90-04-17080-3, Leiden

Tsuda, Y., OH, J.-M. & Kuwahara, R. (2009). Dendronized Polyimides Bearing Long-chain Alkyl Groups and Their Application for Vertically Aligned Nematic Liquid Crystal Displays. *International Journal of Molecular Sciences.* Vol. 10 (November 2009), pp. 5031-5053, ISSN 1422-0067

Tsuda, Y., Nakamura, R., Osajima, S. & Matsuda, T. (2010). Surface Wettability Controllable Polyimides Bearing Long-chain Alkyl Groups by UV Light Irradiation. *PMSE Preprints (ACS Division Proceeding Online),* Vol. 239, ISSN 1550-6703, San Francisco, April 2010

Tsuda, Y., Hashimoto & Matsuda, T. (2011a). Surface Wettability Controllable Polyimides Bearing Long-chain Alkyl Groups by UV Light Irradiation. *Kobunshi Ronbunshu (Japanese),* Vol. 68 (January 2011), pp. 24-32, ISSN 0386-2186

Tsuda, Y. (2011b). Surface Wettability Controllable Polyimides Bearing Long-chain Alkyl Groups by UV Light Irradiation. Proceedings of International Conference on Materials for Advanced Technologies, ISBN 978-981-08-8878-7, SUNTEC Singapore, June 2011.

2

Transparent ZnO Electrode for Liquid Crystal Displays

Naoki Yamamoto, Hisao Makino and Tetsuya Yamamoto
Research Institute, Kochi University of Technology
Japan

1. Introduction

Recently, the scarcity and toxicity of indium, a major constituent element of ITO, has become a concern. Indium is a rare element that ranks 61st in abundance in the Earth's crust (Kempthorne & Myers 2007). In addition, the major amounts of indium consumed by the industries produceing the electronic devices such as liquid crystal displays (LCDs), touch-screens and solar cell systems are supplied by only a few countries. Furthermore, indium has also been suspected to induce lung disease, and particularly indium-related pulmonary fibrosis should be paid attention (Homma et al., 2005).

Transparent conductive oxides have become the focus of attention as a substitute material for ITO currently used for optically transparent electrodes in electronic devices. In particular, transparent conductive ZnO films are expected to be suitable materials to achieve such purposes because, in contrast with indium as a major constituent element of ITO, Zn is an element that the human body requires and is a component of some marketed beverages, in addition to having being used for years in cosmetics and as a vulcanization accelerator for rubber products such as tires. Furthermore, conductive and transparent ZnO films have low electrical resistance and high optical transmittance comparable with those of ITO films reported by some authors (Wakeham et al., 2009; Shin et al., 1999). We have developed the technology to form transparent conductive ZnO films with low resistance (2.4 $\mu\Omega$ m for a 100 nm thick film (Yamada, et al., 2007)), optical transmittance exceeding 95% (film-only transmittance without that of the glass substrate) and high heat-resistance (thermally stable until 300-450 °C (Yamamoto, N. et al., 2010)). The technology of transparent conductive ZnO films applied as alternatives to ITO electrodes for LCD panels is described in this chapter.

2. Preparation of transparent and conductive ZnO film

Ga-doped ZnO (GZO) and Al-doped (AZO) films have been widely studied as the most promising transparent conductive films as alternatives to ITO films used in electronics devices such as LCDs, LEDs and solar cells.

2.1 Magnetron sputtering system

Conventional magnetron sputtering systems, planar- and cylindrical-types (Carousel-type), were used to form transparent ZnO thin films. A schematic diagram of the cylindrical-type magnetron sputtering system is shown in Fig. 1 (a). In the cylindrical-type magnetron sputtering system, a drum with samples set on its surface is rotated concentrically in the chamber with the sputtering target set on the inside wall. A film can be formed by sputtering with dc power (noted as dc MS) and radio frequency power combined with dc power (noted as rf+dc MS) applied to the sputtering target.

2.2 Reactive plasma deposition system

Figure 1 (b) shows a schematic diagram of the reactive plasma deposition system (RPD) (Yamamoto, T. et al., 2008)., which is a type of ion-plating method. An Ar plasma stream is generated by a pressure gradient arc plasma source (Uramoto gun) at the cathode is introduced by control of the electric and the magnetic field to the evaporation source tablet inset in the hearth at the anode. The particles evaporated from the source are deposited onto the substrate set on the tray traveling in front of the heater.

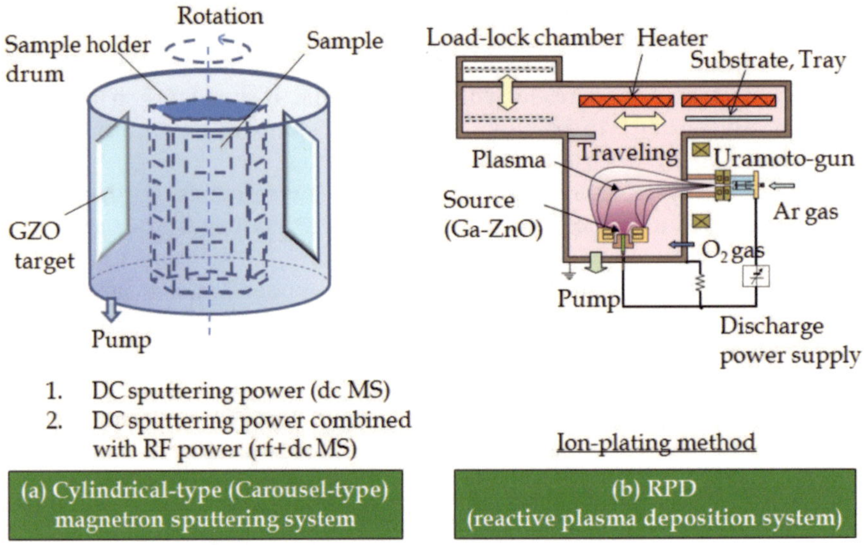

Fig. 1. Schematic diagrams of the deposition systems for the transparent conductive ZnO films.

The specifications for the formation of conductive transparent ZnO film using the magnetron sputtering systems and the RPD system are summarized in Table 1

	Magnetron Sputtering (MS)		RPD
	dc MS	rf+dc MS	
Ga_2O_3/Al_2O_3 content in ZnO source (wt%)	Ga_2O_3: 3.0 - 6.0		Ga_2O_3: 3.0 - 5.0
	Al_2O_3: 2.0 - 5.0		
Power (kW)	0.1 - 2.0	rf: 0.1 - 1.5, dc: 0.1 - 1.5 rf/dc = 0.5 - 2.0	discharge current: 140 - 150 (A)
Operation pressure (Pa)	0.1 - 0.8	0.1 - 0.8	0.4 - 0.6
Operation temperature (°C)	25 - 350	25 - 350	25 - 250

Table 1. Specifications for the formation of GZO or AZO films (Yamamoto. N. et al., 2011a & 2011c).

An Ar plasma stream is generated by a pressure gradient arc plasma source (Uramoto gun) at the cathode is introduced by control of the electric and the magnetic field to the evaporation source tablet inset in the hearth at the anode. The particles evaporated from the source are deposited onto the substrate set on the tray traveling in front of the heater.
The specifications for the formation of conductive transparent ZnO film using the magnetron sputtering systems and the RPD system are summarized in Table 1

3. Basic characteristics of transparent conductive ZnO film

The fundamental characteristics of transparent conductive ZnO films for application to LCD panels are discussed in this section.

3.1 Crystalline structure of transparent conductive ZnO film

X-ray diffraction (XRD; ATX-G, Rigaku) and transmission electron microscopy (TEM; H-9000UHR; Hitachi High-technologies Co.) were applied for analysis of the crystalline structures of transparent conductive ZnO films.
The crystalline structures and orientations of the GZO films were analyzed using both out-of-plane XRD (widely used X-ray diffraction analysis) and in-plane XRD (grazing-incidence wave-dispersive X-ray analysis (Ofuji et al., 2002)). For measurement using the in-plane XRD technique, a Cu Kα X-ray beam with a wavelength of 0.154184 nm was irradiated at a low angle of incidence to the surface of the sample (0.35°). The incident angle is close to the total reflection angle of X-ray for ZnO.
XRD patterns obtained from the GZO films deposited by dc MS, rf+dc MS or RPD were almost identical and had the wurtzite crystalline structure, as with the ZnO films. A typical XRD pattern obtained from a GZO film is shown in Fig. 2.
The in-plane XRD diffraction pattern shows that no diffraction peaks from the ZnO(00x) crystal planes were evident (Fig. 2(a)). In contrast, the out-of plane XRD pattern shows only the (002) and (004) diffraction peaks of the GZO film (Fig. 2(b)).
The appearance of these diffraction peaks clarified that (1) the GZO polycrystalline film consists of the wurtzite structure. (2) the *c*-axes of the wurtzite structure coincides with the direction normal to the GZO film surface, and (3) the *a*-axes of the wurtzite cell structure coincides with the direction in the plane of the film. The TEM image in Fig. 3(a) and the cell structure shown in Fig. 3(b) explains the structure. Columnar grains comprise the interior of the polycrystalline GZO films (Yamamoto, N. et al., 2008). Such crystalline structures also appeared in films formed in the temperature range of 150-250 °C using dc MS, rf+dc MS and RPD.
The lattice constants for the *c*- and *a*-axes, and the volume of the wurtzite crystalline unit cell in 100 nm thick GZO films prepared at 180 °C by dc MS, rf+dc MS and RPD were derived using the XRD peaks diffracted from the (00x) and (x00) crystalline planes and are compared in Fig. 4 (Yamamoto, N. et al., 2010). The lattice constants of the GZO films prepared by RPD were shorter than those of the films formed by magnetron sputtering (Fig. 4(a)). The *c*-axis of the rf+dc MS film was especially expanded toward the direction normal to the surface of the substrate compared with the other films. The *a*-axis was also expanded toward in the direction of the plane of the film. As a result, the cell volume of the wurtzite structure in the films prepared by rf+dc MS were larger than those formed by dc MS and RPD, as shown in Fig. 4 (b).

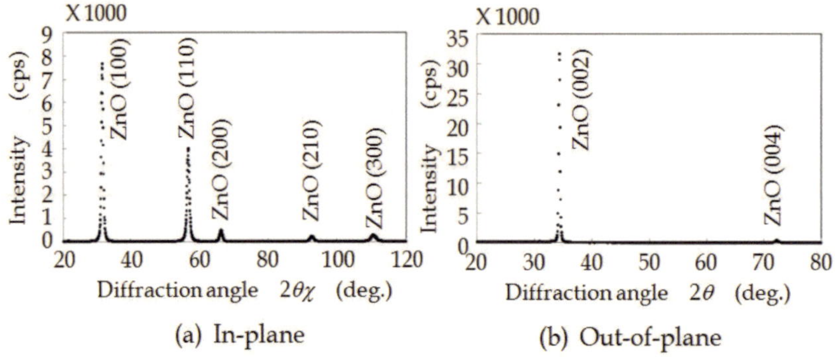

Fig. 2. Typical XRD profile of a 150 nm thick GZO film prepared at 180 °C using dc MS.

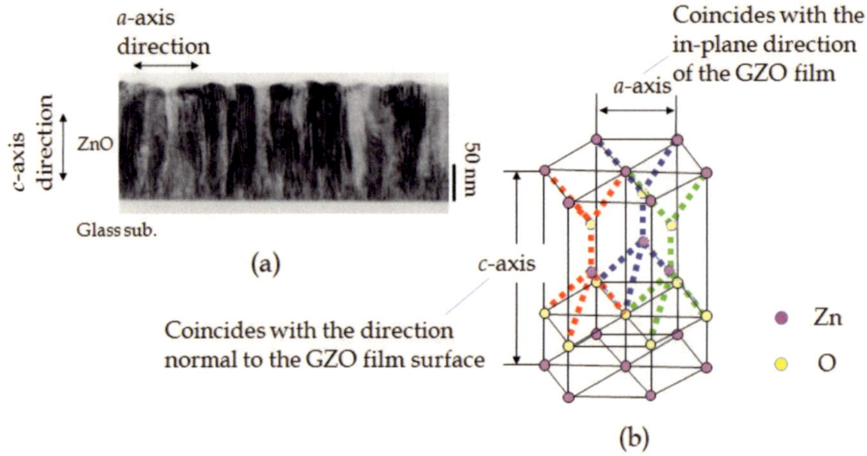

Fig. 3. (a) Cross-sectional TEM image of GZO film formed by dc MS and (b) the wurtzite cell structure.

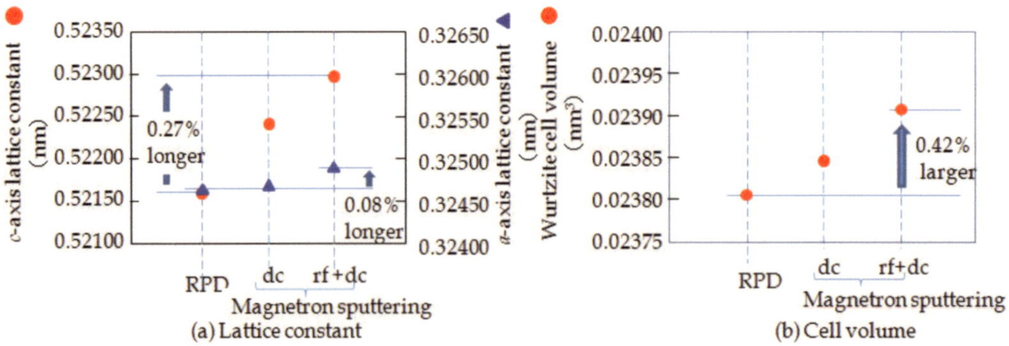

Fig. 4. Comparison of the *a*-axis, *c*-axis lattice constants and the unit cell volumes in crystalline ZnO-based wurtzite structures of films prepared using dc MS, rf+dc MS and RPD (Yamamoto,N. et al 2010).

3.2 Resistance of transparent conductive ZnO film

The resistivity of the film is one of the most important characteristics required for application of alternative transparent electrodes to ITO used in LCDs. The resistivity of ITO transparent films is in the range of 1.3–3.8 μΩm (Wakeham, et al.. 2009; Shin et al., 1999). Therefore, transparent conductive ZnO films as an alternative must have a resistivity of less than ca. 3.8 μΩm.

In the case of thin films, the resistivity is generally derived from the carrier-flow in the plane of the films. The resistivities of even metal films increase with the decrease in film thickness, especially less than ca. 100 nm. Such a phenomenon is caused by the increase in the frequency of collisions or scattering with the carrier–flow and the film surface, the interface with the substrate and irregular crystalline structures in the region of the substrate. The resistivity of the GZO film also shows a similar dependency on the film thickness, as shown in Fig. 5(a).

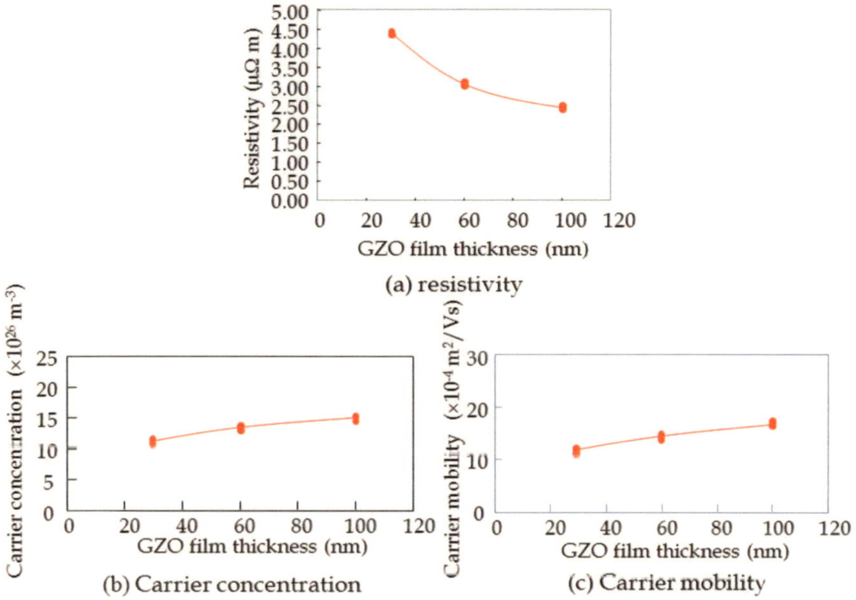

(a) resistivity

(b) Carrier concentration

(c) Carrier mobility

Fig. 5. Electrical characteristics of GZO film as a function of the film thickness.

As the film thickness decreases, the carrier mobility and concentration in the film decreases, as shown in Figs. 5(b) and (c). The data in Fig. 5 was obtained from GZO films prepared using RPD at 180 °C. The electrical characteristics of transparent conductive ZnO films formed by the magnetron sputtering showed similar dependencies on the film thickness, although the values were significantly affected by the formation conditions, i.e., type of dopant and its concentration, the deposition equipment, temperature, pressure and the electrical power supplied to the source during deposition.

Figure 6 shows a comparison of the resistivities of GZO and AZO films formed with an Ar sputtering pressure of 0.66 Pa at room temperature or at 180 °C using the conventional planar magnetron sputtering system. The ZnO sputtering target materials for the deposition of GZO or AZO contained 4 wt% Ga_2O_3 or 2 wt% Al_2O_3, respectively. The total sputtering power used in both cases of dc MS and rf+dc MS was set to 200 W.

The resistivities of the AZO films were 1.5 times or more higher than those of the GZO films. The resistivities of the films formed by rf+dc MS were lower than those of films formed by dc MS (Yamamoto, N. et al., 2011a & 2011c). However, this relationship does not correspond to the lattice constants and cell volumes of the wurtzite structures of the GZO films formed by rf+dc MS, which were larger than those of the films formed by dc MS, as shown in Figs. 4(a) and (b). It is generally considered that the probability of carrier hopping among atoms decreases with distance from each atomic site, and as a result, the resistivity of the film decreases with the length of the lattice constant. The reason for the contradiction between the resistivity and the lattice constant (length) is yet to be clarified.

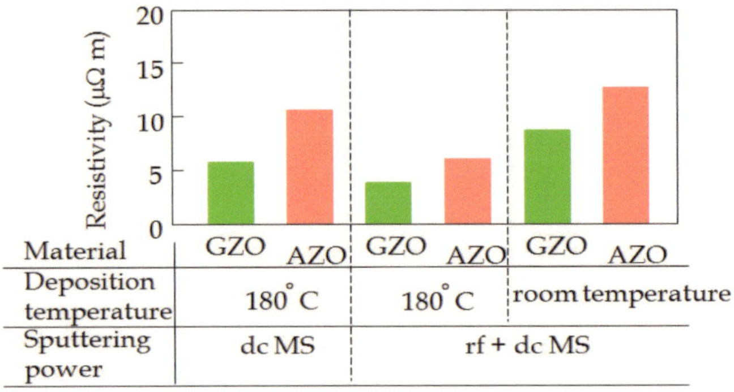

Fig. 6. Comparison of resistivity for ca. 150 nm thick transparent ZnO films formed by dc MS and rf+dc MS

3.3 Visible light transmittance of transparent conductive ZnO film

Alternative materials to the ITO electrodes used for LCDs are required to have transmittance for visible light comparable with that of ITO films. The optical transmittance of the films was analyzed using an optical spectrophotometer (U-4100 UV-Visible-NIR spectrophotometer, Hitachi High-tech. Co. Ltd.). The transmittance of the GZO films in the wavelength range of ultraviolet (UV) to near-infrared (NIR) is shown in Fig. 7(a) The transmittance values presented in this section include the transmittance of the glass substrate. The absorption edges in the spectra are shifted slightly to the shorter side from the wavelength (ca. 370 nm) corresponding to the bandgap of undoped-ZnO according to the Burstein–Moss effect (Moss 1980).

On the other hand, the transmittance in the NIR region was significantly reduced from an increase of reflectance due to the plasma resonance of electron gas in the conduction band in the films with highly density carriers (electrons) of 8×10^{26} to 1.5×10^{27} m^{-3} (Jin et al., 1988; Dong & Fang 2007). The transmittance spectra in the wavelength region of visible light are scaled up in Fig. 7(b). Undulations in the transmittance spectra as a function of wavelength appeared due to optical interference phenomena caused at the surface and at the interface with the glass substrate because the dependency of the 110 nm thick sample on the wavelength differed from those of the ca. 150 nm thick samples. The maximum, minimum and average transmittance of each sample in the range of 400–800 nm were 90.5 - 91.4, 73.1 - 86.4, 87.2 - 87.8, respectively.

The transmittance of a 150 nm thick GZO film formed at 200 °C by conventional dc MS on a 1.1 mm thick Corning #1737 glass substrate was compared with polycrystalline and amorphous ITO films with ca. 150nm thickness formed using a similar sputtering system,

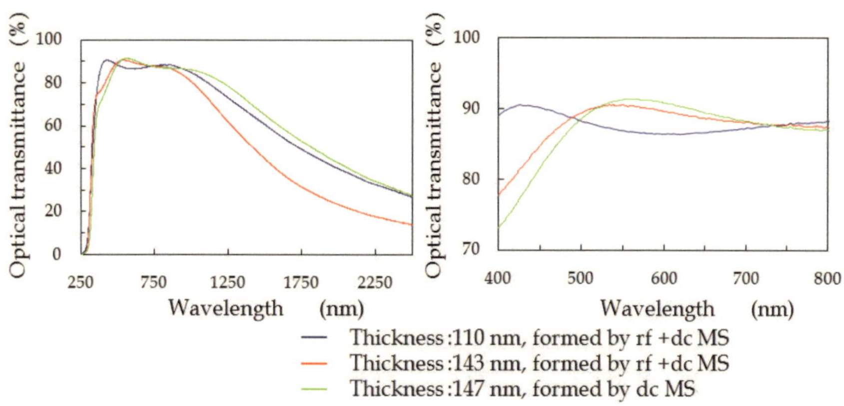

(a) Spectra in the range from UV- to near infrared- wavelength

(b) spectra in the 400 nm -800 nm range of the visible light.

Fig. 7. Comparison of optical transmittance of films formed at 180 °C using rf+dc MS and dc MS on 0.7 mm thick Corning #1737 glass substrates.

and the results are shown in Fig. 8 (Yamamoto, N. et al., 2011a & 2011c). The exact deposition conditions is proprietary and therefore limited to announcement by Geomatec Co., Ltd. Japan. The amorphous and polycrystalline ITO films were formed at approximately 50-70 and 250-300 °C, respectively. The optical transmittance of the GZO film surpassed those of the ITO films in the entire wavelength range of visible light. Visible light in the range of 400-550 nm was transmitted through the GZO film with 2-7% higher transmittance than those of the ITO films.

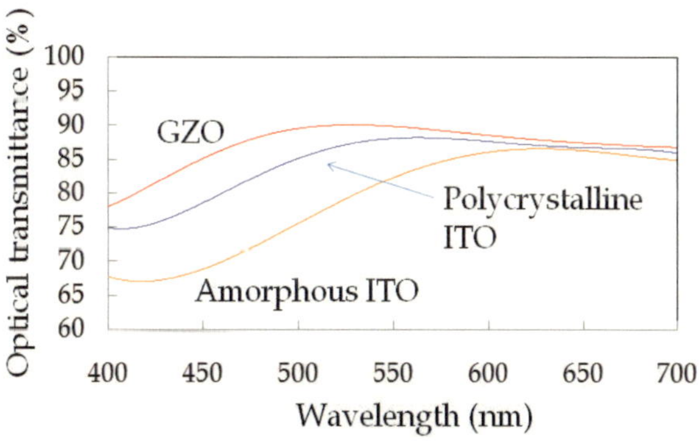

Fig. 8. Comparison of the optical transmittance of transparent GZO and ITO films.

3.4 Residual stress in transparent conductive ZnO film

The development of transparent conductive films that are resistant against external forces and forces induced by thermal processes (maximum temperature: ca. 250 °C) during LCD fabrication requires investigation of the fundamental mechanical characteristics of the films, such as residual stress, thermal stress, strain, Young's modulus, coefficient of thermal expansion, adhesive force and creep. In this section, the residual stress of GZO films is discussed. The conventional optical lever method was applied for residual stress analysis using a HeNe-laser beam at 633 nm (F2300, Flexus Co.). The Si wafer substrate was used as a reflector for the laser beam. The substrate with film was placed on the three projections attached with an equilateral-triangular layout on the surface of the heater plate. The radius of curvature of the Si wafer was measured by detecting the reflected beam. The stress in the film was then derived using the curvature radius, the linear coefficient of thermal expansion (3.34×10^{-6} K^{-1}) and the Young's modulus (160 GPa) of the Si substrate (Wortman & Evans 1965).

The wafer was heated from 25 to 500 °C and subsequently cooled to 25 °C for heat-cycle testing of the film, and the curvature radius of the Si wafer was measured *in situ* during testing. The residual stresses of all the as-deposited films were artificially set to compressive, according to the controlled deposition processing factors. The residual stress in a film is dependent on the degree of energetic particle bombardment, that is, energy striking the condensing film during deposition by magnetron sputtering or RPD with plasma discharge discharge (Thornton, & Hoffman, 1977; Yamamoto, N. et al., 1986).

The residual stress properties of the films were clarified by applying heat-cycle testing. Heat-cycle testings was applied twice to each sample. The samples were heated at a rate of 2.8 °C/min until thermal equilibrium (thermally quasi-static condition) and *in situ* measurements of residual stresses in the films were conducted during heat-cycle testing.

The residual stresses of the films formed at 180 °C using rf+dc MS, dc MS and RPD had following similar behaviors, as shown in Fig. 9 (Yamamoto, N. et al., 2008 & 2010). (I) the residual stress in the films was significantly reduced in the range from 200 to 400 °C during the first heating-up process (step 1); there was an increasing tendency toward strongly compressive residual stress, which changed abruptly to the tensile direction upon heating. This critical temperature of change is dependent on the deposition method of the films. (II) the stress decreased monotonically with cooling from 500 °C (step 2), and (III) the dependence of the stress in each film on the temperature during the second heat-cycle (steps (3) and (4)) almost coincided with that in step (2). These phenomena present evidence that the extrinsic stress components in each film are removed by annealing during the first heating step to 500 °C, and the resulting stress is composed of only thermal stress, i.e., intrinsic stress. The thermal stress is caused by the difference in the linear coefficient of thermal expansion between the GZO film and the Si substrate.

On comparatively thicker GZO films (ca. 500 nm in Fig. 9) formed by dc MS and RPD, the temperature dependencies in step (1) were closer to those in the second cycle testing. The main component of internal stress in such thick films was thermal stress (intrinsic stress), even before heat-cycle testing of the as-deposited film. The GZO films approached the ideal crystalline structure as the thickness increased. The overall temperature dependency change of the residual stresses according with the film thickness corresponds approximately with the increase in crystalline irregularity with the distance from the interface to the substrate. On the other hand, the GZO film thicker than 500 nm formed by rf+dc MS did not provide the same stress behavior in step (1) in the GZO closed to those in steps (2) to (4). High strain or irregular crystalline structures were present until a far distance from the substrate surface of the film.

4. Fabrication of LCD panels with transparent ZnO electrodes on RGB color filters

4.1 Intrinsic weakness of transparent conductive ZnO film in the LCD manufacturing environment

The fundamental properties of GZO films as an alternative to ITO transparent electrodes have been clarified with experimental results. However, the properties required for

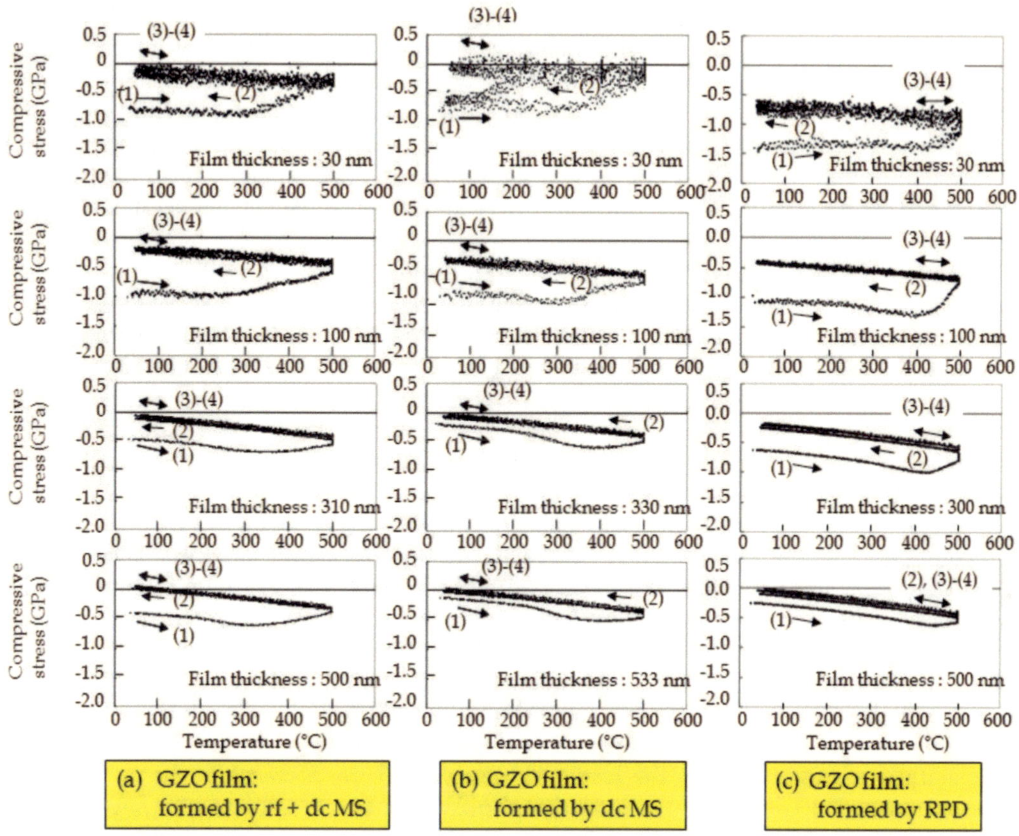

Fig. 9. Comparison of residual stress behavior during heat-cycle testing from room temperature to 500°C in an Ar atmosphere.

application in the production of the LCDs have not yet been discussed thoroughly. ZnO or materials based on ZnO have significant impediments to LCD applications. Such materials have amphoteric properties, so that their films have the extremely low tolerability to the various process conditions and environments used for manufacturing and operating LCD panels, i.e., acidic, basic and highly humid environments. Furthermore, Zn atoms are easily volatilized from materials by heating processes at relatively low temperature such as 300-400 °C. Transparent electrode films are required to endure manufacturing processes at ca. 250 °C for the fabrication of LCDs. A diagrammatic cross-sectional illustration of an LCD is shown in Fig. 10.

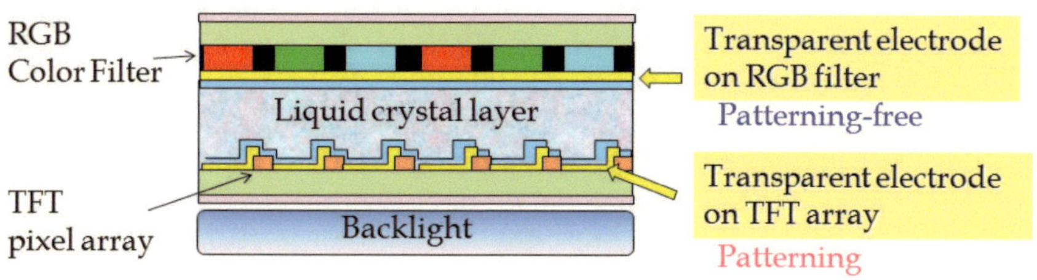

Fig. 10. Schematic cross-sectional view of an LCD.

The transparent conductive films on the RGB color filter is generally used without patterning. On the other hand, the transparent electrode on the TFT pixel array is formed by patterning the film using wet-chemical etching techniques. Therefore, as the first stage of transparent ZnO electrode development, the GZO films were applied as only the transparent electrode on the RGB color filter in an LCD. In this case, the conventional ITO electrode was used on the TFT pixel array. Concurrently, a technique to form the fine-line patterns of the transparent conductive ZnO film was developed using wet-chemical etching.

4.2 Heat resistance against thermal processing during LCD fabrication

Figure 9 should be referred to again for discussion regarding the heat resistance of GZO films, especially with regards to the behavior of 100 nm thick GZO films because this thickness is close to that used for transparent electrodes in LCDs. During the first heating (step (1)) in the heating cycle test, the compressive residual stress started to reduce suddenly at a specific temperature. The temperature range in which a detectable reduction in the compressive stress occurred in each type of film formed by rf+dc MS, dc MS and RPD was 200-250, 250-300 and 350-400 °C, respectively.

The cause of the decrease in the residual stress with heating to higher than the critical temperature was analyzed using thermal desorption spectrometry (TDS; EMD-WA1000S/W, ESCO Co., Ltd.). The dependence of the amount of Zn mass fragment in the TDS mass spectra on the temperature had the highest correlation with stress for various materials subliming from each GZO film, as shown in Fig. 11(a) (Yamamoto. N. et al., 2008). The temperature at which an observable change in the amount of Zn in the three film types due to volatilization was at 200-250, 250-300 and 300-400 °C, respectively (indicated with arrows). Therefore, it was concluded that the volatilization of Zn from the films at these temperatures during step 1 in the heat-cycle test caused the decrease in the compressive residual stress. The correlation indicates that shrinkage of the film volume accompanied the volatilization of Zn, which resulted in a reverse of the temperature-dependence of the residual stress from increasing compressive stress toward tensile stress.

The order among the critical temperatures at which the change of residual stress and the amount of volatilized Zn began to become significant for the three film types has a good correlation with the order among the lengths of the lattice constant or the wurtzite cell volume as shown in Fig. 4. Longer distances between the atoms in the wurtzite structure is likely to cause weakening of the atomic binding forces, which results in lowering of the heat-resistance of the material.

The heat-resistant properties of the three film types derived from these experimental results were in following order: rf+dc MS < dc MS < RPD. The order of the heat-resistance with respect to the electrical characteristics was exactly the same as that with respect to Zn volatilization and the residual stress. The change in the resistivity, carrier concentration and carrier mobility caused by annealing (heat cycle testing) are compared for films formed by rf+dc MS and RPD in Fig. 11(b). The critical temperatures estimated from the electrical characteristics for the three film types were 200-250 °C (rf+dc MS) < 250-300 °C (dc MS) < 350-400 °C (RPD). This order corresponded exactly with the release of compressive residual stress and Zn volatilization in step (1) during the heat-cycle testing (Yamamoto, N. te al 2010).

Therefore, the films formed by dc MS and RPD have heat resistance properties that are suitable for use as an ITO substitute material for optically transparent electrodes in LCDs. In the case of the film formed by rf+dc MS, the heat resistance characteristics were on the borderline for LCD application. It would be necessary to improve the properties of the rf+dc MS film by some process.

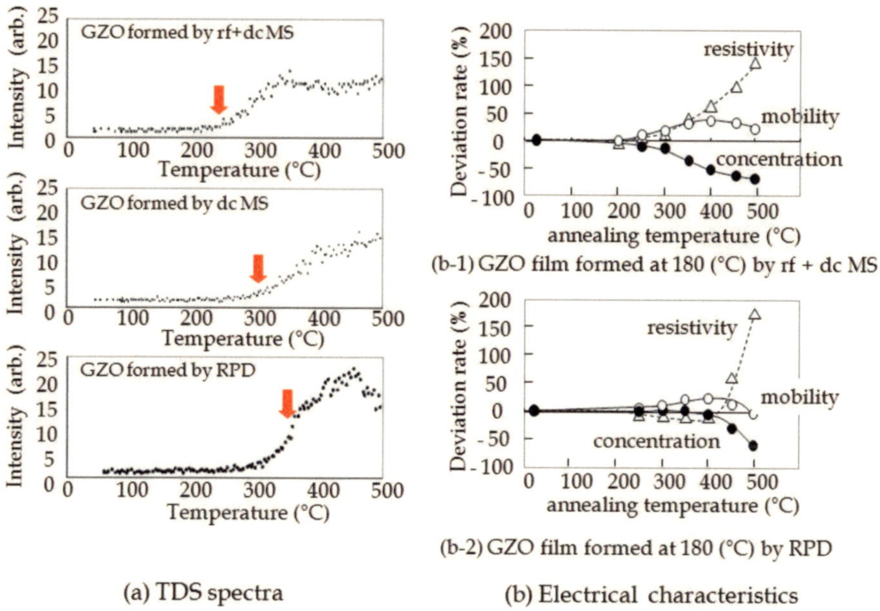

Fig. 11. (a) Comparison of TDS mass spectra for the volatilization of Zn plotted as a function of temperature for 100 nm thick GZO films formed by rf+dc MS, dc MS and RPD at 180 °C. (b) Degradation rates of resistivities, carrier concentrations and carrier mobilities in the films formed by rf+dc MS and RPD at 180 °C as a function of temperature when annealed in flowing Ar (Yamamoto, N. et al., 2010).

4.3 Fabrication of LCDs with transparent ZnO electrode

As the first stage of transparent ZnO electrode development, the GZO electrodes were only applied on the RGB color filter for the manufacture of the LCDs. Conventional ITO films were used as the electrode on the TFT pixel arrays in the LCDs. Among the three film types, the films formed by RPD have the best suited characteristics, such as low electrical resistance and high heat-resistance. However, the current RPD system is not in the usable field for the large size motherglass (1500×1800 mm^2, 2160×2400 mm^2 and 2850×3050 mm^2) used in the production lines of large size LCD TVs, such 6G (generation), 8G and 10G. Therefore, the conventional dc MS technique, which is used for the formation of ITO transparent electrodes used in commercially available LCDs, was used for the fabrication of LCDs with the GZO electrodes on RGB color filters.

150 nm thick transparent GZO films were formed on the RGB color filters at 150 °C using dc MS. The LCDs were fabricated without change to the process flow, except for the GZO deposition. The fabrication process flow is generally kept an industrial secret. Therefore, the order of the most basic process steps for manufacturing the LCDs are shown as a reference in Fig. 12 (Yamamoto, N. et al., 2010).

The fabrication of the RGB color filter side module and the TFT pixel array side module were carried out separately from each other. After the fabrication reached the final step in each side-module process flow, the two side modules are combined with each other. Consequently, the liquid crystal material is poured into the gap between both substrate sides. LCD fabrication is accomplished through hinging the bezel and attaching various

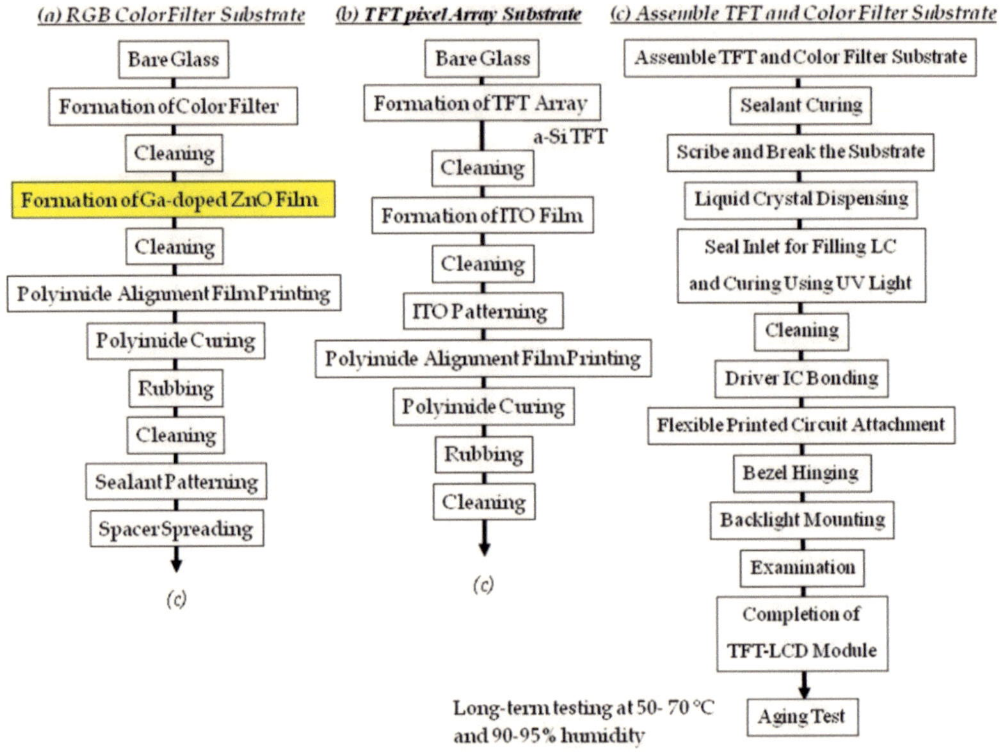

Fig. 12. Basic process flow for manufacture of an LCD with GZO on the RGB color filters.

controlling circuits. Finally the completed LCDs are tested for long term at 50-70 °C in a high humidity (90-95%) environment.

Figure 13 shows the typical display of 3 inch size LCDs and the motherglass before cutting each 3 inch panel part (Yamamoto, N. et al., 2010). The production procedure and process conditions were the same as those for commercially available TFT-LCDs with conventional ITO transparent electrodes, except for the processing step for the formation of transparent conductive GZO films.

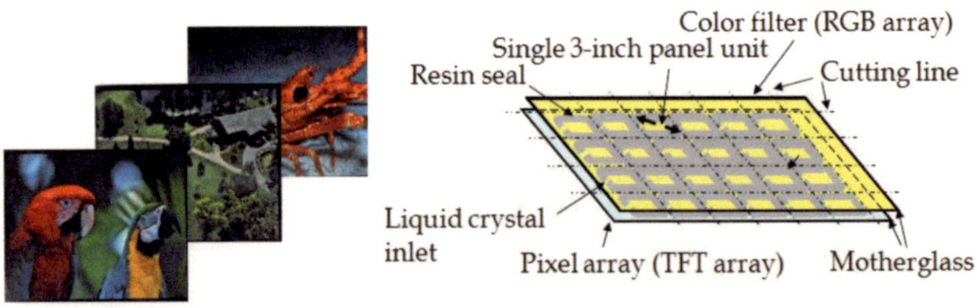

(a) Display pictures of 3-inch LCDs with the GZO transparent electrodes on the RGB color filter

(b) Schematic of 24 sealed 3 in. LCDs sandwiched by the mother glass on both sides.

Fig. 13. (a) Display pictures of 3 inch LCDs with transparent GZO electrodes on the RGB color filters. (b) Schematic of 24 sealed 3 inch LCD sandwiched by the motherglass on both sides.

The displays of 20 inch size LCD TVs with GZO transparent electrodes on the RGB color filters are compared with those of commercially available LCD TVs with ITO electrodes in Fig. 14 (Yamamoto, N. et al., 2011a & 2011c).

Fig. 14. Comparison of the display pictures between 20 inch LCD TVs with GZO and ITO electrodes on the RGB color filter.

The various display properties, such as the contrast ratio of the module, and the chromaticity diagram characteristics, of the LCD TVs with the GZO electrodes were equivalent to those of the LCD TVs with conventional ITO electrodes. The average visible light transmittance of the TV module with the GZO transparent electrode exceeded that of the conventional TV modules with ITO electrodes by 4–5%. These superior characteristics of the GZO module were confirmed by the experimentally measured transmittance of the GZO film in the short wavelength (blue) region of visible light, which exceeded that of polycrystalline and amorphous ITO films (Fig. 8).

The major difference between the 3 inch size LCDs and the 20 inch LCD TVs are the spacers set at the inter-gaps between both side modules. Bead type spacers are distributed between the modules to form a space to inject the liquid crystal, as shown in Fig. 13(b). In the case of the 20 inch LCD TVs, column (rib) type spacers were formed on the plates of the RGB color filters using a conventional photolithography patterning technique. The process steps and the detailed flow for the manufacture of LCD TVs does deviate from that presented in Fig. 12; however, this is proprietary knowledge and the intellectual property of each company.

The display performance of the completed 3 inch LCDs and the 20 inch LCD TVs did not degrade even after long-term operating tests for over 1000 h (on gong) at 50-65 °C in a high humidity (90-95%) environment. Prior to the long-term operation testing, there was some concern regarding degradation of the display performance due to the high humidity environment because the resistance and transmittance of the transparent ZnO film deteriorates under high humidity (Nakagawa, et al., 2009). However, the long-term operation testing results for the LCDs confirmed that transparent ZnO electrodes sealed in the modules of LCDs were not affected by the external environment.

5. Formation of transparent ZnO electrodes with fine-line patterns on TFT pixel arrays

Transparent films formed on the TFT pixel array must be finely patterned using wet-chemical etching techniques. GZO is amphoteric and has low resistance to such agents; therefore, there are no reports on the formation of transparent ZnO patterns with line-widths finer than approximately 10 μm. Thus, the fine-patterning of GZO thin films by wet-chemical etching is a significant developmental challenge.

Fine-patterns of transparent conductive ZnO films were formed by the process flow shown in Fig. 15 (Yamamoto, N. et al., 2011a; 2011b & 2011c). At first, a positive-type novolac-photoresist layer was formed on a transparent conductive ZnO film using a spin-coating applicator for the formation of fine-patterned films. A contact aligner (UV-light exposure system) was used to print the designed mask patterns onto the photoresist layer. The UV-exposed parts in the photoresist layer were removed with a photoresist-developer containing tetramethyl ammonium hydroxide (TMAH, $(CH_3)_4NOH$). The unexposed parts remained as photoresist patterns. Patterning of the GZO films was conducted using a wet-chemical etching technique with organic-acidic etchants; carboxylic acid agents with the photoresist-patterns as the etching mask. The patterning-mask photoresist remaining on the ZnO material patterns was subsequently removed using a photoresist-stripper containing an amine (ELM-R10-F22, produced by Mitsubishi Gas Chemical (MGC) Co., Inc.). The key to realize fine-patterns with widths of a few micrometers is fulfilled by the following three requirements. (1) Development or choice of a suitable developer for the photoresist, i.e., alkaline solution. An aqueous solution prepared by the addition of TMAH to deionized pure water was used as the developer. (2) Development or choice of an acidic solution (etchant) for wet-etching (patterning) of the transparent conductive ZnO films. Solutions based on organic acids and inorganic acid with different pH values were prepared in this work. (3) Optimization of the photolithography and etching processes for GZO films.

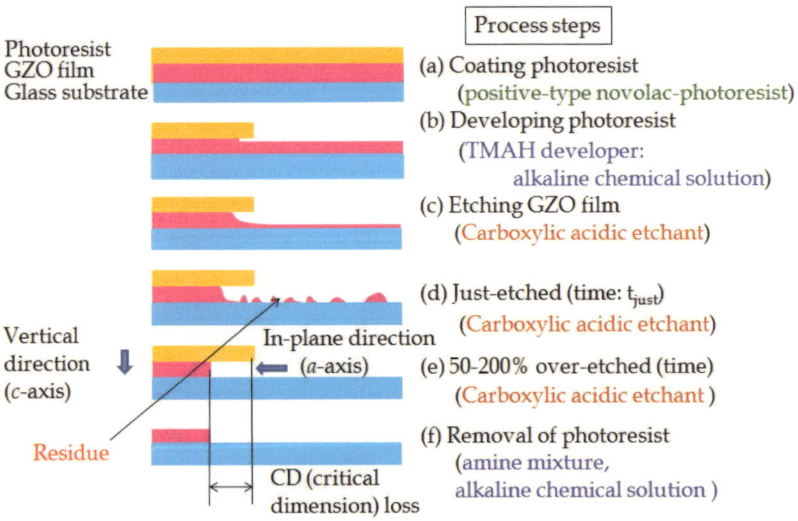

Fig. 15. Process flow for patterning GZO films using wet-chemical etching techniques

The technology for fabricating fine-patterns of transparent conductive ZnO films developed in this work is summarized as follows. The key factors are (1) the development or selection of appropriate chemical agents, and (2) the determination of an appropriate pH range for each agent (Yamamoto, N. et al., 2011a; 2011b & 2011c).

1. Alkaline chemicals for photolithography process
   a. Developer: TMAH aqueous solution pH: 12.0-13.0
   b. Stripper: Amine chemical solution pH: 11.0-12.0 (ELM-R10-F22, MGC Inc.)
2. Acidic agent for wet-chemical etching (patterning) pH: 5.5-6.8
   a. Organic acidic etchants (carboxylic acid series) prepared by MGC Inc. (Chemical composition: nondisclosur. Not for sale at the present stage)
   b. Inorgabic acidic etchants prepared by Naoki Yamamoto
3. Optimization and control of conditions related to lithography process (UV light exposure, baking of photoresist) and etching process (temperature of etchant, rinsing and drying after etching)

Fine-patterns with line and space widths of a few micrometers were successfully formed using this technology, as shown in the optical micrographs in Fig. 16 (Yamamoto, N. et al., 2011a; 2011b & 2011c).

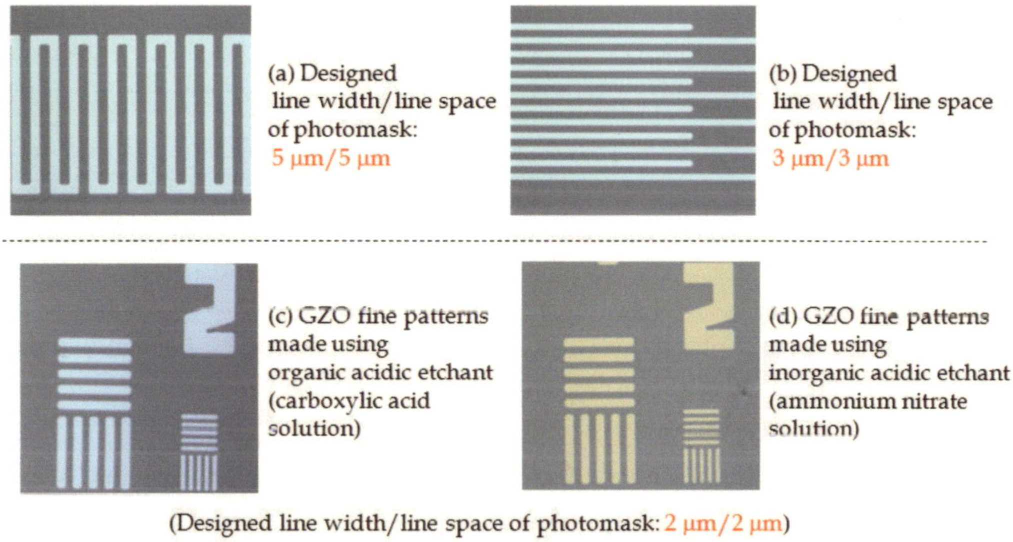

Fig. 16. Typical GZO patterns formed using a weakly acidic etchant (pH 5.5 - 6.8).

Micrographs (a) and (b) show typical meander patterns with line and space widths of 5 μm and comb-like line and space widths of 3 μm, respectively.

Figure 16(c) and (d) shows dense 2 μm line and space patterns of 120 nm- and 150nm- thick GZO film formed successfully using the developed wet-chemical etching technique.

It should be noted here that the patterns shown in Fig. 16 had smooth edges and no residues were observed on the glass substrates after etching of the GZO films. In addition, the 2 μm width line patterns of the GZO film are comparable to the narrowest ITO transparent electrodes formed by wet-etching techniques using leading-edge proximity exposure systems available in commercial LCD production lines.

6. Summary

GZO and AZO films are considered to be the most suitable conductive ZnO films for the current purpose of LCD fabrication. Particular focus was made on the GZO films, because the resistivities of GZO films are lower than those of AZO films.
GZO films have the wurtzite structure as with ZnO films. The resistivity of the GZO films reached approximately 2.4 μΩm, which is in the range of 1.7–3.8 μΩm reported for ITO films formed at temperatures lower than 250 °C.
Conductive ZnO films are amphoteric and are therefore susceptible to acidic and alkaline environments. Therefore, the most important issue for the development of transparent GZO electrodes for LCDs is the development of wet-chemical etching techniques to form fine-patterns with widths of a few micrometers.
The fabrication of LCDs includes process steps that require heating up at 200-250 °C. The thermal resistance of GZO films was investigated from the residual stresses behavior of the films and the volatilization of Zn in the films during heat-cycle testing between room temperature and 500 °C. During the first heating step, the critical temperature where significant Zn volatilization from the GZO film begins coincided with the temperature where significant changes begin to appear in the residual stress and electrical characteristics, that is, resistivity, carrier mobility and carrier concentration. The relationship of these changes with the volatilization of Zn caused by heating the films was clarified. The temperature at which significant change occurs is the critical temperature of the GZO film. The critical temperature was dependent on the method of film formation as follows. The critical temperatures of the films were in the order of 250 °C (rf+dc MS film) < 300 °C (dc MS film) < 400 °C (RPD film). Films formed using dc MS and RPD would therefore be resistant to the thermal process steps conducted at a maximum temperature of 250 °C for the manufacture of LCD panels (TV sets). The transmittance of the GZO film exceeded that of the conventional ITO electrode by ca. 2-7% in the short wavelength region of visible light. As the first development stage of the transparent ZnO electrode, ca. 150 nm thick GZO films were used as only the transparent electrodes on the RGB color filters in 3 inch LCDs and 20 inch LCD TVs. Conventional ITO electrodes were used as the transparent electrodes on the TFT pixel arrays. The initial display performance of both the 3 inch LCDs and the 20 inch LCD TVs was maintained even after long-term operation testing (>2000 h) at 50-65 °C in a high humidity (90-95%) environment.
A wet-chemical etching technique was developed concurrently for the formation of fine-line dense patterns with widths of a few micrometer using the amphoteric GZO films as the 2nd stage of transparent ZnO electrode development. This involved (1) development or selection of appropriate alkaline chemicals for photolithographic processing, (2) development of acidic chemicals for the etching of transparent ZnO films, and (3) optimization of the photolithography and etching processes. The suitable pH range of each chemical agent was determined experimentally, and included the following: (1) TMAH photoresist-developer: pH 12.0–13.2, (2) an etchant for patterning the transparent ZnO film: pH 5.1-6.8, and (3) photoresist remover (stripper): pH 11.0–12.0. Fine line dense patterns with 2 μm line/2 μm space widths were successfully fabricated from 50–150 nm GZO films using these selected reagents and the developed technique. AZO fine-line patterns with the same widths could be also formed using the pattering techniques, because the CD loss (the critical dimension loss is defined as the distance encroached in the ZnO conductive films by acidic etchant from the edge of the patterns under the photoresist layer. reffer Fig.15) dependency rates of AZO films on the over-etched times processed using the acidic etchant were the same as those of the GZO films.

7. Future perspective of transparent ZnO electrode technology

It is expected that LCD TVs manufactured in near future will have the transparent ITO electrodes replaced with transparent ZnO electrodes. Furthermore, the successful development of fine-patterning technology for transparent ZnO films should realize the application of transparent ZnO electrodes to not only LCD TVs that are operated by the VA (vertical alignment) method, but also LCD TVs driven by the IPS (in-plane switching) method. In addition, it is expected that this fine-patterning technology will enable the application of transparent ZnO electrodes to not only LCDs, but also organic LEDs (OLEDs), touch panels LEDs and solar batteries. This technology may also contribute to the development of discrete electronic devices and integrated circuits consisting of ZnO or derivative semiconductors.

8. Acknowledgement

We would like to thank Mr. Ujihara, A., Dr. Ito, T. and co-researchers of Geomatec Co., Ltd. for preparation of the GZO films using a magnetron sputtering system, Mr. Maruyama, T. and co-researchers at Mitsubishi Gas Chemical Co., Inc. for preparation of the chemicals for patterning the GZO films, and Mr. Hokari, H. of Ortus Technology Co., Ltd., for fabrication of the 3 inch LCD panels. We are also grateful to Mr. Morisawa, K. and Mr. Osone, S. of Kochi university of Technology for their support, which aided in the development of the technique for froming the fine-patterns of the ZnO transparent films.
This work was made possible by grants for the Development of Indium Substitute Materials for a Transparent Conducting Electrode in the Rare Metal Substituted Materials Development Project from the Ministry of Economy, Trade and Industry and the New Energy and Industrial Technology Development Organization of Japan.

9. References

Dong, B.-Z.; Fang, G.-J.; Wang,, J.-F.; Guan, W.-J.; & Xing-Zhong Zhao (2007), Effect of thickness on structural, electrical, and optical properties of ZnO: Al films deposited by pulsed laser deposition, *Journal of Applied Physics*, Vol. 101, 033713-1 - 033713-7, ISSN: 0021-8979.

Homma, S.; Miyamoto, A.; Sakamoto, S.; Kishi, K.; Motoi, N. & Yoshimura, K. (2005), Pulmonary fibrosis in an individual occupationally exposed to inhaled indium-tin oxide, Europian *Recipiratory Journal*, Vol. 25, No. 1, pp. 200–204, ISSN: 0903-1936.

Jin, Z. C.; Hamberg, I. & Granqvist, C. G. (1988), Optical properties of sputter-deposited ZnO:Al thin films, *Journal of Applied Physics*, Vol. 64, No. 10, pp. 5117-5131, doi:10.1063/1.342419, ISSN: 0021-8979.

Kempthorne, D. & Myers, M. D. (January 12, 2007), *Mineral Commodity Summaries 2007*, United States Government Printing Office, pp. 78-79. prepared by Carlin, J. F. Jr. U.S. Geological Survey, ISBN-10: 0160778956, ISBN-13: 978-0160778957. http://minerals.usgs.gov/minerals/pubs/mcs/2007/mcs2007.pdf

Moss, T. S., (1980), Theory of intensity dependence of refractive index, *Physica Status Solidi (b)*, Vol. 101, No. 2, pp. 555-561, ISSN: 0370-1972

Nakagawara, O.; Kishimoto, Y.; Seto, H.; Koshido, Y.; Yoshino, Y.; Makino, T. (2009), Moisture-resistant ZnO transparent conductive films with Ga heavy doping, *Applied Physics Letters*, Vol.89, No.9, pp. 091904 - 091904-3, ISSN: 0003-6951.

Ofuji, M.; Inaba, K.; Omote, K.; Hoshi, H.; Takanishi, Y.; Ishikawa, K. & Takezoe, H. (2002), Grazing incidence in-plane X-ray diffraction study on oriented copper phthalocyanine thin films", Japanese Journal of Applied Physics, Part 1, vol. 41, No. 8, pp. 5467–5471, ISSN: 0021- 4922.

Shin, S. H.; Shin, J. H.; Park, K. J.; Ishida, O. & Kim, H. H. (1999), Low resistivity indium tin oxide films deposited by unbalanced DC magnetron sputtering, *Thin Solid Films*, Vol. 341, issues 1-2, pp. 225 - 229, ISSN: 0040-6090.

Thornton, J. A. & Hoffman, D. W. (1977), Internal stresses in titanium, nickel, molybdenum, and tantalum films deposited by cylindrical magnetron sputtering, *Journal of vacuum science and technology*, Vol. 14, pp. 164- 168, ISSN 0022-5355.

Wakeham, S. J.; Thwaites,M. J.; Holton, B. W.; Tsakonas, C.; Cranton, W. M; Koutsogeorgis, D. C. & R. Ranson (2009), Low temperature remote plasma sputtering of indium tin oxide for flexible display applications, *Thin Solid Films* doi:10.1016/j.tsf.2009.04.072, ISSN: 0040-6090.

Wortman J. J. & Evans, R. A. (1965), Young's Modulus, Shear Modulus, and Poisson's Ratio in Silicon and Germanium, *Japanese Journal of Applied Physics*, Vol. 36, pp. 153 - 156, doi:10.1063/1.1713863, ISSN: 0021-8979

Yamada, T.; Miyake, A.; Kishimoto, S.; Makino, H.; Yamamoto, N. & Yamamoto, T. (2007) Low resistivity Ga-doped ZnO thin films of less than 100 nm thickness prepared by ion plating with direct current arc discharge, *Applied Physics Letters*, Vol. 91, 051915, doi:10.1063/1.2767213 (*3 pages*). ISSN : 0003-6951.

Yamamoto, N.; Kume, K.; Iwata, S.; Yagi, K. & Kobayashi, N. (1986), Fabrication of highly reliable tungsten gate MOS VLSI's, *Journal of the Electrochemical Society*, Vol. 133, No.2, pp. 401-407, ISSN: 0013-4651.

Yamamoto, N.; Yamada, T.; Miyake, A.; Makino, H.; Kishimoto, S. & Yamamoto, T. (2008), Reletionship between residual stress and crystallographic structure in Ga-doped ZnO film, *Journal of the Electrochemical Society*, Vol. 155, No. 9, pp. J221-J225, ISSN: 0013-4651.

Yamamoto, N.; Makino, H.; Yamada, T.; Hirashima, Y.; Iwaoka, H.; Ito, T.; Ujihara, A.; Hokari, H.; Morita, H. & Yamamotoa, T. (2010), Heat resistance of Ga-doped ZnO thin films for application as transparent electrodes in liquid crystal displays, *Journal of the Electrochemical Society*, Vol. 157, No. 2, pp. J13-J20, ISSN: 0013-4651.

Yamamoto, N.; Makino, H.; Osone, S.; Ujihara, A.; Ito, T.; Hokari, H.; Maruyama, T. & Yamamoto, T. (2011a), Development of Ga-doped ZnO transparent electrodes for liquid crystal display panels, , *Thin Solid Films*, doi: 10.1016/j.tsf.2011.04.067, Article in press, ISSN: 0040-6090.

Yamamoto, N.; Makino, H.; Sato, Y. & Yamamoto, T, (2011b), Controlled formation of ZnO fine-pattern transparent electrodes by wet-chemical etching, ECS Transactions, Vol. 35, Issue 8, pp. 165 - 172, ISSN: 1938-5862.

Yamamoto, N.; Makino, H.; Sato, Y. & Yamamoto, T (2011c), Development of Ga-doped ZnO Transparent Electrodes as Alternatives for ITO Electrodes in Liquid Crystal Displays, the *SID 2011 digest of technical papers*, pp. 1375-1378, ISSN: 0097-966X.

Yamamoto, T.; Yamada, T.; Miyake, A.; Makino, H. & Yamamoto, N. (2008), Ga-doped zinc oxide: An attractive potential substitute for ITO, large-area coating, and control of electrical and optical properties on glass and polymer substrates, *Journal of the Society for Information Display*, Vol. 16/7, pp. 713-719, ISSN: 1071-0922.

3

Inkjet Printing of Microcomponents: Theory, Design, Characteristics and Applications

Chin-Tai Chen
Department of Mechanical Engineering,
National Kaohsiung University of Applied Sciences, Kaohsiung
Taiwan, ROC

1. Introduction

Inkjet printing technology has been invented and used in the form of continuous jets for typewriting and recording over 50 years. However, this kind of printing technique becomes truly popular within twenty years for the public at homes and offices greatly thanks to the consumer products of desktop inkjet printers developed and marketed in the middle of the 1980s. Those leading manufacturing companies including Hewlett-Packard (HP), Canon, and Seiko Epson *et al.* have been celebrated around the world for the evolution of the modern digital printers since then. Compared to the mechanical typewriters in the past, the new inkjet printers perform the printing by dye or pigment-based ink droplets jetted through micro-electro-mechanical actuators, which feature non-impact and digital-control merits, thereby generating less noise and power consumption. Furthermore, as operating in called drop-on-demand (DOD) mode, their droplets with sub-nanoliter volume form the dot-matrix patterns onto medium surfaces, rendering static images for human sight with resolution from low to more than 1200 dot per inch (dpi).

To achieve high quality ink printing on various substrates, dispensing the micro droplets precisely from nozzles to underlying medium propose some technical difficulties concerning fluidic issues such as the variations in droplet volume, flight direction, and deposition morphologies. In terms of microfluidic flow, a stream of micro droplets significantly undergo high speed flying in air, impacting on a solid substrate, forming and drying above a substrate surface. Through the whole process, the issues about transition and stability of flow from liquid channels to individual droplets, and vice versa, will be frequently encountered in theory, which need to be delicately dealt with in advance of various applications. For instance, undesirable satellites from main droplets may be generated due to Rayleigh instability; similarly, it is possible to form irregular bulges within a lengthy liquid channel on a non-penetrable substrate (*e.g.*, glass substrate) instead of porous paper. Moreover, nonuniform evaporation can be caused in nature for volatile sessile droplets placed on flat surfaces, in which the called coffee-ring effect often seen for diluted fluids (Deegan *et al.*, 1998) should be avoided for the requirements of uniform thickness. Besides the conventional dye and pigment inks, almost the materials in solution involve the complexity of evaporable droplets, when applying to the inkjet printing processes that rely heavily on the full understanding of droplet behaviors on various surfaces from wet to dry stages.

Nevertheless, in the recent years, much attention of studies in academics and industries as well is paid to the uses of various materials in the inkjet printing processes, including nanoparticle metals (Szczech *et al.*, 2002; Kang *et al.*, 2010), bio-chemicals (Busato *et al.*, 2007; Fuchs *et al.*, 2011), colloidal polymers (MacFarlane *et al.*, 1994; Biehl *et al.*, 1998; Jeon *et al.*, 2005; Chen *et al.*, 2008), coloring and light emitting materials (Bale *et al.*, 2006; Chen *et al.*, 2010; Chang *et al.*, 2011), transistors and semiconductors (Sirringhaus, et al, 2000; Han *et al.*, 2009; Hinemawari *et al.*, 2011), and so forth. Compared to conventional two deposition techniques, *i.e.* physical vapor deposition (PVD) and chemical vapor deposition (CVD), this current inkjet printing has been considered as alternative third one that can be termed here droplet vaporization deposition (DVD). The DVD method (Chen *et al.*, 2011) further provides capability of direct patterning on substrates without the employment of photolithography, and therefore becomes a very promising green technology that will consume fewer amounts of material and power with minor waste compared to both the formers as mentioned earlier. Therefore, using the inkjet printing as a digital microfabrication tool, many associated manufactures in the world have developed the inkjet printing platforms and equipments from the laboratory to commercial grade, for example MicroFab Inc. (www.microfab.com) and Dimatix Inc. (www.dimatix.com), in order to provide the required technology services for research and pilot production.

With the above inkjet printing materials and equipments required, many different processes and strategies have consequently been followed to directly print specific products by applying this tooling method. For example, the manufacturing apparatus of inkjet-printed color filter that has been one of key components for liquid crystal display (LCD) was first proposed and patented by Canon Company (Satoi, 2001). Primary three colors of red (R), green (G), and blue (B) can be firmly formed on the receptor layer of a substrate through a series of RGB coloring, ink fixing, and curing of receptor layer, where the substrate requires film coating for absorbing the color inks prior to inkjet printing. Based on the this kind of application, more persisting efforts were made within recent years about the improvement of ink formulas (Kim *et al.*, 2009; Chang *et al.*, 2011), surface treatments and characteristics (Koo *et al.*, 2005; Chen *et al.*, 2010), and large-area and active lighting (Bale *et al.*, 2006). In a similar development way, more extensive studies for inkjet printing applications were also directed to other fields including biochips (Xu *et al.*, 2005; Gutmann *et al.*, 2005), optical microlenses (Nallani *et al.*, 2006; Chen *et al.*, 2009), microelectromechanical systems (Fuller *et al.*, 2002; Cho *et al.*, 2006; Alfeeli *et al.*, 2008), and electronic components (Lee *et al.*, 2005; Scandurra *et al.*, 2010; Perelaer *et al.*, 2010). Among them, many of the fundamental processes (*e.g.*, surface treatments and droplet depositions) and components (*e.g.*, inkjet-printed polymers and conductors) are in common with the present display applications, in particular the polymer light-emitting and liquid crystal displays (Katayama, 1999; Mentley, 2002; van der Vaart *et al.*, 2005). Therefore, full understanding of inkjet printing microcomponents is a substantial route to produce the key components for the applications of display based on the inkjet printing techniques.

This chapter will comprehensively report the microfabrication techniques and applications for various microcomponents via inkjet-printing processes over decades, especially used for the present and future liquid crystal displays. It is organized briefly as follows. The background of current research and development about the inkjet printing technique is introduced here in Section 1, followed by the technical strategy of inkjet printing processes detailed with the concentrated respects of critical requirements for materials and facilities in Section 2. Section 3 addresses the critical design issues of microfabrication such as the position accuracy and morphology formations required to be satisfactory for realizing mass product. Section 4 describes the discovered characterization of droplet depositions involved mostly with liquid evaporation, solute deposit, and patterns in geometries. Section 5 presents the concept-proof applications of the inkjet-printing processes, elaborating their present technical progresses and challenges in the future, which are intimately related to the display components including color filters, polymer light emitting diodes, microlenses and backlight planes, conductive lines and electrodes, transistors and integrated circuits. Finally, we draw a conclusion in Section 6.

2. Strategy of inkjet printing processes

To fulfill the specific requirements of applications, basic inkjet printing processes should be comprehensively designed and realized with four major elements of system implementation: ink materials, substrate properties, droplet generation, electromechanical platforms and printing algorithm. As shown in Figure 1, each element involves different technical considerations and plays a critical role of a whole inkjet printing system for realizing the process. Those key elements of fulfilling the inkjet printing components will be thoroughly described and discussed in the next paragraphs.

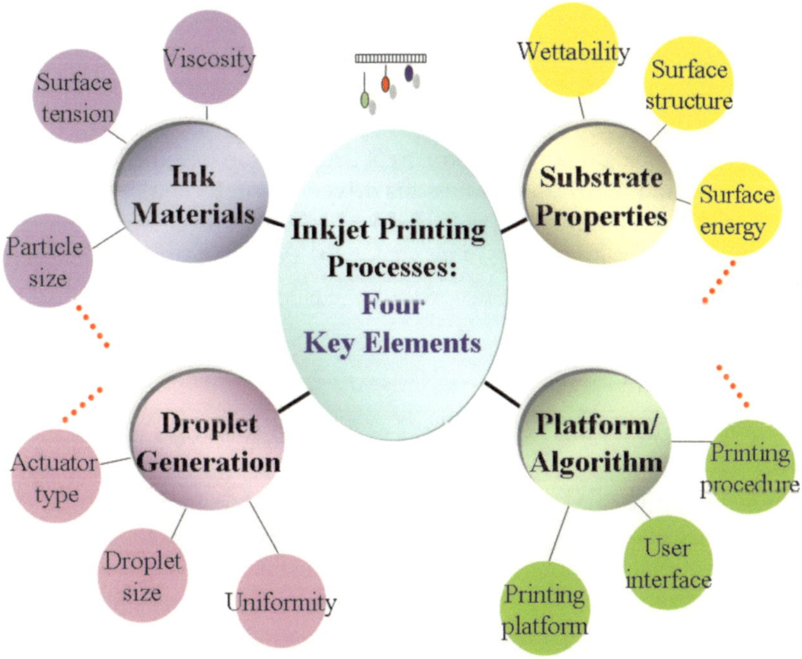

Fig. 1. Major elements of system implementation in strategy of inkjet printing processes including ink materials, substrate properties, droplet generation, and electromechanical platform combined with printing algorithm.

2.1 Ink materials

For any application, appropriate selection of inks is firstly important as the primary consumable during inkjet printing, which is a liquid solution generally composed of solvent and solute. To be stored stability within a printhead reservoir, the solid contents (*i.e.*, solute ~ 5-25 wt%) of ink materials have to be uniformly dispersed in the solvent by additional suitable surfactant, in which no sedimentation is required for the long run duration (de Gans *et al.*, 2004). And the ink viscosity needs to be low sufficiently for jettability of droplet actuators as well. Hence, the surface energy (γ) together with dynamic viscosity (μ) of ink solutions depending on various printheads are typically in the range of 25-55 mN/m and 3-10 mNs/m² (cp), respectively, where pure water has γ of ~72 mN/m and μ of 1cp at 20 °C. The sub-micrometer size of suspended particles, and in particular pigment and metal particles, is also limited to extent that is determined by some thresholds (*e.g.*, filter cavity and nozzle neck) inside the micro liquid channels of printheads.

Besides, heating issues of ink materials should be considered when thermal bubble printheads are used instead of dominate piezoelectric ones. For instance, some biological materials (*e.g.*, DNA and proteins) and polymers (*e.g.*, UV-curable photoresists) may be sensitive to the elevated temperature (up to over 200°C) during tens of microseconds of heating pulses, although a few studies on epinephrine, collagens (Chiu *et al.*, 2006) and UV color-resists (Chang *et al.*, 2004) were demonstrated successfully by bubble inkjet printing. More different types of ink materials, for example, the conductive nanoparticle metals or polymers in bubble jetting (Shaw & Seidler, 2001), are still unclear for the heating sustainability when curing and sintering at higher temperature as required. Since thermal printheads are much less expensive than piezoelectric ones (Tseng *et al.*, 2002), it deserves to be further investigated for versatile materials, in view of the low-cost disposable circumstances.

2.2 Droplet generation

Filled with inks of suitable properties, a stream of droplets can be expelled from tiny nozzles of reservoirs on demand when the printhead is driven by electric pulses. Based on different DOD types of printheads (Le, 1998), such as thermal, piezoelectric, electrostatic, and acoustic ones, those droplets generally undergo morphological transition due to interplay of viscous force and surface tension, from initial to necking stages as demonstrated in Figure 2. Thus the volume of formed droplet is estimated as

$$V = \pi d^3/12 + \pi d^2 h/12 \tag{1}$$

where d is the diameter and h is the height for the drop, respectively. In principle, the larger surface energy liquid possesses during ejection, the smaller drop (V) is formed by shorter height (h), thus producing finer width of a line on the substrate. In the practical case of h=6d, for example, the final spherical diameter of droplet becomes 1.5 times the original size of main droplet, which is determined largely by the nozzle diameter. In general, the size of nozzle in diameter can range from 10 ~ 100 μm, thereby corresponding to the droplet volume with one to hundreds of picoliters (pl=10^{-12} liter). Obviously, higher resolution of the inkjet printing is therefore yielded by reducing the printhead nozzle with smaller diameter.

In fact, except for printing resolution, the uniformity of individual droplets in directionality and sizing is another considerable issue for precise manufacturing of microcomponents (Chen *et al.*, 2001). Positioning accuracy together with morphological formation of droplet deposition that will be discussed in next Section 3 can be significantly influenced by the deviation from the targeted values. And this technical difficulty, in some situations permitted (Chen *et al.*, 2010), may be compensated in terms of tuning substrate properties and printing algorithms as follows.

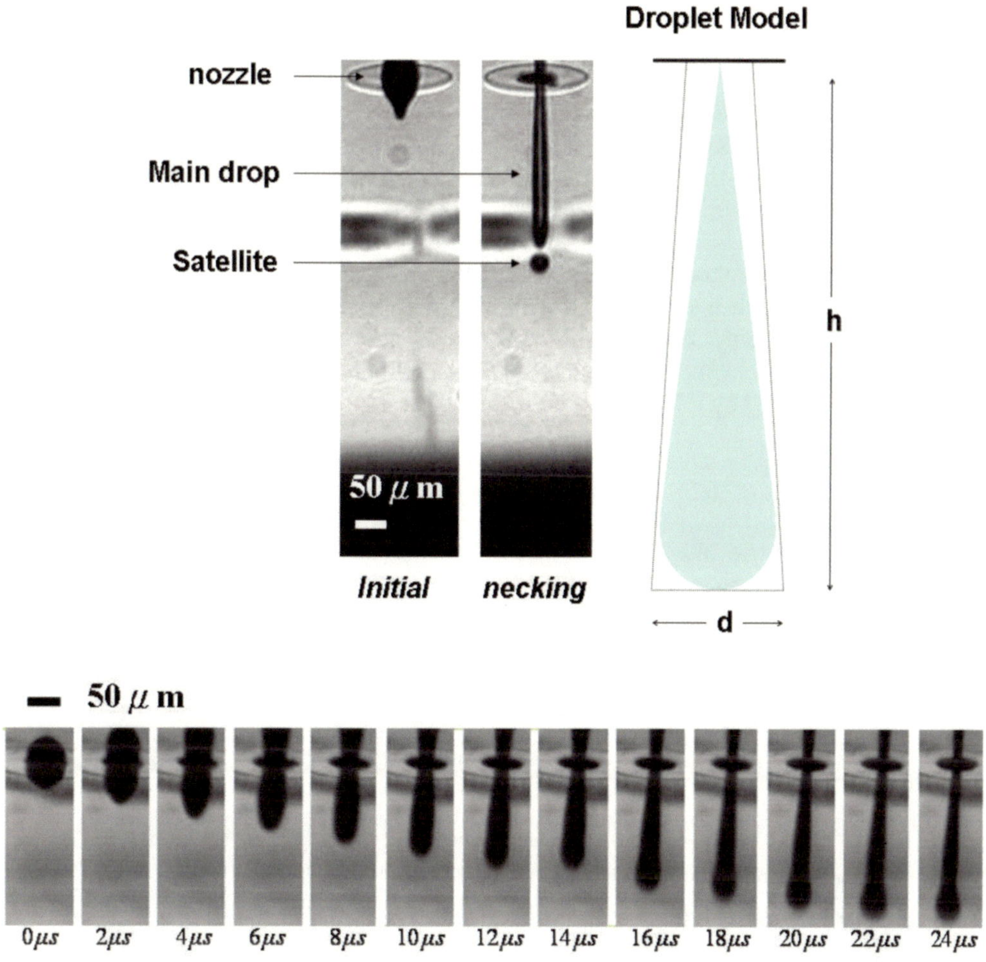

Fig. 2. (Top) Typical jetting process of a micro droplet generated through a tiny nozzle, from initial to necking stages; (Bottom) dynamic evolution of droplet formation in tens of microseconds.

2.3 Substrate properties

The conventional inkjet printers use regular paper made of fibers with porous top surface to absorb ink droplets. However, as illustrated in Figure 3(a), a glass surface can generate different deposition patterns from that on paper for the same ink, since free surface fluid flow is induced on such a solid surface. For a homogeneous surface, a deposited sessile droplet obeys a classic Young-Laplace (Y-L) relation as below (Pujado *et al.*, 1972; Chen *et al.*, 2008):

$$\gamma_{sg} - \gamma_{sl} = cos\theta_c \times \gamma_{gl} = cos\theta_c \times \frac{\Delta P}{\left(R_1^{-1} + R_2^{-1}\right)} \quad (2)$$

where γ_{sg} , γ_{sl} , and γ_{gl} are the surface tensions of solid-gas (*s-g*), solid-liquid (*s-l*), and gas-liquid (*g-l*) interfaces, respectively; θ_c is the contact angle of the *s-l* interface, ΔP is the pressure difference across the *g-l* interface, and R_1 , R_2 are the two principal radii of the droplet curvature.

However, a different story can be given by many research groups for droplets on heterogeneous surfaces (Kim *et al.*, 1995; Gau *et al.*, 1999; Anton *et al.*, 2000; Lenz *et al.*, 2001; Bao *et al.*, 2002), where the use of minimal free energy for self-assembly to form stable shapes was investigated in theory and experiments. For example, as demonstrated in Figure 3(b), pure water droplets spontaneously formed (self-assembled) the circular and stripe shapes on hydrophilic (glass) domains only of the structured surfaces, on which the glass were pre-patterned with hydrophobic (Teflon) domain by photolithography. Based on this concept of self-assembly on structured surfaces, stable solid formations of droplet deposition for striped channel (Chen *et al.*, 2007) and circular shapes (Chen *et al.*, 2009) were experimentally realized by inkjet printing of colloidal inks.

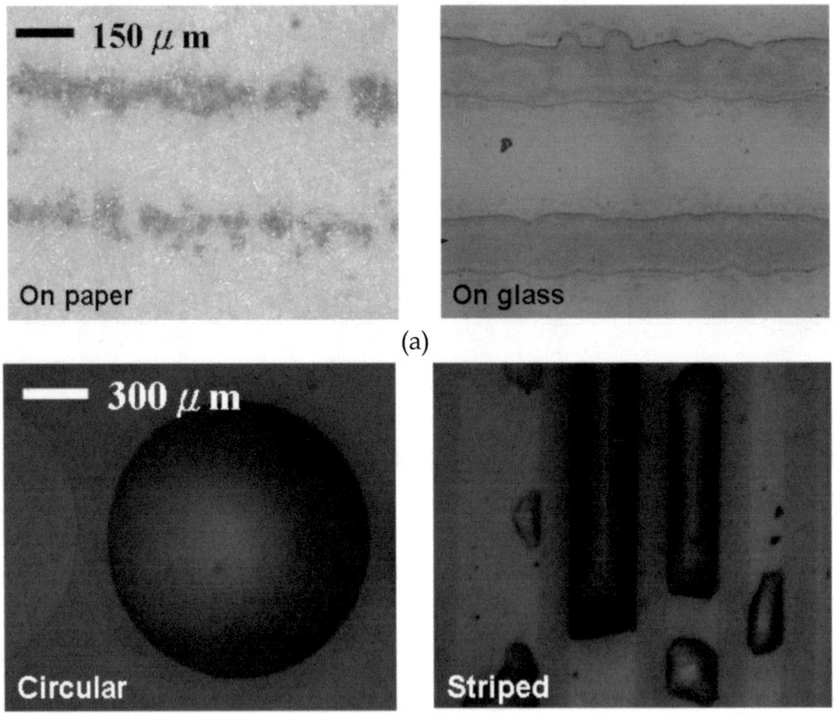

(b)

Fig. 3. (a) Dried formations of color inkjet printing on (porous: left) paper and (solid: right) glass, respectively; (b) wetting topography of pure water droplet formation on circular (left) and striped (right) Teflon-patterned glass surfaces.

2.4 Inkjet printing platform and printing algorithms

In recent years, there have been some microfabrication platforms particularly developed for the inkjet printing processes as mentioned above. As can seen in Figure 4, an inkjet printing platform prototype, laboratory-designed and developed for inkjet printing of color

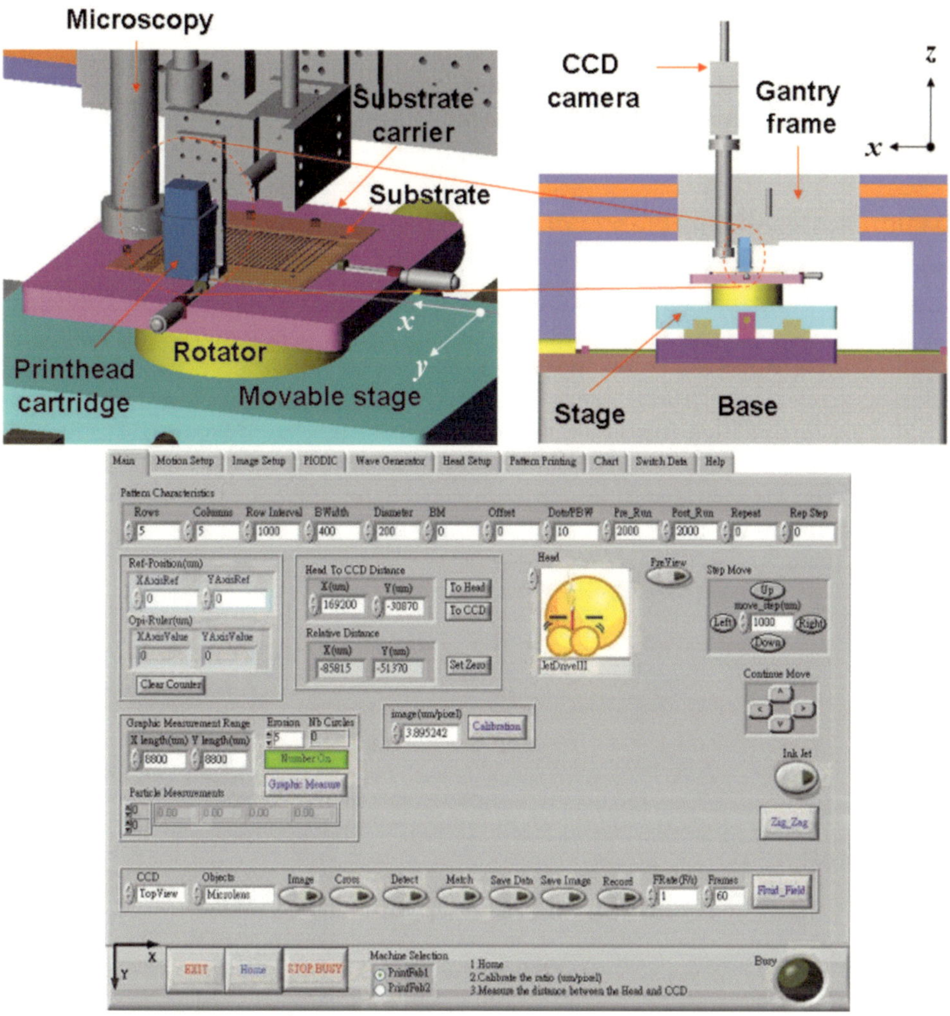

Fig. 4. (Right) Illustration of inkjet printing platform mainly composed of the printhead cartridge, movable stage, CCD camera, base, gantry frame, etc.; (Left) enlarged portion of the platform for highlight of the inkjet printing area; (Bottom) User interface window built by LabView in the controlling computer.

filters (Chen *et al.*, 2010) by Industrial Technology Research Institute (www.itri.org.tw) in Taiwan, illustrates a common configuration primarily comprising the printhead cartridge, substrate carrier, movable stage and rotator, CCD camera and microscopy, base and gantry frame. With the printhead cartridge fixed on the gantry frame, this platform can perform DOD inkjet printing algorithm by moving the motored stage to deliver the substrate in *x*-*y* coordinates so that the imaging pattern is online input and completed through a friendly user interface (PC-based LabView).

To achieve high quality inkjet printing, however, additional pre-printing procedures should be carried out in advance of actual inking the underlying substrates, which includes the *z*-axis gap tuning between the nozzle plate to substrate surface, cleaning of printhead nozzles, alignment and calibration of homing coordinates in *x*-*y* axes (Huang *et al.*, 2009), whereas special care is also paid for rotational registration if a large-area substrate (*e.g.*, LCD color filter) is to be printed here.

After the full preparation of inkjet printing processes, two major design issues of microfabrication associated with precise DVD implementation will be further addressed and explained below.

3. Design issues of microfabrication

3.1 Positioning accuracy

Because of the jet instability in microfluidic nature, as demonstrated in Figures 2 and 3, most of droplets jetted from the nozzles exhibit slightly uncertain deviation of angle, $\Delta\theta$ (*e.g.*, ~1°) from the normal direction of the horizontal plate. This uncertainty poses the issue of positioning precision for droplet deposition, which results in the inaccuracy of the location and width for the thin films formed on the substrates, as depicted in Figure 5. The rectilinear displacement on the surface nearly equals to $H\Delta\theta$, where H (typically, ~ 0.5-1 mm) is the distance between the nozzle and the substrate surface (for example, it can amount to about 17 μm comparable to the desired film width, *i.e.*, 100 μm).

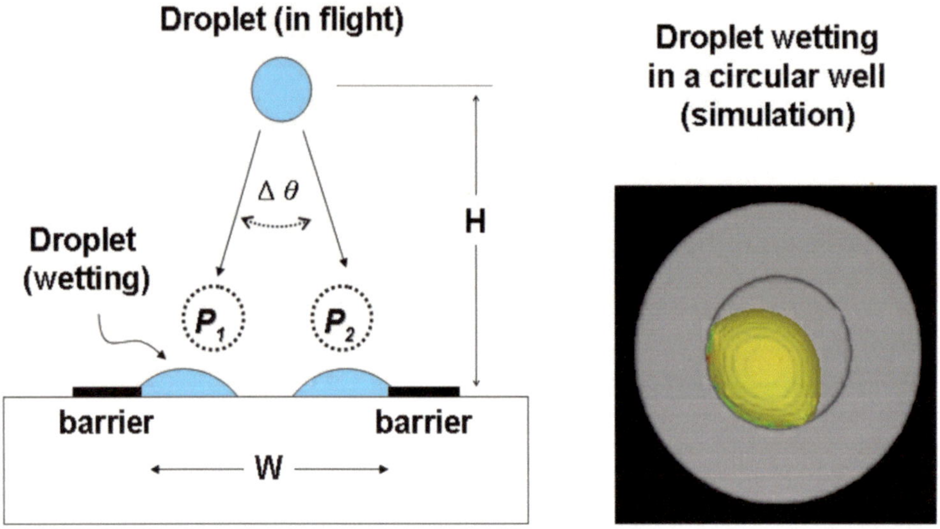

Fig. 5. (Left) Uncertain deviation angle of individual jetted droplet from the normal direction of the horizontal plate; (Right) microfluidic simulation of droplet wetting in a circular well.

Concerning the uneven width as well as positioning uncertainty, those difficulties can be surmounted using heterogeneous (structured) surfaces, as aforementioned previously. They have demonstrated the remarkable effect of registration on wetting and dried positions for droplets on heterogeneous glass substrates (*e.g.*, Teflon coated and patterned on the surfaces). Those wetting droplets substantially exhibit minimum surface tension in the hydrophobic domain by repelling the other ones, leading to self-align along the surrounding rim (Joshi & Sun, 2010). In fact, the wetting rim acts as a "virtual barrier" for droplets to resist flowing across the hydrophobic regimes. However, this energy-patterning strategy using thin-film coating technique suffers from the instability of liquid morphology imposing the limit of liquid volume to the droplets and thus making the thickness of the dried deposition film insufficient and nonuniform. Hence, another approach applying a concept of 'physical barrier' (Chen *et al.*, 2010) was proposed to deal with the above constraints without losing the any positioning accuracy. In this case, as demonstrated in Figure 5, the simulated droplet was capable of self-aligning along the surrounding sidewall to prevent flowing over a circular well. Therefore, for a number of droplets, their allowable collective deviation of $H\Delta\theta$ can be raised to be W at maximum (Chen, 2004).

3.2 Morphology formation

Another concern with the inkjet printed microcomponents involves the control of the morphology of the droplet deposition that is typically complex and varying in different situations. A deposited liquid droplet with volume of V_1 obeying the Y-L relation will simply form a hemispherical shape on homogeneous surface, with characteristic base radius R_b of footprint as

$$R_b = \frac{3cot(\theta_c / 2)}{2\left[3 + tan^2(\theta_c / 2)\right]} \times V_1^{1/3} \tag{3}$$

where θ_c is the contact angle as defined previously. Namely, the droplet footprint radius R_b is proportional to its volume V_1 with scaling exponent of 1/3; additionally, the smaller the contact angle θ_c is, the larger the footprint radius R_b will be.
Furthermore, linear morphology from a number of droplets show more complicated than that of single dot, including individual-drop, scalloped, uniform, bulging, and stacked-coins formations, which are controlled by the delay and drop spacing as well (Soltman & Subramanian, 2008). As the evaporation and curing temperature involved (Biswas *et al.*, 2010; Scandurra *et al.*, 2010), their morphology formations will change dramatically with more complexity in geometry and structure that will be further discussed in next Section 4.

4. Characterization of droplet deposition

Two key dimensionless parameters describe the hydrodynamics of droplet deposition: the Reynolds number (*Re*) and Weber number (*We*). Typically, supposed the values of U ranging from one to ten meters per second (*i.e.*, 1-10 m/s), the Reynolds number (*Re*, a ratio of inertia force to viscous force) gives corresponding values of 2 to 277, which is small sufficiently to render the laminar flow (typical requirement of less than 2300). Also, the Weber number (*We*, a ratio of inertia force to surface-tension force) yields the corresponding values of 0.36 to 320 ensuring the final formation of droplet (Liou *et al.*, 2008). Moreover, the droplet on substrate surface dynamically evolves into three distinct stages in succession: impacting, spreading (and wetting), and drying, as shown in Figure 6. As a result, the droplet deposition of interest for practical applications can be further discussed and characterized in three respects in the following.

4.1 Evaporation deposit

Over the last decades, evaporation kinematics of a pure droplet, without solid content involved, on homogeneous surface were thoroughly investigated, in theoretical and experimental ways, for various conditions (*e.g.*, droplet and substrate materials), in which all mostly featured highly nonlinear (hysteresis) behaviors for the rates of contact angle, base

Fig. 6. Hydrodynamic evolution of a droplet on substrate surface through three distinct stages in succession: (I) impacting (II) spreading, and (III) drying.

radius and height (Bourges-Monnier & Shanaha, 1995; Decker & Garoff, 1997; Erbil *et al.*, 2002; Hu & Larson, 2002; Chen *et al.*, 2006). A the same time, solution droplets that contain either suspended particles or colloidal polymers exhibit more complex fluidic properties (induced flows) due to such non-uniform evaporation (Adachi *et al.*, 1995; Parisse & Allain, 1997; Conway, *et al.*, 1997; *Gorand et al.*, 2004). One significant breakthrough in theories and experiments for evaporation deposit was disclosed by Deegan *et al.* (Deegan *et al.*, 1997), with a derived expression of evaporative flux $J(r)$ under a small contact angle as

$$J(r) \propto \frac{1}{(R-r)^{1/2}} \tag{4}$$

where R is the droplet base radius with contact line fixed on surface, and r is the radial distance from the center of the droplet. Radial liquid flow towards the droplet side is induced during evaporation, thereby carrying the suspended particles within the droplet to its surrounding that was termed coffee-ring (CR) effect. As can be seen in Figure 7, an aqueous PVA (Polyvinyl Alcohol) 20% droplet formed non-uniform surface profile after drying, due to this remarkable CR effect, showing a characteristic concave shape that the perimeter region was much thicker than the center one over three times (*i.e.*, 12 μm/4 μm =3).

Some research efforts to avoid the non-uniform droplet deposit were recently reported (Chang *et al.*, 2004; Chen *et al.*, 2004; Weon & Je, 2010), since the deposit thickness is important for many applications such as biochips, LCD color filters, and light-emitting displays. Among them, special treatment on either homogeneous or heterogeneous surfaces plays a critical role on controlling the final deposit formations during evaporation, because of pinning or de-pinning condition as boundary constraints (Chen *et al.*, 2009).

4.2 Deposit patterns and properties

As a whole, deposit patterns that fulfill the duplication from virtual (digital) codes in computers to real (printed) formations on substrates can be rendered and featured in geometry, including two-dimensional (planar) dot matrix, one-dimensional (linear) stripes, and arbitrary images. Those digital patterns can be dealt with in various formats: either text (*e.g.*, location coordinates) or drawing ones (*e.g.*, bmp, jpg). For example, as shown in Figure 8(a), a typical dot-matrix (150×200) covering a rectangular region can be formed by PU (Polyurethane) 15% droplets on the hydrophobic (Teflon-coated) substrate, in which each individual 173 μm-diameter dot with spacing of 450 μm was inkjet-printed to exhibit uniformly hemispherical. Rather, on hydrophilic glass surface, as demonstrated in Figure

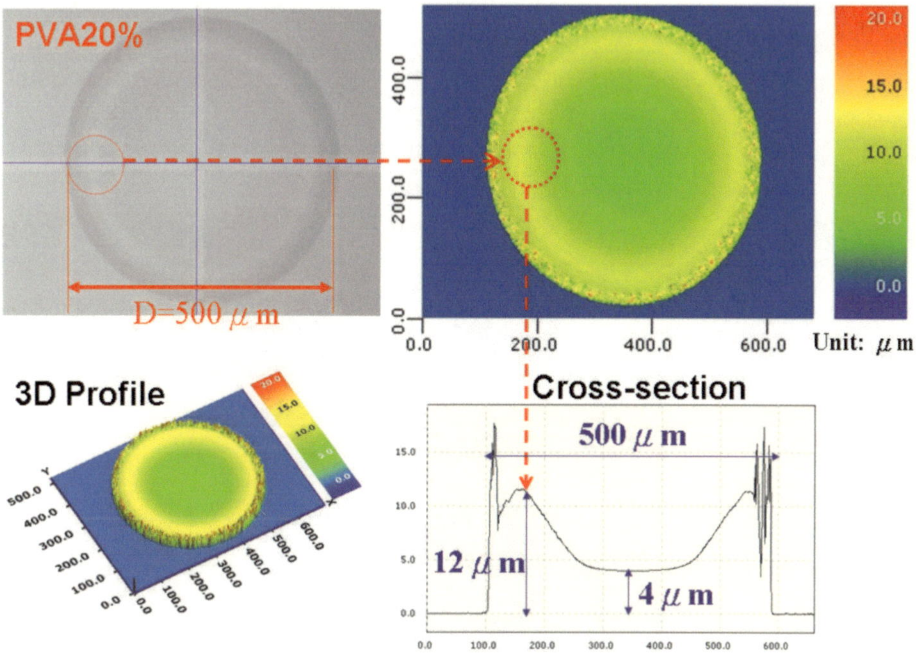

Fig. 7. Evaporation deposit of an aqueous PVA 20% droplet on homogeneous glass characterized with apparent coffee-ring effect on nonuniform surface profile.

8(b), simple straight lines were self-formed by Ag (silver) nanoparticle inks when the smaller dot spacing of 5 μm was used. As further proceeding, any arbitrary images, like cartoon Doraemon as depicted in Figure 8(c), were carried out with ease demonstrating versatile capabilities of image processing in inkjet printing.

As a matter of fact, this allowable versatility of deposit patterns exactly offer such a unique advantage of material and time saving as a cost-efficient technique compared to the conventional others. Hence, different evaporation depositions and patterns can be selected for specific applications. Also, their corresponding properties such as optical, mechanical, and electronic performances depend solely on the technical requirements of specifications in commercial products. For instance, the dot-matrix as shown in Figure 8(a) can be used a microlens array such that optical transparency is dominant, whileas the electric conductivity should be emphasized for the straight lines in Figure 8(b) being used as the conductors in circuitry.

Therefore, typical inkjet printing applications, insofar as potentially useful candidates for electric display fields, will be described and explained in the next paragraph.

5. Applications

5.1 Color filters

Generally, LCD color filters (CFs) feature a dot-matrix with primary red (R), green (G), blue (B) colors. Each color dot presents a tiny pixel of the full-color display with characteristic size ranging from tens to hundreds of micrometers, which match the droplet size if a high-resolution inkjet printing process is applied. Thus, much research has been done in the development of the inkjet-printed color filters, including the suitable UV-curable inks and

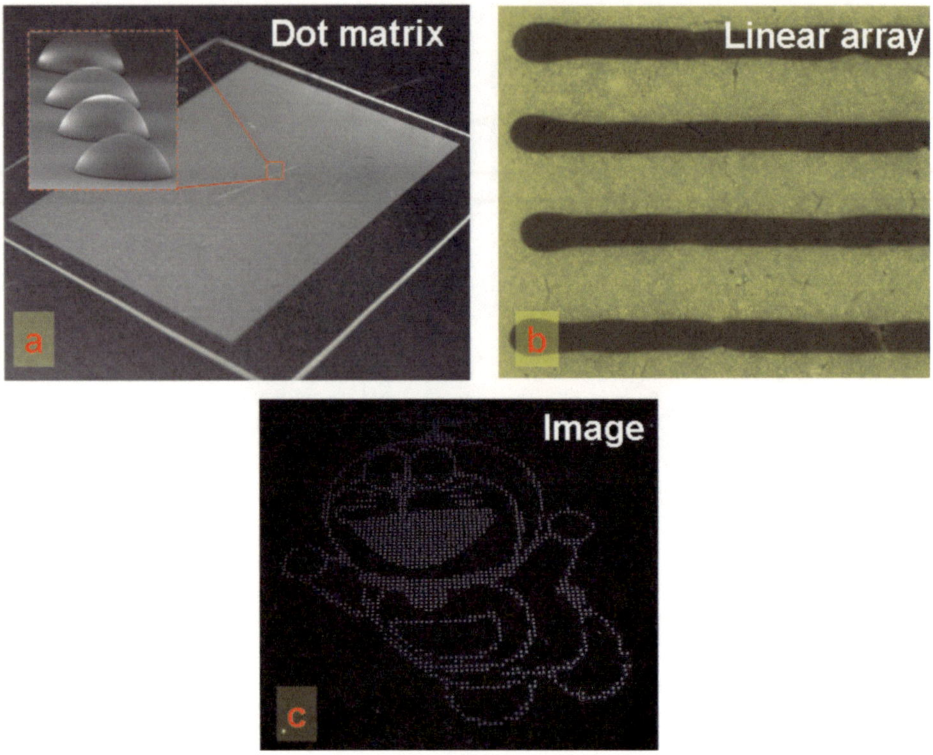

Fig. 8. (a) Individual convex 173 μm-diameter PU deposits inkjet-printed in a 150×200 matrix on a 10 ×10 cm^2 Teflon-coated glass, (b) linear Ag-nanoparticle deposits of ~200 μm-width inkjet-printed on glass surface, (c) inkjet-printed cartoon Doraemon on glass surface.

novel printing platforms (Satoi, 2001; Chang *et al.*, 2005; Koo *et al.*, 2006; Chen *et al.*, 2010), to replace the conventional techniques based on photolithography. Figure 9 shows an inkjet-printed stripe-type color filter with RGB thin-film layers built on the underlying black-matrix (BM) glass, where the sidewalls were pre-patterned by photolithography to prevent the overflows between different color inks (Chen *et al.*, 2010).

Although great success of inkjet-printed color filters was achieved in some respects, there are challenging issues, including higher color density, reliability and yield rate, to be further resolved in future mass production. At the same time, the similar inkjet printing processes have been adopted for active-lighting components, polymer light-emitting-diode (LED) displays, which are described as below.

5.2 Polymer light emitting diodes

Instead of performing light-filter in CFs, polymer light emitting diodes (LEDs) serve as the active-matrix components for lighting without back light required for CFs. As conjugated polymer materials used for electroluminescence that are commercially available (www.cdtltd.co.uk), the polymer LEDs can be directly applied for full-color displays using the inkjet printing technique (van der Vaart *et al.*, 2005; Bale *et al.*, 2006). Since the LED materials are sensitive and degenerative via chemical reactions (*e.g.*, for water H_2O and oxygen O_2), their productions through inkjet printing processes require delicate control of background environment when the droplet depositions of conjugated polymer materials are

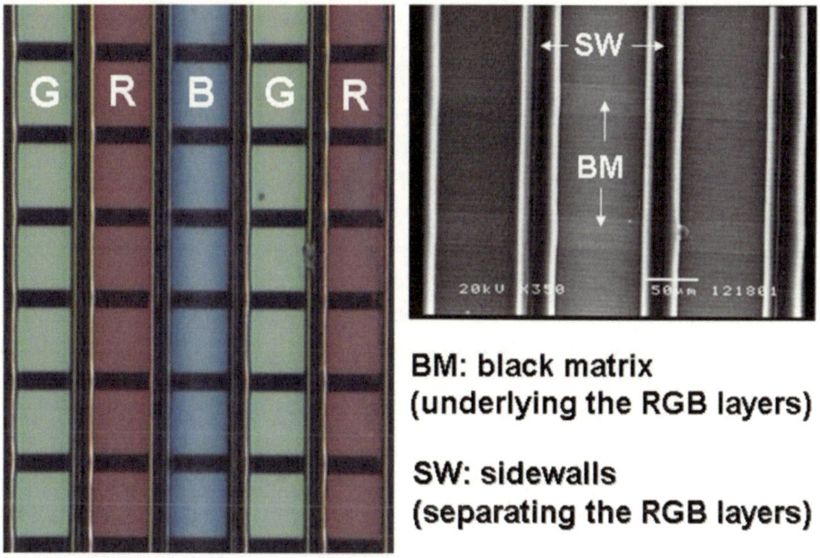

Fig. 9. One inkjet-printed stripe-type LCD color filter with primary colors of red (R), green (G), and blue (B) built on the underlying black-matrix (BM) glass.

being performed. With high flexibility and light weight, the polymer LED display is one of promising candidates for low power consumption in the near future, particularly in the applications of portable consumer devices (*e.g.*, mobile phones and electronic books).

Besides, the LEDs can be enhanced in brightness together with the microlens embedded on top. Figure 10 demonstrates such a lens-cap effect on LEDs, in which the polymer microlenses were deposited to introduce more illumination out of the lighting plane that will be further explained below.

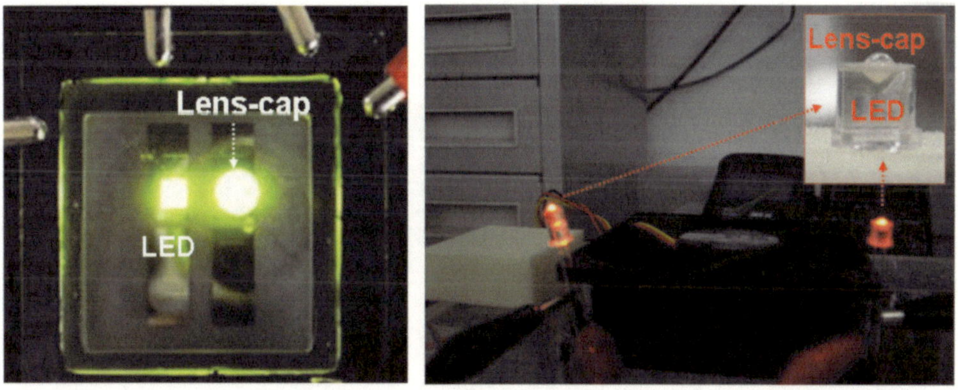

Fig. 10. (Left) Comparison between lens-less and lens-cap green light-emitting diodes (LEDs) showing the lens effect of brightness enhancement; (Right) lens-cap red LEDs.

5.3 Microlenses and back light planes

The inkjet-printed microlenses was introduced in 1994 when MacFarlane *et al.* published their works on microjet fabrication of microlens arrays for collimating light beam

(MacFarlane *et al.*, 1994). Since then, the refractive microlenses were widely investigated by direct inkjet printing for more functionality with incorporation of other devices such as LEDs and VCSEL (Jeon *et al.*, 2005; Nallani *et al.*, 2006). As shown previously in Figure 8, the microlenses feature three-dimensional (3D) curvatures of hemispherical shapes, significantly different from those thin-film layers for CFs and LEDs. As evaporative inks used herein for polymer lenses, the CR effect should be treated in inkjet printing by modifying the substrate surface energy (Chen *et al.*, 2008).

In addition, one potential application for microlenses is associated to the back light plane that transports light of source from the back (side) to front surface of plane by virtue of lens curvature. Nevertheless, compared to conventional techniques of fabrication such as molding and injection, this application is limited to hemispherical profile of a lens, and therefore suffers significantly from low coverage of inkjet printing on plane surface that needs to be further improved in the future.

5.4 Conductive lines and electrodes

Besides the light emitting or transport in CFs, LEDs, and microlenses, both the conductive lines and electrodes are basic elements in electricity delivery for electronic devices. Mostly, with synthesis of nanoparticle metals instead of polymers for inks, the electrical properties of inkjet-printed conductors have been investigated recently in many researches (Fuller *et al.*, 2002; Lee *et al.*, 2005; Kang *et al.*, 2010; Scandurra *et al.* 2010). Because of the need for fusing the nanoparticles, those inkjet print of metal inks typically feature a sintering process at elevated temperature (> 100 °C) to reduce their porous portions of structure, in which the resistivity of printed materials can be as low as 5-7×10^{-6} Ωcm (Scandurra *et al.* 2010).

Furthermore, this type of conductive elements can be commonly applied in flexible microelectronics that has been attracting many efforts in recent years (Perelaer & Schubert, 2010). As demonstrated in Figure 11, the conductive Ag (silver) lines and electrodes can be directly inkjet printed and sintered on a flexible PET (Polyethylene terephthalate) substrate using a commercial Dimatix material printer (DMP 2800). Similarly, electric transistors and integrated circuits can be fulfilled as below.

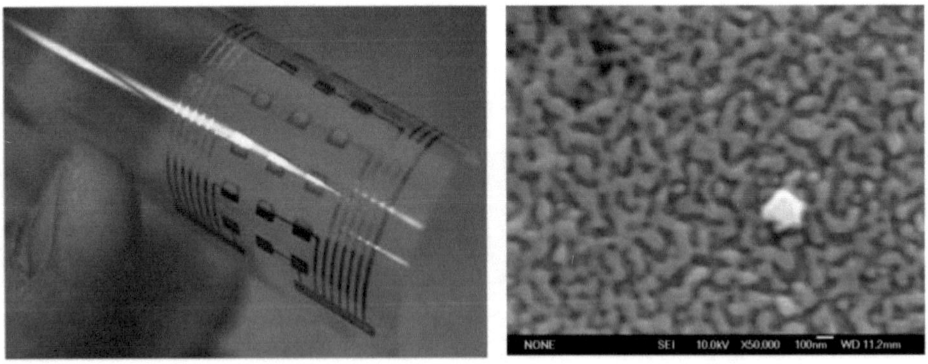

Fig. 11. (Left) Electronic conductors inkjet-printed on a highly flexible PET substrate using Ag-nanoparticle solutions; (Right) the surface morphology of the conductors after sintering at 250 °C.

5.5 Transistors and integrated circuits

Ultimate aim in the field of the inkjet-printed microelectronics is no doubt led to fully fabricate the transistors and integrated circuits that is still at early stage of development in scientific researches (Sirringhaus *et al.*, 2000; Han *et al.*, 2009; Lim *et al.*, 2010; Hinemawari *et al.*, 2011). This revolutionary development, in science and technique as well, can be eventually conducted into the many applications including the thin-film transistor liquid crystal display (TFT-LCD).

Interestingly, more other technical disciplines and ideas, such as soft-lithography and self-assembly (Bruzewicz *et al.*, 2008; Chen *et al.*, 2011), are being gradually blended into inkjet printing of microcomponents, whereby perhaps generating a novel phase for microfabrication in the future (see Figure 12).

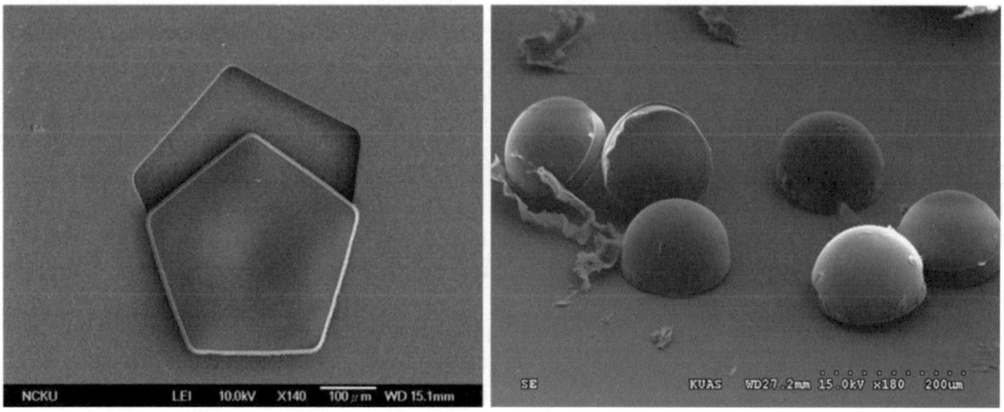

Fig. 12. (Left) one 5-sided regular polygon inkjet-printed and self-formed from a micro cavity; (Right) multiple hemispherical polymer microstructures inkjet-printed and self-leased from their corresponding master molds.

6. Concluding remarks

Indeed, the DOD inkjet printing technology has proved, in recent decades, a powerful tool for digital microfabrication. Key success elements for fulfilling quality inkjet printing involve availabilities and selections of ink materials, substrates, droplet generation, platform and algorithm. Technical issues such as positioning accuracy and morphology formation should be well dealt with in good design, which strongly rely on the full understanding of fundamental fluidics and mechanics.

Droplet depositions, including evaporation deposit and pattern, will eventually find most suitable applications, in which LCD color filters, polymer LEDs, microlenses, conductors, transistors and integrated circuits have been demonstrated using the inkjet printing technique. In the future, this developing technique fused with other disciplines may open novel routes to fabricate more versatile microcomponents.

7. Acknowledgements

The author thanks research grants for this work partially by the National Science Council (NSC) under NSC-99-2221-E-151-034 and NSC-100-2221-E-151-042, Taiwan, ROC.

8. References

Adachi, E.; Dimitrov, A. S. & Nagayama, K. (1995). Stripe patterns formed on a glass surface during droplet evaporation. *Langmuir*, 11, 1057-1060.

Alfeeli, B.; Cho, D.; Ashraf-Khorassani, M.; Taylor, L. T. & Agah, M. (2008). MEMS-based multi-inlet/outlet preconcentrator coated by inkjet printing of polymer adsorbents. *Sens. Actuators B*, 133, 24-32.

Bale, M.; Carter, J. C.; Creighton, C. J.; Gregory, H. J.; Lyon, P. H.; Ng, P.; Webb, L. & Wehrum, A. (2006). Ink-jet printing: the route to production of full-color P-OLED displays. *J. Soc. Inf. Display*, 15, 453-459.

Bao, Z.; Chen, L.; Weldon, E. C.; Cherniavskaya, O.; Dai, Y. & Tok, J. B.-H. (2002). Toward controllable self-assembly of microstructures: selective functionalization and fabrication of patterned spheres. Chem. Mater., 14, 24-26.

Biehl, S.; Danzebrink, R.; Oliveira, P. & Aegerter, M. A. (1998). Refractive microlens fabrication by inkjet-jet process. *J. Sol-Gel Sci. Tech.*, 13, 177-182.

Biswas, S.; Gawande, S.; Bromberg V. & Sun, Y. (2010). Effects of particle size and substrate surface properties on deposition dynamics of inkjet-printed colloidal drops for printable photovoltaics fabrication. *J. Sol. Energy Eng.*, 132, 021010 (7 pages).

Bourges-Monnier, C. & Shanaha, M. E. R. (1995). Influence of evaporation on contact angle. *Langmuir*, 11, 2820-2829.

Bruzewicz, D. A.; Reches, M. & Whitesides, G. M. (2008). Low-cost printing of poly(dimethylsiloxane) barriers to define microchannels in paper. *Anal. Chem.*, 80, 3387-3392.

Busato, S.; Belloli, A. & Ermanni, P. (2007). Inkjet printing of palladium catalyst patterns on polyimide film for electroless copper plating. *Sens. Actuators B*, 123, 840-846.

Chang, C.-J.; Chang, S.-J. ; Wu, F.-M. ; Hsu, M.-W. ; Chiu, W. W. W. & Chen, K. (2004). Effect of compositions and surface treatment on the jetting stability and color uniformity of ink-jet printed color filter. *Jpn. J. Appl. Phys.*, 43, 8227-8233.

Chang, C.-J.; Lin, Y.-H. & Tsai, H.-Y. (2011). Synthesis and properties of UV-curable hyperbranched polymers for ink-jet printing of color micropatterns on glass. *Thin Solid Films*, 519, 5243-5248.

Chen, C.-T. & Yang, T.-Y. (2001). Flow behavior of micro inkjet drop on the layer surface of micro color device. *Proc. of NIP*, 17, 776-779.

Chen, C.-T. (2004). Experimental and numerical study of thin-film formation by microfluidic deposition method. *Proc. of SPIE*, 5519, 255-261.

Chen, C.-T.; Tseng, F.-G.; Chieng, C.-C. (2006). Evaporation evolution of volatile liquid droplets in nanoliter wells. *Sens. Actuators A*, 130-131, 12-19.

Chen, C.-T.; Hsu, C.-Y. & Chiu, C.-L. (2007). Striped droplet deposition on patterned surfaces using inkjet-printing method. *Tamkang J. of Sci. Eng.*, 10, 107-112.

Chen, C.-T.; Chiu, C.-L.; Tseng, Z.-F. & Chuang, C.-T. (2008). Dynamic evolvement and formation of refractive evaporative polyurethane droplets. *Sens. Actuators A*, 147, 369-377.

Chen, C.-T.; Tseng, Z.-F.; Chiu, C.-L.; Hsu, C.-Y. & Chung, C.-T. (2009). Self-aligned hemispherical formation of microlenses from colloidal droplets on heterogeneous surfaces. *J. Micromech. Microeng.*, 19, 025002 (9pp).

Chen, C.-T.; Wu, K.-H.; Lu, C.-F. & Shieh, F. (2010). An inkjet printed stripe-type color filter of liquid crystal display. *J. Micromech. Microeng.*, 20, 005004 (11pp).

Chen, C.-T.; Chiu, C.-L.; Hsu, C.-Y.; Tseng, Z.-F. & Chuang, C.-T. (2011). Inkjet-printed polymeric microstructures in n-sided regular polygonal cavities. *J. Microelectromech. Syst.* , 20, 1001-1009.

Chiu, C.-L. & Chen, C.-T. (2006). The concept and prototype system of medicine-jet capsule endoscope. *Opt. Quantum Electron.*, 37, 1447-1456.

Cho, H.; Parameswaran, M. & Yu, H.-Z. (2007). Fabrication of microsensors using unmodified office inkjet printers. *Sens. Actuators B*, 123, 749-756.

Conway, J.; Korns, H. & Fisch, M. R. (1997). Evaporation kinematics of polystyrene bead suspensions. *Langmuir*, 13, 426-431.

Darhuber, A. A.; Troian, S. M.; Miller, S. M. & Wager, S. (2000). Morphology of liquid microstructures on chemically patterned surfaces. *J. Appl. Phys.*, 87, 7768-7775.

Decker, E. L. & Garoff, S. (1997). Contact line structure and dynamics on surfaces with contact angle hysteresis. *Langmuir*, 13, 6321-6332.

Deegan, R. D.; Bakajin, O.; Dupont, T. F.; Huber G.; Nagel, S. R. & Witten T. A. (1997). Capillary flow as the cause of ring stains from dried liquid drops. *Nature*, 389, 827-829.

de Gans B.-J.; Duineveld, P. C. & Schubert, U. S. (2004). Inkjet printing of polymers: state of the art and future developments. *Adv. Mater.*, 16, 203-213.

Erbil, H. Y.; McHale, G. & Newton, M. I. (2002). Drop evaporation on solid surfaces: constant angle mode. *Langmuir*, 18, 2636-2641.

Fuchs, G.; Diges, C.; Kohlstaedt, L. A.; Wehner, K. A. & Sarnow, P. (2011). Proteomic analysis of ribosomes: translational control of mRNA populations by glycogen synthase GYS1. *J. Mol. Biol.*, 410, 118-130.

Fuller, S. B.; Wilhelm, E. J. & Jacobson, J. M. (2002). Ink-jet printed nanoparticle microelectromechanical systems. *J. Microelectromech. Syst.*, 11, 54-60.

Gau, H.; Herminghaus, S.; Lenz, P. & Lipowsky, R. (1999). Liquid morphologies on structured surfaces: from microchannels to microchips. *Science*, 283, 46-49.

Gorand, Y.; Pauchard, L.; Calligari, G.; Hulin, J. P. & Allain, C. Mechanical instability induced by the desiccation of sessile drops. *Langmuir*, 20, 5138-5140.

Gutmann, O.; Kuehlewein, R.; Reinbold, S.; Niekrawietz, R.; Steinert, C. P.; de Heij, B.; Zengerle, R. & Daub, M. (2005). Fast and reliable protein microarray production by a new drop-in-drop technique. *Lab Chip*, 5, 675-681.

Han, S.-Y.; Lee, D.-H.; Herman, G. S. & Chang C.-H. (2009). Inkjet-printed high mobility transparent-oxide semiconductors. *J. Display Tech.*, 5, 520-524.

Hinemawari, H.; Yamada, T.; Matsui, H.; Tsutsumi, J.; Haas, S.; Ryosuke, C. & Hasegawa, T. (2011). Inkjet printing of single-crystal films. *Nature*, 475, 364-367.

Hu, H. & Larson, R. G. (2002). Evaporation of a sessile droplet on a substrate. *J. Phys. Chem.*, 106, 1334-1344.

Jeon, C. W.; Gu, E.; Liu, C.; Girkin, J. M. & Dawson, M. D. (2005). Polymer microlens arrays applicable to AlInGaN ultraviolet micro-light-emitting diodes. *IEEE Photo. Technol. Lett.*, 17, 1887-1889.

Joshi, A. S. & Sun Y. (2010). Numerical simulation of colloidal drop deposition dynamics on patterned substrates for printable electronics fabrication. *J. Display Tech.*, 6, 579-585.

Kang, J. S.; Kim, H. S.; Ryu, J.; Hahn, H. T.; Jang, S. & Joung, J. W. (2010). Inkjet printed electronics using copper nanoparticle ink, *J. Mater. Sci: Mater. Electron.*, 21, 1213-1220.

Katayama, M. (1999). TFT-LCD technology. *Thin Solid Films*, 341, 140-147.

Kim, E. & Whitesides, G. M. (1995). Use of minimal free energy and self-assembly to form shapes. *Chem. Mater.*, 7, 1257-1264.

Kim, Y. D.; Kim, J. P.; Kwon, O. S. & Cho, I. H. (2009). The synthesis and application of thermally stable dyes for ink-jet printed LCD color filters. *Dyes Pigments*, 81, 45-52.

Koo, H. S.; Chen, M.; Pan., P. C.; Chou, L. T.; Wu, F. M.; Chang, S. J. & Kawai, T. (2006). Fabrication and chromatic characteristics of the greenish LCD colour-filter layer with nano-particle ink using inkjet printing technique. *Displays*, 27, 124-129.

Le, H. P. (1998). Progress and trends in ink-jet printing technology. *J. Imaging Sci. Technol.*, 42, 49-62.

Lee, H.-H.; Chou, K.-S. & Huang, K.-C. (2005). Inkjet printing of nanosized silver colloids. *Nanotechnology*, 16, 2436-2441.

Lenz, P.; Fenzl, W. & Lipowsky, R. (2011). Wetting of ring-shaped surface domains. *Europhys. Lett.*, 53, 618-624.

Lim, J. A.; Kim, J.-H.; Qiu, L.; Lee W. H.; Lee, H. S.; Kwak, D. & Cho, K. (2010). Inkjet-printed single-droplet organic transistors based on semiconductor nanowires embedded in insulating polymers. *Adv. Func. Mater.*, 20, 3292-3297.

Liou, T.-M.; Chan, C.-Y.; Fu, C.-C. & Shih, K.-C. (2008). Effects of impact inertia and surface characteristics on deposited polymer droplets in microcavities. *J. Microelectromech. Syst.*, 17, 278-287.

MacFarlane, D. L.; Narayan, V.; Tatum, J. A.; Cox, W. R.; Chen, T. & Hayes, D. J. (1994). Microjet fabrication of microlens arrays. IEEE Photonics Tech. Lett., 6, 1112-1114.

Mentley, D. E. (2002). State of flat-panel display technology and future trends. *Proc. of the IEEE*, 90, 453-459.

Nallani, A. K.; Chen, T.; Hayes, D. J.; Che, W.-S. & Lee, J.-B. (2006). A method for improved VCSEL packaging using MEMS and ink-jet technologies. *J. Lightw. Technol.*, 24, 1504-1512.

Parisse, F. & Allain, C. (1997). Drying of colloidal suspension droplets: experimental study and profile renormalization. *Langmuir*, 13, 3598-3602.

Perelaer, J. & Schubert, U. S. (2010). *Inkjet printing and alternative sintering of narrow conductive tracks on flexible substrates for plastic electronic applications*, INTECH, ISBN 978-953-7619-72-5, Book Chapter 16.

Satoi, T. (2001). Color filter manufacturing apparatus. US Patent 6,331,384 B1.

Scandurra, A.; Indelli, G. F.; Sparta, N. G.; Galliano, F.; Ravesi, S. & Pignatro S. (2010). Low-temperature sintered conductive silver patterns obtained by inkjet printing for plastic electronics. Surf. Interface Annal., 42, 1163-1167.

Shaw, J. M. & Seidler, P. F. (2001). Organic electronics: introduction. *IBM J. Res. & Dev.*, 45, 3-9.

Sirringhaus, H.; Kawase, T.; Friend, R. H.; Shimoda, T.; Inbasekaran, M.; Wu. W. & Woo, E. P. (2000). High-resolution inkjet printing of all-polymer transistor circuits. *Science*, 290, 2123-2126.

Soltman, D. & Subramanian, V. (2008). Inkjet-printed line morphologies and temperature control of the coffee ring effect. *Langmuir*, 24, 2224-2231.

Szczech B. J.; Megaridis, C. M.; Gamota, D. R. & Zhang, J. (2002). Fine-line conductor manufacturing using drop-on-demand PZT printing technology. *IEEE Trans. Electron. Packag. Manufact.*, 25, 26-33.

Tseng, F.-G.; Kim, C.-J.; Ho, C.-M. (2002). A high-resolution high-frequency monolithic top-shooting microinjector free of satellite drops - part I: concept, design, and model. *J. Microelectromech. Syst.* , 11, 427-436.

van der Vaart, N. C.; Lifka, H.; Budzelaar, F. P. M.; Rubingh, J. E. J. M.; Hoppenbrouwers, J. J. L.; Dijksman, J. F.; Verbeek, R. G. F. A.; van Woudenberg, R.; Vossen, F. J.; Hiddink, M. G. H.; Rosink, J. J. W. M.; Bernards, T. N. M.; Giraldo, A.; Young, N. D.; Fish, D. A.; Childs, M. J.; Steer, W. A. & George, D. S. (2005). Towards large-area full-color active-matrix printed polymer OLED television. *J. Soc. Inf. Display*, 13/1, 9-16.

Weon, B. M. & Je, J. H. (2010). Capillary force repels coffee-ring effect. *Phys. Rev. E*, 82, 015305(R) (4 pages).

Xu, T.; Jin, J.; Gregory, C.; Hickman, J. J. & Boland, T. (2005). Inkjet printing of viable mammalian cells. Biomaterials, 26, 93-99.

Part 2

Technical Schemes and Processes

4

Electromagnetic Formalisms for Optical Propagation in Three-Dimensional Periodic Liquid-Crystal Microstructures

I-Lin Ho and Yia-Chung Chang
Research Center for Applied Sciences, Academia Sinica, Taipei, Taiwan 115, R.O.C. Taiwan

1. Introduction

Nanoscale structures have achieved novel functions in liquid crystal devices such as liquid crystal displays, optical filters, optical modulators, phase conjugated systems, optical attenuators, beam amplifiers, tunable lasers, holographic data storage and even as parts for optical logic systems over the last decades (Blinov et al. (2006; 2007); Sutkowski et al. (2006)). Many theoretical works also have been reported on liquid crystal (LC) optics. Jones method (Jones (1941)) is first proposed for an easy calculation, which stratifies the media along the cell normal while remains the transverse LC orientation uniform, and hence supplies a straightforward way to analyze the forward propagation at normal incidence. This was later followed by the extended Jones method (Lien (1997)), which allows to trace the forward waves at an oblique incidence. The Berreman method (Berreman (1972)) then provides an alternative process to include forward and backward waves.
A further step in LC optics is to consider rigorously the LC variation both along the cell normal and along a single transverse direction, leading to a two-dimensional treatment of light propagation. This step is fulfilled by implementing the finite-difference time-domain method (Kriezis et al. (2000a); Witzigmann et al. (1998)), the vector beam propagation method (Kriezisa & Elston (1999); Kriezis & Elston (2000b)), coupled-wave theory (Galatola et al. (1994); Rokushima & Yamakita (1983)), and an extension of the Berreman approach (Zhang & Sheng (2003)), and has proven to be successful in demonstrating the strong scattering and diffractive effects on the structures with transverse LC variation lasting over the optical-wavelength scale.
For three-dimensional LC medium with arbitrary normal and transverse LC variations, Kriezis et al. (2002) proposed a composite scheme based on the finite-difference time-domain method and the plane-wave expansion method to evaluate the light propagation in periodic liquid-crystal microstructures. Olivero & Oldano (2003) applied numerical calculations by a standard spectral method and the finite-difference frequency-domain method for electromagnetic propagation in LC cells. Glytsis & Gaylord (1987) gave three-dimensional coupled-wave diffraction algorithms via the field decomposition into ordinary and extraordinary waves, although the transverse variation of the ordinary/extraordinary axis raises the complexity. Alternatively, this work neglects the multiple reflections and gives a coupling-matrix algorithm that is much easier to manipulate algebraically for three-dimensional LC media, yet accounts for the effects of the Fresnel refraction and

the single reflection at the surfaces of the media. The detailed derivations are described in appendix A. Furthermore, analogous with the Berreman approach (Berreman (1972)) to consider the multiple reflections for one-dimensional layered media (i.e. stratifying the media along the cell normal while remaining the transverse LC orientation uniform), another supplementary formulae including the influences of multiple reflections for three-dimensional media (i.e. stratifying the media along the cell normal and simultaneously including the varying LC orientation along the transverse) are also addressed in the appendix A. The program code of wolfram mathematica for coupling-matrix method is appended in appendix B for references.

2. Extended Jones matrix method revisited

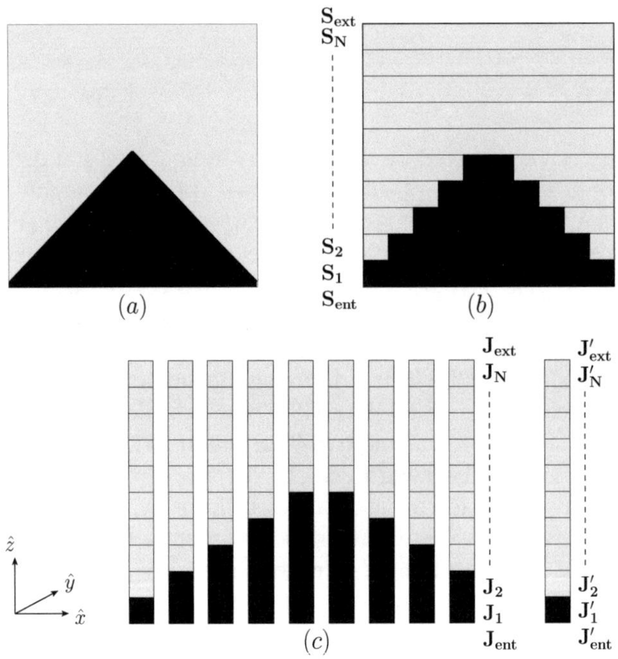

Fig. 1. (a) Schematic depiction of one unit cell of the periodic LC structures. (b) Stratification of the cell along the cell normal $\hat{z}$ with remaining the real transverse $\hat{x}(\hat{y})$ profile as in coupling-matrix method.(c) Decompose the cell along the transverse direction $\hat{x}(\hat{y})$ into independent strips, and treat the stratification of each stripe with uniform transverse profiles, as in (extended) Jones matrix method.

In this section, extended Jones matrix method is revisited first due to its similar underlying concepts can supply an accessibility to understand the coupling-matrix method. In the extended Jones matrix method, the liquid crystal cell (Figure 1(a)) is decomposed into multiple one-dimensional (z) independent stripes (Figure 1(c)), treating the transverse LC orientation uniform within each stripes and being irrelevant each other. Each stripe is further divided into N layers along the z direction, including two separate polarizer and analyzer layers. In the layer, there are four eigen-mode waves: two transmitted and two reflected waves; while at the interface of the layer, the boundary condition is that the tangential components of the electric field are continuous. Without loss of generality, considering the propagation of waves in the xz plane at angle angle θ related to z axis, it specifies $\vec{k} = (k_0 sin\theta, 0, k_0 cos\theta)$, extended Jones Matrix can relate the electric fields at the bottom of the ℓ_{th} layer to the fields at the top of the ℓ_{th} layer of each strip by:

$$\begin{bmatrix} E_x \\ E_y \end{bmatrix}_{\ell,dz_\ell} = \mathbf{J}_\ell \begin{bmatrix} E_x \\ E_y \end{bmatrix}_{\ell,0} ; \ \mathbf{J}_\ell = \mathbf{A}_\ell \Xi_\ell \mathbf{A}_\ell^{-1} \tag{1}$$

with

$$\Xi_\ell = \begin{bmatrix} \exp(ik_{z1}dz_\ell) & 0 \\ 0 & \exp(ik_{z2}dz_\ell) \end{bmatrix}; \ \mathbf{A}_\ell = \begin{bmatrix} e_{x1} & e_{x2} \\ e_{y1} & e_{y2} \end{bmatrix} \tag{2}$$

$$\frac{k_{z1}}{k_0} = \left(n_0^2 - \frac{k_x^2}{k_0^2} \right)^{1/2} ; \tag{3}$$

$$\frac{k_{z2}}{k_0} = -\frac{\varepsilon_{xz}}{\varepsilon_{zz}} \frac{k_x}{k_0} + \frac{n_o n_e}{\varepsilon_{zz}} \left(\varepsilon_{zz} - \left(1 - \frac{n_e^2 - n_o^2}{n_e^2} \cos^2\theta_o \sin^2\phi_o \right) \frac{k_x^2}{k_0^2} \right)^{1/2} \tag{4}$$

$$e_{x1} = \left(\frac{k_{z1}^2}{k_0^2} + \frac{k_x^2}{k_0^2} - \varepsilon_{yy} \right) \left(\frac{k_x^2}{k_0^2} - \varepsilon_{zz} \right) - \varepsilon_{yz}\varepsilon_{zy} \tag{5}$$

$$e_{y1} = \left(\frac{k_x^2}{k_0^2} - \varepsilon_{zz} \right) \varepsilon_{yx} + \left(\frac{k_x}{k_0} \frac{k_{z1}}{k_0} + \varepsilon_{zx} \right) \varepsilon_{yz} \tag{6}$$

$$e_{x2} = \left(-\frac{k_x^2}{k_0^2} + \varepsilon_{zz} \right) \varepsilon_{xy} - \left(\frac{k_x}{k_0} \frac{k_{z2}}{k_0} + \varepsilon_{xz} \right) \varepsilon_{zy} \tag{7}$$

$$e_{y2} = \left(-\frac{k_{z2}^2}{k_0^2} + \varepsilon_{xx} \right) \left(\frac{k_x^2}{k_0^2} - \varepsilon_{zz} \right) + \left(\frac{k_x}{k_0} \frac{k_{z2}}{k_0} + \varepsilon_{zx} \right) \left(\frac{k_x}{k_0} \frac{k_{z2}}{k_0} + \varepsilon_{xz} \right) \tag{8}$$

Here, $k_0 = \omega/c = 2\pi/\lambda$ with λ the wavelength of the incident light in free space. dz_ℓ is the thickness of the the ℓ_{th} layer. θ_o and ϕ_o are the orientation angles of the LC director defined in the spherical coordinate. $\varepsilon_{i,j\in\{x,y,z\}}$ is the dielectric tensors defined in appendix A. Equation (1) can be understood as follow. $\mathbf{A}_\ell^{-1}$ transforms the electric fields at the bottom of the ℓ_{th} layer into the eigen-mode fields. Ξ_ℓ then propagates the eigen-mode fields from the bottom of the ℓ_{th} layer to the top of the ℓ_{th} layer through the distance dz_ℓ. Finally, $\mathbf{A}_\ell$ transform the eigen-mode fields at the top of the the ℓ_{th} layer back into the electric fields at the top of the ℓ_{th} layer, which is equal to the electric fields at the bottom of the $(\ell+1)_{th}$ layer by boundary condition. Grouping all layers, the extended Jones matrix formula that relates the incident electric fields ($\ell = 0$) and the emitted electric fields ($\ell = N+1$) is given by

$$\begin{bmatrix} E_x \\ E_y \end{bmatrix}_{N+1} = \mathbf{J}_{ext}\mathbf{J}_N\mathbf{J}_{N-1} \cdots \mathbf{J}_2\mathbf{J}_1\mathbf{J}_{ent} \begin{bmatrix} E_x \\ E_y \end{bmatrix}_0 \tag{9}$$

$$\mathbf{J}_{ent} = \begin{bmatrix} \frac{2\cos\theta_p}{\cos\theta_p + n_p\cos\theta} & 0 \\ 0 & \frac{2\cos\theta}{\cos\theta + n_p\cos\theta_p} \end{bmatrix} \tag{10}$$

$$\mathbf{J}_{ext} = \begin{bmatrix} \frac{2n_p\cos\theta}{\cos\theta_p + n_p\cos\theta} & 0 \\ 0 & \frac{2n_p\cos\theta_p}{\cos\theta + n_p\cos\theta_p} \end{bmatrix} \tag{11}$$

with $\theta_p = \sin^{-1}(\sin\theta/\Re(n_p))$ in which $\Re(n_p)$ stands for the average of the real parts of the two indices of refraction (n_e and n_o) of the polarizer. The total transmission for the stripe is calculated by

$$trans. = \frac{\left|E_{x,N+1}\right|^2 + \cos^2\theta \cdot \left|E_{y,N+1}\right|^2}{\left|E_{x,0}\right|^2 + \cos^2\theta \cdot \left|E_{y,0}\right|^2} \tag{12}$$

The total transmission of the three-dimensional LC media then can be evaluated by summing up the contributions from the individual stripe.

3. Coupling matrix method

Parallel to the equation (9) by one-dimensional treatments for strips, an analogous coupling-matrix formulae for the propagations of waves through the three-dimensional periodic microstructures can be given as:

$$\begin{bmatrix} \vec{E}^{+}_{q,N+1} \\ \vec{M}^{+}_{q,N+1} \\ \vec{E}^{-}_{q,N+1} \\ \vec{M}^{-}_{q,N+1} \end{bmatrix} = \mathbf{S}_{ext}\mathbf{S}_{N}...\mathbf{S}_2\mathbf{S}_1\mathbf{S}_{ent} \begin{bmatrix} \vec{E}^{+}_{q,0} \\ \vec{M}^{+}_{q,0} \\ \vec{E}^{-}_{q,0} \\ \vec{M}^{-}_{q,0} \end{bmatrix} \tag{13}$$

Here, $\vec{E}^{+}_{q,\ell}$ and $\vec{M}^{+}_{q,\ell}$ ($\vec{E}^{-}_{q,\ell}$ and $\vec{M}^{-}_{q,\ell}$) represent the physical forward (backward) TE and TM fields, i.e. transverse electric and transverse magnetic fields corresponding to the planes of the diffraction waves in the incident ($\ell = 0$) and emitted ($\ell = N+1$) regions. In which the components of the vectors $\vec{E}^{+}_{q,\ell}$, $\vec{M}^{+}_{q,\ell}$, $\vec{E}^{-}_{q,\ell}$, or $\vec{M}^{-}_{q,\ell}$ define the diffraction waves along the direction $\mathbf{n}_{gh} = n_{xg}\hat{\imath} + n_{yh}\hat{\jmath} + \xi_{gh}\hat{k}$:

$$n_{xg} = n_I \sin\theta\cos\phi - g\frac{\lambda}{\Lambda_x} \tag{14}$$

$$n_{yh} = n_I \sin\theta\sin\phi - h\frac{\lambda}{\Lambda_y} \tag{15}$$

$$\xi_{gh} = \sqrt{\varepsilon_{I(E)} - n_{yh}n_{yh} - n_{xg}n_{xg}} \tag{16}$$

with $\varepsilon_I = n_I^2$ ($\varepsilon_E = n_E^2$) being the dielectric coefficient in the incident (emitted) region. Note that the components with imaginary ξ_{gh} values are ignored for studied cases due to the decaying natures along the electromagnetic propagations parallel to the z direction. Λ_x (Λ_y) is the periodicity of the LC structure along the x (y) direction. $\mathbf{S}_{\ell\in\{1\sim N\}}$ is the matrix representing the propagations of waves through the ℓ_{th} structured layer. It consists of the matrix $\mathbf{T}^{(a)}_{\ell}$, which is the (column) eigen-vector matrix of the characteristic matrix $\mathbf{G}_\ell$ for the ℓ_{th} layer , and the diagonal matrix $exp\left[i\kappa^{(a)}_{\ell}\overline{dz}_\ell\right]$ relates to the eigen-value $\kappa^{(a)}_{\ell}$ of $\mathbf{G}_\ell$ with dimensionless $\overline{dz}_\ell = dz_\ell k_0$:

$$\mathbf{S}_\ell = \mathbf{T}^{(a)}_{\ell} exp\left[i\kappa^{(a)}_{\ell}\overline{dz}_\ell\right] (\mathbf{T}^{(a)}_{\ell})^{-1} \tag{17}$$

$$\mathbf{G}_\ell = \begin{bmatrix} \tilde{n}_x\tilde{\varepsilon}_{zz}^{-1}\tilde{\varepsilon}_{zx} & \tilde{n}_x\tilde{\varepsilon}_{zz}^{-1}\tilde{n}_x - 1 & \tilde{n}_x\tilde{\varepsilon}_{zz}^{-1}\tilde{\varepsilon}_{zy} & -\tilde{n}_x\tilde{\varepsilon}_{zz}^{-1}\tilde{n}_y \\ \tilde{\varepsilon}_{xz}\tilde{\varepsilon}_{zz}^{-1}\tilde{\varepsilon}_{zx} - \tilde{\varepsilon}_{xx} + \tilde{n}_y\tilde{n}_y & \tilde{\varepsilon}_{xz}\tilde{\varepsilon}_{zz}^{-1}\tilde{n}_x & \tilde{\varepsilon}_{xz}\tilde{\varepsilon}_{zz}^{-1}\tilde{\varepsilon}_{zy} - \tilde{\varepsilon}_{xy} - \tilde{n}_y\tilde{n}_x & -\tilde{\varepsilon}_{xz}\tilde{\varepsilon}_{zz}^{-1}\tilde{n}_y \\ \tilde{n}_y\tilde{\varepsilon}_{zz}^{-1}\tilde{\varepsilon}_{zx} & \tilde{n}_y\tilde{\varepsilon}_{zz}^{-1}\tilde{n}_x & \tilde{n}_y\tilde{\varepsilon}_{zz}^{-1}\tilde{\varepsilon}_{zy} & -\tilde{n}_y\tilde{\varepsilon}_{zz}^{-1}\tilde{n}_y + 1 \\ -\tilde{\varepsilon}_{yz}\tilde{\varepsilon}_{zz}^{-1}\tilde{\varepsilon}_{zx} + \tilde{\varepsilon}_{yx} + \tilde{n}_x\tilde{n}_y & -\tilde{\varepsilon}_{yz}\tilde{\varepsilon}_{zz}^{-1}\tilde{n}_x & -\tilde{\varepsilon}_{yz}\tilde{\varepsilon}_{zz}^{-1}\tilde{\varepsilon}_{zy} + \tilde{\varepsilon}_{yy} - \tilde{n}_x\tilde{n}_x & \tilde{\varepsilon}_{yz}\tilde{\varepsilon}_{zz}^{-1}\tilde{n}_y \end{bmatrix} \tag{18}$$

In this context, the notation $\vec{E}$ (or $\vec{M}$) denotes the $N_gN_h \times 1$ vector with components E_{gh} (or M_{gh}) describing the wave along $\mathbf{n}_{gh}$. $\tilde{n}_x$ ($\tilde{n}_y$) are $N_gN_h \times N_gN_h$ diagonal matrices with N_gN_h diagonal elements n_{xg} (n_{yh}) being the same (g,h) sequence as that of $\vec{E}$ and $\vec{M}$, and are calculated by Equations (14-15). $\tilde{\varepsilon}_{ij\in\{x,y,z\}}$ are $N_gN_h \times N_gN_h$ matrices with elements $\varepsilon_{ij,\alpha\beta}$ being the Fourier transform of the spatial dielectric coefficients $\varepsilon_{ij}(x,y;z)$, in which the indexes α,β are arranged by the relation $\vec{M} \sim \tilde{\varepsilon}_{ij}\vec{E}$, i.e. $M_{gh} \sim \sum_{g'h'}\varepsilon_{ij,(g-g')(h-h')}E_{g'h'}$ (derived in appendix A). Above $N_{g(h)}$ define the number of considered total Fourier orders g (h) in the x (y) direction. 1 represents the $N_gN_h \times N_gN_h$ identity matrix. One may understand the Equation (17) for the ℓ_{th} layer by the similar way as described in extended Jones method: the $(\mathbf{T}_\ell^{(a)})^{-1}$ term represents the coordinate transformation from the spatial tangential components of fields $\mathbf{f}_{\hat{t},\ell} = [\vec{e}_{x,\ell}\ \vec{h}_{y,\ell}\ \vec{e}_{y,\ell}\ \vec{h}_{x,\ell}]^t$ denoted by Equations (46)-(47) at ℓ_{th} interface into the orthogonal components of the eigen-modes in the ℓ_{th} layer; the $exp\left[i\kappa_\ell^a dz_\ell\right]$ term describes eigen-mode propagation over the distance dz_ℓ (thickness of the ℓ_{th} layer); the $\mathbf{T}_\ell^{(a)}$ term then is the inversely coordinate transformation from the eigen-mode components back to the spatial tangential components of fields at the next interface. Considering the continuum of tangential fields on interfaces, these fields emitted from the ℓ_{th} layer hence can be straightforwardly treated as the incident fields $\mathbf{f}_{\hat{t},\ell+1}$ for the $(\ell+1)_{th}$ layer, and allow to follow the next transfer matrix $\mathbf{S}_{\ell+1}$ to describe the sequential propagations of fields through the $(\ell+1)_{th}$ layer as in Equation (13).

For the matrices $\mathbf{S}_{ent}$ and $\mathbf{S}_{ext}$ defined for the (isotropic) uniform incident ($\ell = 0$) and emitted ($\ell = N+1$) regions, respectively, the eigen-modes are specially chosen (and symbolized) as $\vec{E}_q^+$ and $\vec{M}_q^+$ ($\vec{E}_q^-$ and $\vec{M}_q^-$) (Ho et al. (2011); Rokushima & Yamakita (1983)), representing the physical forward (backward) *TE* and *TM* waves as the above-mentioned. In which the transform matrix $\mathbf{T}_{\varepsilon_I}^{(i)}$ between the eigen-mode components and the tangential components $\mathbf{f}_{\hat{t},0} = [\vec{e}_{x,0}\ \vec{h}_{y,0}\ \vec{e}_{y,0}\ \vec{h}_{x,0}]^t$ for the isotropic incident region ($\ell = 0$) is given as:

$$\begin{bmatrix} \vec{e}_{x,0} \\ \vec{h}_{y,0} \\ \vec{e}_{y,0} \\ \vec{h}_{x,0} \end{bmatrix} = \begin{bmatrix} \dot{\mathbf{n}}_y & \dot{\mathbf{n}}_x & \dot{\mathbf{n}}_y & \dot{\mathbf{n}}_x \\ \dot{\mathbf{n}}_y\xi & \varepsilon_I\dot{\mathbf{n}}_x\xi^{-1} & -\dot{\mathbf{n}}_y\xi & -\varepsilon_I\dot{\mathbf{n}}_x\xi^{-1} \\ -\dot{\mathbf{n}}_x & \dot{\mathbf{n}}_y & -\dot{\mathbf{n}}_x & \dot{\mathbf{n}}_y \\ \dot{\mathbf{n}}_x\xi & -\varepsilon_I\dot{\mathbf{n}}_y\xi^{-1} & -\dot{\mathbf{n}}_x\xi & \varepsilon_I\dot{\mathbf{n}}_y\xi^{-1} \end{bmatrix} \begin{bmatrix} \vec{E}_{q,0}^+ \\ \vec{M}_{q,0}^+ \\ \vec{E}_{q,0}^- \\ \vec{M}_{q,0}^- \end{bmatrix}$$

$$\equiv \mathbf{T}_{\varepsilon_I}^{(i)} \begin{bmatrix} \vec{E}_{q,0}^+ \\ \vec{M}_{q,0}^+ \\ \vec{E}_{q,0}^- \\ \vec{M}_{q,0}^- \end{bmatrix} \tag{19}$$

Here, $\dot{\mathbf{n}}_y$ and $\dot{\mathbf{n}}_x$ are $N_gN_h \times N_gN_h$ diagonal matrices with normalized elements $\frac{n_{yh}}{m_{gh}}$ and $\frac{n_{xg}}{m_{gh}}$ respectively. ξ^{-1} is the diagonal matrix with elements $1/\xi_{gh}$ (not the inverse of the matrix ξ), in which $m_{gh} = (n_{yh}n_{yh} + n_{xg}n_{xg})^{1/2}$, $\zeta_{gh} = (\varepsilon_I - n_{yh}n_{yh} - n_{xg}n_{xg})^{1/2}$, and $\varepsilon_I = n_I^2$ have been applied for the incident region. A similar transform for $\mathbf{f}_{\hat{t},N+1}$ in the emitted region can be derived straightforwardly by replacing all the ε_I in Equation (19) with ε_E and can be denoted as $\mathbf{f}_{\hat{t},N+1} = \mathbf{T}^{(i)}_{\varepsilon_E}[\vec{E}^{+}_{q,N+1}\ \vec{M}^{+}_{q,N+1}\ \vec{E}^{-}_{q,N+1}\ \vec{M}^{-}_{q,N+1}]^t$, with $\zeta_{gh} = (\varepsilon_E - n_{yh}n_{yh} - n_{xg}n_{xg})^{1/2}$, and $\varepsilon_E = n_E^2$.

$\mathbf{S}_{ent}$ is the matrix representing the light propagation from the incident region into the medium, and indicates the essential refraction and the reflection at the first interface of the medium. To consider these effects in a simple way, a virtual (isotropic) uniform layer, which has zero thickness and effective dielectric coefficient $\varepsilon_a = n^2_{avg}$, e.g. $n_{avg} = (n_e + n_o)/2$ for the liquid-crystal grating, is assumed to exist between the incident region and the 1_{st} layer. $\mathbf{S}_{ent}$ thereby can be approximately evaluated as:

$$\mathbf{S}_{ent} = \mathbf{T}^{(i)}_{\varepsilon_a}\begin{bmatrix}\mathbf{W}'^{-1}_1 & \mathbf{0}\\ \mathbf{0} & \mathbf{0}\end{bmatrix} \tag{20}$$

$$\begin{bmatrix}\mathbf{W}'_1 & \mathbf{W}'_2\\ \mathbf{W}'_3 & \mathbf{W}'_4\end{bmatrix} = \left[(\mathbf{T}^{(i)}_{\varepsilon_a})^{-1}\mathbf{T}^{(i)}_{\varepsilon_I}\right]^{-1} \tag{21}$$

Here, $\mathbf{T}^{(i)}_{\varepsilon_a}$ is formulated as equation (19) with the replacements of ε_I by ε_a, $\zeta_{gh} = (\varepsilon_a - n_{yh}n_{yh} - n_{xg}n_{xg})^{1/2}$, and $\varepsilon_a = n^2_{avg}$. Similar to the argument of $\mathbf{S}_{ent}$, another virtual (isotropic) uniform layer is included between the emitted region and the N_{th} layer to consider the effects of refraction and the reflection at the last interface. Here, $\mathbf{S}_{ext}$ is approximated as:

$$\mathbf{S}_{ext} = \begin{bmatrix}\mathbf{W}''^{-1}_1 & \mathbf{0}\\ \mathbf{0} & \mathbf{0}\end{bmatrix}(\mathbf{T}^{(i)}_{\varepsilon_a})^{-1} \tag{22}$$

$$\begin{bmatrix}\mathbf{W}''_1 & \mathbf{W}''_2\\ \mathbf{W}''_3 & \mathbf{W}''_4\end{bmatrix} = \left[(\mathbf{T}^{(i)}_{\varepsilon_E})^{-1}\mathbf{T}^{(i)}_{\varepsilon_a}\right]^{-1} \tag{23}$$

Put everything together, and the propagation of fields through three-dimensional periodic microstructures hence can be evaluated as in Equation (13).

4. Numerical analyses

In this section, a simple case is applied to demonstrate the algorithms and is verified by finite-difference time-domain (FDTD) method. Consider a one-layer film ($N = 1$) with liquid-crystal orientation $\theta_o = \pi x/\Lambda_x = \lambda\overline{x}/2\Lambda_x$, $\phi_o = \pi/2$. By the Fourier transform defined in equations (42-45), the non-zero Fourier components for the dielectric elements $\bar{\varepsilon}_{ij,gh}$ are: $\bar{\varepsilon}_{xx,00} = n_o^2$, $\bar{\varepsilon}_{yy,00} = \left(n_o^2 + n_e^2\right)/2$, $\bar{\varepsilon}_{yy,\pm 10} = \left(n_o^2 - n_e^2\right)/4$, $\bar{\varepsilon}_{yz,\pm 10} = \pm i\left(n_o^2 - n_e^2\right)/4$, $\bar{\varepsilon}_{zz,00} = \left(n_o^2 + n_e^2\right)/2$, $\bar{\varepsilon}_{zz,\pm 10} = \left(n_e^2 - n_o^2\right)/4$. For simplicity, we only consider three Fourier components of fields, i.e. $(g,h) = (\pm 1,0)$ and $(0,0)$, for this case. The corresponding transfer-matrix formula in equation (13) are then given as:

$$\begin{bmatrix}\vec{E}^{+}_{q,N+1}\\ \vec{M}^{+}_{q,N+1}\\ \vec{E}^{-}_{q,N+1}\\ \vec{M}^{-}_{q,N+1}\end{bmatrix} = \mathbf{S}_{ext}\mathbf{S}_1\mathbf{S}_{ent}\begin{bmatrix}\vec{E}^{+}_{q,0}\\ \vec{M}^{+}_{q,0}\\ \vec{E}^{-}_{q,0}\\ \vec{M}^{-}_{q,0}\end{bmatrix} \tag{24}$$

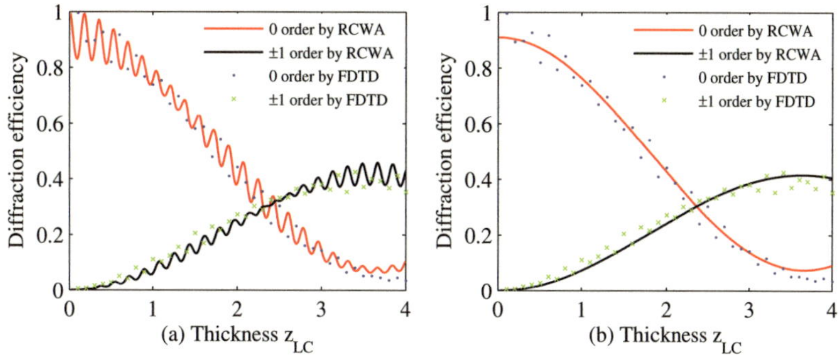

Fig. 2. Diffraction efficiency for the periodic LC structures at thickness $z_{LC} = 0 - 4\mu m$, in which (a) the solid line indicates numerical results by RCWA with considering multiple reflections as in the appendix (equations 89-92), and are in agreement with those (dotted line) from the FDTD method, while (b) the solid line indicates numerical results by the RCWA with ignoring multiple reflections, yet accounting for the effects of the Fresnel refraction and the single reflection at the surfaces of the media as in equation (13), showing comparable results.

which relates to the eigen-vector matrices $\mathbf{T}_0/\mathbf{T}_2$ for the isotropic incident/emitted layer in the equation (19), and the eigen-values $\kappa_1^{(a)}$ and eigen-vector $\mathbf{T}_1^{(a)}$ matrices of $\mathbf{G}$ in equation (18) for the liquid-crystal film. Here, $z_{lc} = \bar{z}_{lc}/k_0$ is the thickness of the liquid-crystal film. In this case, we simply choose the unit-amplitude normal TE incidence with respect to the xz incident plane, i.e. $\vec{E}^+_{q,0} = \left[E^+_{q,0,-10}\ E^+_{q,0,00}\ E^+_{q,0,10}\right]^t = [0\ 1\ 0]^t$ and $\vec{M}^+_{q,0} = \left[M^+_{q,0,-10}\ M^+_{q,0,00}\ M^+_{q,0,10}\right]^t = [0\ 0\ 0]^t$. For the isotropic incident/emitted air layer ($\varepsilon = 1$), the associated $\dot{\mathbf{n}}_x$, $\dot{\mathbf{n}}_y$, ε, and ξ in $\mathbf{T}_0/\mathbf{T}_2$ are referred to equations (14,15,16), and are given as:

$$\dot{\mathbf{n}}_x = \begin{bmatrix} 1 & 0 & 0 \\ 0 & 1 & 0 \\ 0 & 0 & -1 \end{bmatrix}; \dot{\mathbf{n}}_y = \begin{bmatrix} 0 & 0 & 0 \\ 0 & 0 & 0 \\ 0 & 0 & 0 \end{bmatrix}; \varepsilon = \begin{bmatrix} 1 & 0 & 0 \\ 0 & 1 & 0 \\ 0 & 0 & 1 \end{bmatrix} \tag{25}$$

$$\xi = \begin{bmatrix} \sqrt{1-\lambda^2/\Lambda_x^2} & 0 & 0 \\ 0 & 1 & 0 \\ 0 & 0 & \sqrt{1-\lambda^2/\Lambda_x^2} \end{bmatrix} \tag{26}$$

$$\xi^{-1} = \begin{bmatrix} 1/\sqrt{1-\lambda^2/\Lambda_x^2} & 0 & 0 \\ 0 & 1 & 0 \\ 0 & 0 & 1/\sqrt{1-\lambda^2/\Lambda_x^2} \end{bmatrix} \tag{27}$$

Note we have used a small incident angle ($\theta = 10^{-5}$, $\phi = 0$) to avoid the numerical instability at $\theta = 0$. For the layer of liquid-crystal film, the associated $\tilde{\mathbf{n}}_x$, $\tilde{\mathbf{n}}_y$ and $\varepsilon_{ij\in\{x,y,z\}}$ in $\mathbf{G}$ in equation (18) are written out as below:

$$\tilde{\mathbf{n}}_x = \begin{bmatrix} \lambda/\Lambda_x & 0 & 0 \\ 0 & 1 & 0 \\ 0 & 0 & -\lambda/\Lambda_x \end{bmatrix}; \tilde{\mathbf{n}}_y = \begin{bmatrix} 0 & 0 & 0 \\ 0 & 0 & 0 \\ 0 & 0 & 0 \end{bmatrix} \tag{28}$$

$$\varepsilon_{xx} = \begin{bmatrix} \varepsilon_{xx,00} & \varepsilon_{xx,-10} & \varepsilon_{xx,-20} \\ \varepsilon_{xx,10} & \varepsilon_{xx,00} & \varepsilon_{xx,-10} \\ \varepsilon_{xx,20} & \varepsilon_{xx,10} & \varepsilon_{xx,00} \end{bmatrix} = \begin{bmatrix} n_o^2 & 0 & 0 \\ 0 & n_o^2 & 0 \\ 0 & 0 & n_o^2 \end{bmatrix} \tag{29}$$

$$\varepsilon_{yy} = \begin{bmatrix} \frac{n_o^2+n_e^2}{2} & \frac{n_o^2-n_e^2}{4} & 0 \\ \frac{n_o^2-n_e^2}{4} & \frac{n_o^2+n_e^2}{2} & \frac{n_o^2-n_e^2}{4} \\ 0 & \frac{n_o^2-n_e^2}{4} & \frac{n_o^2+n_e^2}{2} \end{bmatrix} \tag{30}$$

$$\varepsilon_{zz} = \begin{bmatrix} \frac{n_o^2+n_e^2}{2} & \frac{n_e^2-n_o^2}{4} & 0 \\ \frac{n_e^2-n_o^2}{4} & \frac{n_o^2+n_e^2}{2} & \frac{n_e^2-n_o^2}{4} \\ 0 & \frac{n_e^2-n_o^2}{4} & \frac{n_o^2+n_e^2}{2} \end{bmatrix} \tag{31}$$

$$\varepsilon_{yz} = \begin{bmatrix} 0 & \frac{-i\left(n_o^2-n_e^2\right)}{4} & 0 \\ \frac{i\left(n_o^2-n_e^2\right)}{4} & 0 & \frac{-i\left(n_o^2-n_e^2\right)}{4} \\ 0 & \frac{i\left(n_o^2-n_e^2\right)}{4} & 0 \end{bmatrix} \tag{32}$$

$$\varepsilon_{xy} = 0, \quad \varepsilon_{xz} = 0 \tag{33}$$

Consequently, the eigen-values $\kappa_1^{(a)}$ and eigen-vector $\mathbf{T}_1^{(a)}$ matrices of $\mathbf{G}$ can be numerically evaluated and a similar process for $\mathbf{S}_{ent}$ and $\mathbf{S}_{ext}$ can be followed straightforwardly. Together with all these definitions of matrixes in equation (24), the transmittance fields $\vec{E}^{+}_{q,2}$ and $\vec{M}^{+}_{q,2}$ then can be decided. Here, we set $\lambda = 0.55um$, $\Lambda_x = 2.0um$, $n_o = 1.5$, and $n_e = 1.6$. The numerical results of far-field diffractions for this case by RCWA ignoring the influences of the multiple reflections are shown in figure 2(b), and are in agreement with these obtained by FDTD. Besides, an alternative consideration described by the equations (89-92) in appendix A, which includes the multiple reflections, is shown in figure 2(a), and clarifies the effectiveness of the easy-manipulated algorithm in equation (13) for the three-dimensional periodic LC media.

A. Derivation of the coupling matrix

In this appendix, detailed derivations of the coupling matrix method are demonstrated for references.

A.1 Maxwell's equations in spatial-space descriptions

Without charges and currents, Maxwell's equations can be read as:

$$\nabla \cdot \mathbf{E} = 0 \tag{34}$$

$$\nabla \cdot \mathbf{B} = 0 \tag{35}$$

$$\nabla \times \mathbf{E} = -\frac{\partial \mathbf{B}}{\partial t} \tag{36}$$

$$\nabla \times \mathbf{B} = \mu\mu_0\varepsilon\varepsilon_0 \frac{\partial \mathbf{E}}{\partial t} \tag{37}$$

Define variables $k_0 = \omega\sqrt{\mu_0\varepsilon_0} = \frac{2\pi}{\lambda}$, $Y_0 = \frac{1}{Z_0} = \sqrt{\frac{\varepsilon_0}{\mu_0}}$, $\overline{r} = k_0 r$, $\overline{x} = k_0 x$, $\overline{y} = k_0 y$, $\overline{z} = k_0 z$, and $\overline{\nabla}_i = \partial/\partial\overline{r}_i = \partial/\partial r_i k_0 = \nabla_i/k_0$, and the equations can be derived as:

$$\overline{\nabla}\cdot\mathbf{E} = 0 \tag{38}$$

$$\overline{\nabla}\cdot\mathbf{B} = 0 \tag{39}$$

$$\overline{\nabla}\times\sqrt{Y_0}\mathbf{E} = -i\sqrt{Z_0}\mathbf{H} \tag{40}$$

$$\begin{aligned}\overline{\nabla}\times\sqrt{Z_0}\mathbf{H} &= i\overline{\varepsilon}\left(\overline{r}\right)\sqrt{Y_0}\mathbf{E}\\ &= i\begin{bmatrix}\overline{\varepsilon}_{xx}\left(\overline{r}\right) & \overline{\varepsilon}_{xy}\left(\overline{r}\right) & \overline{\varepsilon}_{xz}\left(\overline{r}\right)\\ \overline{\varepsilon}_{yx}\left(\overline{r}\right) & \overline{\varepsilon}_{yy}\left(\overline{r}\right) & \overline{\varepsilon}_{yz}\left(\overline{r}\right)\\ \overline{\varepsilon}_{zx}\left(\overline{r}\right) & \overline{\varepsilon}_{zy}\left(\overline{r}\right) & \overline{\varepsilon}_{zz}\left(\overline{r}\right)\end{bmatrix}\sqrt{Y_0}\mathbf{E}\end{aligned} \tag{41}$$

Here, all the field components are assumed to have time dependence of $\exp\left(i\omega t\right)$ and are omitted everywhere. The relative permeability of the medium is assumed to be $\mu = 1$. Note that $\varepsilon_{ij\in\{x,y,z\}}$ are defined as functions of position (x, y, z) and $\overline{\varepsilon}_{ij}$ are of $(\overline{x}, \overline{y}, \overline{z})$. λ is the vacuum wavelength of the incident wave. Variables x, y, z generally represent spatial positions while these appeared in suffix, e.g $\varepsilon_{ij\in\{x,y,z\}}$, denote the orientations along the directions $\hat{x}, \hat{y}, \hat{z}$. Moreover, the variable i is the imaginary constant number $i = \sqrt{-1}$ and that appeared in suffix, e.g. dz_i, is an integer indexing number. For liquid crystals, the dielectric matrix ε is associated with the orientation of director (θ_o, ϕ_o):

$$\varepsilon = \begin{bmatrix}\varepsilon_{xx} & \varepsilon_{xy} & \varepsilon_{xz}\\ \varepsilon_{yx} & \varepsilon_{yy} & \varepsilon_{yz}\\ \varepsilon_{zx} & \varepsilon_{zy} & \varepsilon_{zz}\end{bmatrix} \tag{42}$$

with

$$\begin{aligned}
\varepsilon_{xx} &= n_o^2 + \left(n_e^2 - n_o^2\right)\sin^2\theta_o\cos^2\phi_o,\\
\varepsilon_{xy} &= \varepsilon_{yx} = \left(n_e^2 - n_o^2\right)\sin^2\theta_o\sin\phi_o\cos\phi_o,\\
\varepsilon_{xz} &= \varepsilon_{zx} = \left(n_e^2 - n_o^2\right)\sin\theta_o\cos\theta_o\cos\phi_o,\\
\varepsilon_{yy} &= n_o^2 + \left(n_e^2 - n_o^2\right)\sin^2\theta_o\sin^2\phi_o,\\
\varepsilon_{yz} &= \varepsilon_{zy} = \left(n_e^2 - n_o^2\right)\sin\theta_o\cos\theta_o\sin\phi_o,\\
\varepsilon_{zz} &= n_o^2 + \left(n_e^2 - n_o^2\right)\cos^2\theta_o,
\end{aligned} \tag{43}$$

where n_e and n_o are extraordinary and ordinary indices of refraction of the birefringent liquid crystal, respectively, θ_o is the angle between the director and the z axis, and ϕ_o is the angle between the projection of the director on the xy plane and x axis.

A.2 Maxwell's equations in k-space descriptions

Consider the general geometry illustrated in Figure 3 of stacked multi-layer two-dimensional periodic microstructures. To apply the rigorous coupled-wave theory to the stack, all of the layers have to define the same periodicity: Λ_x along the x direction and Λ_y along the y

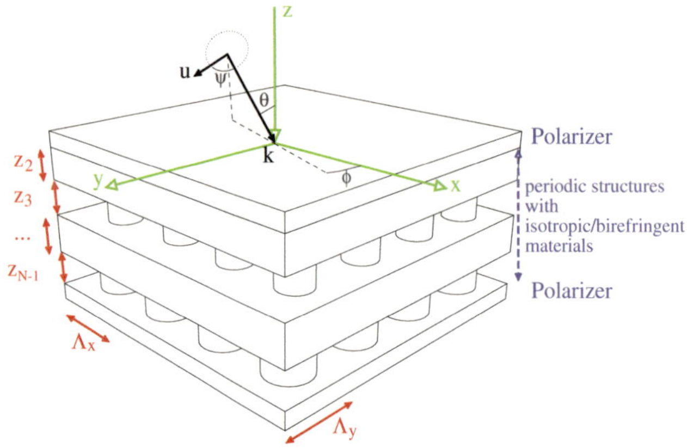

Fig. 3. Geometry of three-dimensional RCWA algorithm for a multi-layer stack with two-dimensional periodic microstructures in arbitrary isotropic and birefringent material arrangement.

direction. The thickness for the ℓ_{th} layer is dz_ℓ, and these layers contribute to a total thickness of the stack $Z_N = \sum_{\ell=1}^{N} dz_\ell$. The periodic permittivity of an individual layer in the stack can be expanded in Fourier series of the spatial harmonics as:

$$\bar{\varepsilon}_{ij}\left(\bar{x},\bar{y};\bar{z}_\ell\right) = \sum_{g,h} \bar{\varepsilon}_{ij,gh}\left(\bar{z}_\ell\right) \exp\left(i\frac{g\lambda\bar{x}}{\Lambda_x} + i\frac{h\lambda\bar{y}}{\Lambda_y}\right) \tag{44}$$

$$\bar{\varepsilon}_{ij,gh}\left(\bar{z}_\ell\right) = \frac{\lambda}{2\pi\Lambda_x}\frac{\lambda}{2\pi\Lambda_y}\int_0^{\frac{2\pi\Lambda_x}{\lambda}}\int_0^{\frac{2\pi\Lambda_y}{\lambda}} \bar{\varepsilon}_{ij}\left(\bar{x},\bar{y};\bar{z}_\ell\right)\exp\left(-i\frac{g\lambda\bar{x}}{\Lambda_x} - i\frac{h\lambda\bar{y}}{\Lambda_y}\right) d\bar{x}d\bar{y} \tag{45}$$

A similar transform for the fields in the stack can be expressed in terms of Rayleigh expansions:

$$\sqrt{Y_0}\mathbf{E}\left(\bar{x},\bar{y};\bar{z}_\ell\right) = \sum_{g,h} \mathbf{e}_{gh}\left(\bar{z}_\ell\right)\exp\left[-i\left(n_{xg}\bar{x} + n_{yh}\bar{y}\right)\right] \tag{46}$$

$$\sqrt{Z_0}\mathbf{H}\left(\bar{x},\bar{y};\bar{z}_\ell\right) = \sum_{g,h} \mathbf{h}_{gh}\left(\bar{z}_\ell\right)\exp\left[-i\left(n_{xg}\bar{x} + n_{yh}\bar{y}\right)\right] \tag{47}$$

$$n_{xg} = n_I \sin\theta\cos\phi - g\frac{\lambda}{\Lambda_x} \tag{48}$$

$$n_{yh} = n_I \sin\theta\sin\phi - h\frac{\lambda}{\Lambda_y} \tag{49}$$

where n_I (n_E) is the refraction index for the isotropic incident (emitted) region. θ, ϕ are the incident angles defined in sphere coordinates, and z is the normal direction for the xy plane of periodic structures. Here, the electric field of an incident unit-amplitude wave has been introduced by $\mathbf{E}_{inc} = \mathbf{u} \times exp\left(-i\mathbf{k}\cdot\mathbf{r}\right)$ as illustrated in figure 3, in which the wave vector $\mathbf{k}$ as well as the unit polarization vector $\mathbf{u}$ are given by:

$$\mathbf{k} = k_0 n_I \left(\sin\theta\cos\phi\hat{x} + \sin\theta\sin\phi\hat{y} + \cos\theta\hat{z}\right) \tag{50}$$

$$\begin{aligned}\mathbf{u} &= u_x\hat{x} + u_y\hat{y} + u_z\hat{z} = \left(\cos\Psi\cos\theta\cos\phi - \sin\Psi\sin\phi\right)\hat{x} \\ &\quad + \left(\cos\Psi\cos\theta\sin\phi + \sin\Psi\cos\phi\right)\hat{y} - \left(\cos\Psi\sin\theta\right)\hat{z}\end{aligned} \tag{51}$$

with the Ψ angle between the electric field vector and the incident plane.
Now we can express Maxwell's equations by the (g,h) Fourier components in k-space descriptions. For simplicity, we introduce the definitions of the tangential and normal fields at the interfaces as

$$\mathbf{f}_{\hat{t}} = \begin{bmatrix} \vec{e}_x \\ \vec{h}_y \\ \vec{e}_y \\ \vec{h}_x \end{bmatrix}, \quad \mathbf{f}_{\hat{n}} = \begin{bmatrix} \vec{e}_z \\ \vec{h}_z \end{bmatrix} \tag{52}$$

Here, $\vec{e}_{i\in\{x,y,z\}} = \vec{e}_i\left(\overline{z}_\ell\right)$ and $\vec{h}_{i\in\{x,y,z\}} = \vec{h}_i\left(\overline{z}_\ell\right)$ are column matrices with Fourier components $e_{i,gh}\left(\overline{z}_\ell\right)$ and $h_{i,gh}\left(\overline{z}_\ell\right)$, respectively. In the following context, a straightforward calculation to obtain the infinite set of coupled-wave equations corresponding to the infinite Fourier components is fulfilled. First, we express Maxwell's curl equations (40)-(41) in terms of the spatial x, y, z components:

$$\begin{aligned}
\overline{\nabla} \times \sqrt{Y_0}\mathbf{E} &= \left[\partial_{\overline{x}}\sqrt{Y_0}E_y - \partial_{\overline{y}}\sqrt{Y_0}E_x\right]\hat{\mathbf{z}} + \left[\partial_{\overline{y}}\sqrt{Y_0}E_z - \partial_{\overline{z}}\sqrt{Y_0}E_y\right]\hat{\mathbf{x}} \\
&\quad + \left[\partial_{\overline{z}}\sqrt{Y_0}E_x - \partial_{\overline{x}}\sqrt{Y_0}E_z\right]\hat{\mathbf{y}} \\
&= -i\sqrt{Z_0}H_z\hat{\mathbf{z}} - i\sqrt{Z_0}H_x\hat{\mathbf{x}} - i\sqrt{Z_0}H_y\hat{\mathbf{y}}
\end{aligned} \tag{53}$$

$$\begin{aligned}
\overline{\nabla} \times \sqrt{Z_0}\mathbf{H} &= \left[\partial_{\overline{x}}\sqrt{Z_0}H_y - \partial_{\overline{y}}\sqrt{Z_0}H_x\right]\hat{\mathbf{z}} + \left[\partial_{\overline{y}}\sqrt{Z_0}H_z - \partial_{\overline{z}}\sqrt{Z_0}H_y\right]\hat{\mathbf{x}} \\
&\quad + \left[\partial_{\overline{z}}\sqrt{Z_0}H_x - \partial_{\overline{x}}\sqrt{Z_0}H_z\right]\hat{\mathbf{y}} \\
&= i\sqrt{Y_0}\left[\overline{\varepsilon}\left(\overline{r}\right)\mathbf{E}\right]_z\hat{\mathbf{z}} + i\sqrt{Y_0}\left[\overline{\varepsilon}\left(\overline{r}\right)\mathbf{E}\right]_x\hat{\mathbf{x}} + i\sqrt{Y_0}\left[\overline{\varepsilon}\left(\overline{r}\right)\mathbf{E}\right]_y\hat{\mathbf{y}}
\end{aligned} \tag{54}$$

Next, we introduce the Fourier representations of $\mathbf{E}$, $\mathbf{H}$, and $\overline{\varepsilon}\left(\overline{r}\right)$ as defined in Equations (44)-(47). Maxwell's curl equations (53)-(54) can thereby be regrouped by the components of $\mathbf{f}_{\hat{t}}$ and $\mathbf{f}_{\hat{n}}$. For the component $h_{z,gh}\left(\overline{z}_\ell\right)$, the equation can be derived as:

$$\begin{aligned}
\partial_{\overline{x}}\sqrt{Y_0}E_y - \partial_{\overline{y}}\sqrt{Y_0}E_x &= \sum_{gh} -in_{xg}e_{y,gh}\left(\overline{z}_\ell\right)\exp\left[-i\left(n_{xg}\overline{x} + n_{yh}\overline{y}\right)\right] \\
&\quad - \sum_{gh} -in_{yh}e_{x,gh}\left(\overline{z}_\ell\right)\exp\left[-i\left(n_{xg}\overline{x} + n_{yh}\overline{y}\right)\right] \\
&= -i\sqrt{Z_0}H_z = -i\sum_{gh} h_{z,gh}\left(\overline{z}_\ell\right)\exp\left[-i\left(n_{xg}\overline{x} + n_{yh}\overline{y}\right)\right]
\end{aligned} \tag{55}$$

It is simplified to be:

$$h_{z,gh}\left(\overline{z}_\ell\right) = n_{xg}e_{y,gh}\left(\overline{z}_\ell\right) - n_{yh}e_{x,gh}\left(\overline{z}_\ell\right) \tag{56}$$

For the component $\frac{\partial e_{y,gh}\left(\overline{z}_\ell\right)}{\partial_{\overline{z}}}$, the equation can be derived as:

$$\begin{aligned}
\partial_{\overline{y}}\sqrt{Y_0}E_z - \partial_{\overline{z}}\sqrt{Y_0}E_y &= \sum_{gh} -in_{yh}e_{z,gh}\left(\overline{z}_\ell\right)\exp\left[-i\left(n_{xg}\overline{x} + n_{yh}\overline{y}\right)\right] \\
&\quad - \sum_{gh} \frac{\partial e_{y,gh}\left(\overline{z}_\ell\right)}{\partial_{\overline{z}}}\exp\left[-i\left(n_{xg}\overline{x} + n_{yh}\overline{y}\right)\right] \\
&= -i\sqrt{Z_0}H_x = -i\sum_{gh} h_{x,gh}\left(\overline{z}_\ell\right)\exp\left[-i\left(n_{xg}\overline{x} + n_{yh}\overline{y}\right)\right]
\end{aligned} \tag{57}$$

and is simplified to be:

$$\frac{\partial e_{y,gh}\left(\overline{z}_{\ell}\right)}{\partial_{\overline{z}}} = -in_{yh}e_{z,gh}\left(\overline{z}_{\ell}\right) + ih_{x,gh}\left(\overline{z}_{\ell}\right) \tag{58}$$

For the component $\frac{\partial e_{x,gh}\left(\overline{z}_{\ell}\right)}{\partial_{\overline{z}}}$, the equation can be derived as:

$$\begin{aligned}
\partial_{\overline{z}}\sqrt{Y_0}E_x - \partial_{\overline{x}}\sqrt{Y_0}E_z &= \sum_{m}\frac{\partial e_{x,gh}\left(\overline{z}_{\ell}\right)}{\partial_{\overline{z}}}\exp\left[-i\left(n_{xg}\overline{x}+n_{yh}\overline{y}\right)\right] \\
&\quad -\sum_{gh}-in_{xg}e_{z,gh}\left(\overline{z}_{\ell}\right)\exp\left[-i\left(n_{xg}\overline{x}+n_{yh}\overline{y}\right)\right] \\
&= -i\sqrt{Z_0}H_y \\
&= -i\sum_{gh}h_{y,gh}\left(\overline{z}_{\ell}\right)\exp\left[-i\left(n_{xg}\overline{x}+n_{yh}\overline{y}\right)\right]
\end{aligned} \tag{59}$$

and is simplified to be:

$$\frac{\partial e_{x,gh}\left(\overline{z}_{\ell}\right)}{\partial_{\overline{z}}} = -in_{xg}e_{z,gh}\left(\overline{z}_{\ell}\right) - ih_{y,gh}\left(\overline{z}_{\ell}\right) \tag{60}$$

For the component $e_{z,gh}$, the equation can be derived as:

$$\begin{aligned}
\partial_{\overline{x}}\sqrt{Z_0}H_y - \partial_{\overline{y}}\sqrt{Z_0}H_x &= \sum_{gh}-in_{xg}h_{y,gh}\left(\overline{z}_{\ell}\right)\exp\left[-i\left(n_{xg}\overline{x}+n_{yh}\overline{y}\right)\right] \\
&\quad -\sum_{gh}-in_{yh}h_{x,gh}\left(\overline{z}_{\ell}\right)\exp\left[-i\left(n_{xg}\overline{x}+n_{yh}\overline{y}\right)\right] \\
&= i\sqrt{Y_0}\left[\overline{\varepsilon}\left(\overline{r}\right)\mathbf{E}\right]_z \\
&= i\sum_{ghuv}\overline{\varepsilon}_{zx,uv}e_{x,gh}\left(\overline{z}_{\ell}\right)\exp\left[-i\left(n_{x(g+u)}\overline{x}+n_{y(h+v)}\overline{y}\right)\right] \\
&\quad +i\sum_{ghuv}\overline{\varepsilon}_{zy,uv}e_{y,gh}\left(\overline{z}_{\ell}\right)\exp\left[-i\left(n_{x(g+u)}\overline{x}+n_{y(h+v)}\overline{y}\right)\right] \\
&\quad +i\sum_{ghuv}\overline{\varepsilon}_{zz,uv}e_{z,gh}\left(\overline{z}_{\ell}\right)\exp\left[-i\left(n_{x(g+u)}\overline{x}+n_{y(h+v)}\overline{y}\right)\right]
\end{aligned} \tag{61}$$

and is simplified to be:

$$\begin{aligned}
n_{yh}h_{x,gh}\left(\overline{z}_{\ell}\right) - n_{xg}h_{y,gh}\left(\overline{z}_{\ell}\right) &= \sum_{u'v'}\overline{\varepsilon}_{zx,(g-u')(h-v')}e_{x,u'v'}\left(\overline{z}_{\ell}\right) \\
&\quad +\sum_{u'v'}\overline{\varepsilon}_{zy,(g-u')(h-v')}e_{y,u'v'}\left(\overline{z}_{\ell}\right) \\
&\quad +\sum_{u'v'}\overline{\varepsilon}_{zz,(g-u')(h-v')}e_{z,u'v'}\left(\overline{z}_{\ell}\right)
\end{aligned} \tag{62}$$

For the component $\frac{\partial h_{y,gh}(\overline{z}_\ell)}{\partial_{\overline{z}}}$, the equation can be derived as:

$$\begin{aligned}
\partial_{\overline{y}}\sqrt{Z_0}H_z - \partial_{\overline{z}}\sqrt{Z_0}H_y &= \sum_{gh} -in_{yh}h_{z,gh}\left(\overline{z}_\ell\right)\exp\left[-i\left(n_{xg}\overline{x}+n_{yh}\overline{y}\right)\right] \\
&\quad -\sum_{m}\frac{\partial h_{y,gh}\left(\overline{z}_\ell\right)}{\partial_{\overline{z}}}\exp\left[-i\left(n_{xg}\overline{x}+n_{yh}\overline{y}\right)\right] \\
&= i\sqrt{Y_0}\left[\overline{\varepsilon}\left(\overline{r}\right)\mathbf{E}\right]_x \\
&= i\sum_{ghuv}\overline{\varepsilon}_{xx,uv}e_{x,gh}\left(\overline{z}_\ell\right)\exp\left[-i\left(n_{x(g+u)}\overline{x}+n_{y(h+v)}\overline{y}\right)\right] \\
&\quad +i\sum_{ghuv}\overline{\varepsilon}_{xy,uv}e_{y,gh}\left(\overline{z}_\ell\right)\exp\left[-i\left(n_{x(g+u)}\overline{x}+n_{y(h+v)}\overline{y}\right)\right] \\
&\quad +i\sum_{ghuv}\overline{\varepsilon}_{xz,uv}e_{z,gh}\left(\overline{z}_\ell\right)\exp\left[-i\left(n_{x(g+u)}\overline{x}+n_{y(h+v)}\overline{y}\right)\right] \qquad (63)
\end{aligned}$$

and is simplified to be:

$$\begin{aligned}
\frac{\partial h_{y,gh}\left(\overline{z}_\ell\right)}{\partial_{\overline{z}}} &= -in_{yh}h_{z,gh}\left(\overline{z}_\ell\right) - i\sum_{u'v'}\overline{\varepsilon}_{xx,(g-u')(h-v')}e_{x,u'v'}\left(\overline{z}_\ell\right) \\
&\quad -i\sum_{u'v'}\overline{\varepsilon}_{xy,(g-u')(h-v')}e_{y,u'v'}\left(\overline{z}_\ell\right) - i\sum_{u'v'}\overline{\varepsilon}_{xz,(g-u')(h-v')}e_{z,u'v'}\left(\overline{z}_\ell\right) \qquad (64)
\end{aligned}$$

For the component $\frac{\partial h_{x,gh}(\overline{z}_\ell)}{\partial_{\overline{z}}}$, the equation can be derived as:

$$\begin{aligned}
\partial_{\overline{z}}\sqrt{Z_0}H_x - \partial_{\overline{x}}\sqrt{Z_0}H_z &= \sum_{gh}\frac{\partial h_{x,gh}\left(\overline{z}_\ell\right)}{\partial_{\overline{z}}}\exp\left[-i\left(n_{xg}\overline{x}+n_{yh}\overline{y}\right)\right] \\
&\quad -\sum_{gh} -in_{xh}h_{z,gh}\left(\overline{z}_\ell\right)\exp\left[-i\left(n_{xg}\overline{x}+n_{yh}\overline{y}\right)\right] \\
&= i\sqrt{Y_0}\left[\overline{\varepsilon}\left(\overline{r}\right)\mathbf{E}\right]_y \\
&= i\sum_{ghuv}\overline{\varepsilon}_{yx,uv}e_{x,gh}\left(\overline{z}_\ell\right)\exp\left[-i\left(n_{x(g+u)}\overline{x}+n_{y(h+v)}\overline{y}\right)\right] \\
&\quad +i\sum_{ghuv}\overline{\varepsilon}_{yy,uv}e_{y,gh}\left(\overline{z}_\ell\right)\exp\left[-i\left(n_{x(g+u)}\overline{x}+n_{y(h+v)}\overline{y}\right)\right] \\
&\quad +i\sum_{ghuv}\overline{\varepsilon}_{yz,uv}e_{z,gh}\left(\overline{z}_\ell\right)\exp\left[-i\left(n_{x(g+u)}\overline{x}+n_{y(h+v)}\overline{y}\right)\right] \qquad (65)
\end{aligned}$$

and is simplified to be:

$$\begin{aligned}
\frac{\partial h_{x,gh}\left(\overline{z}_\ell\right)}{\partial_{\overline{z}}} &= -in_{xh}h_{z,gh}\left(\overline{z}_\ell\right) + i\sum_{u'v'}\overline{\varepsilon}_{yx,(g-u')(h-v')}e_{x,u'v'}\left(\overline{z}_\ell\right) \\
&\quad +i\sum_{u'v'}\overline{\varepsilon}_{yy,(g-u')(h-v')}e_{y,u'v'}\left(\overline{z}_\ell\right) + i\sum_{u'v'}\overline{\varepsilon}_{yz,(g-u')(h-v')}e_{z,u'v'}\left(\overline{z}_\ell\right) \qquad (66)
\end{aligned}$$

To solve the fields systematically, these equations are reformulated in terms of the full fields $\mathbf{f}_{\hat{t}}$ and $\mathbf{f}_{\hat{n}}$ in the following context, and show an eigen-system problem for studied periodic structures.

A.3 Derive the coupled-wave equation of the normal field $\mathbf{f}_{\hat{n}}$

To obtain the coupled-wave equations of the normal field $\mathbf{f}_{\hat{n}}$, we consider the above-mentioned formulas for its components $h_{z,gh}(\overline{z}_\ell)$ and $e_{z,gh}(\overline{z}_\ell)$ in Equations (56) and (62), respectively, i.e.:

$$h_{z,gh}(\overline{z}_\ell) = n_{xg}e_{y,gh}(\overline{z}_\ell) - n_{yh}e_{x,gh}(\overline{z}_\ell) \tag{56}$$

$$\begin{aligned} n_{yh}h_{x,gh}(\overline{z}_\ell) - n_{xg}h_{y,gh}(\overline{z}_\ell) &= \sum_{u'v'} \overline{\varepsilon}_{zx,(g-u')(h-v')}e_{x,u'v'}(\overline{z}_\ell) \\ &+ \sum_{u'v'} \overline{\varepsilon}_{zy,(g-u')(h-v')}e_{y,u'v'}(\overline{z}_\ell) \\ &+ \sum_{u'v'} \overline{\varepsilon}_{zz,(g-u')(h-v')}e_{z,u'v'}(\overline{z}_\ell) \end{aligned} \tag{62}$$

Up to the Fourier order $g,h \in \{0,1\}$, an example corresponding to Equations (56) and (62) can be matrixized:

$$\begin{bmatrix} h_{z,00}(\overline{z}_\ell) \\ h_{z,01}(\overline{z}_\ell) \\ h_{z,10}(\overline{z}_\ell) \\ h_{z,11}(\overline{z}_\ell) \end{bmatrix} = \begin{bmatrix} n_{x0} & 0 & 0 & 0 \\ 0 & n_{x0} & 0 & 0 \\ 0 & 0 & n_{x1} & 0 \\ 0 & 0 & 0 & n_{x1} \end{bmatrix} \begin{bmatrix} e_{y,00}(\overline{z}_\ell) \\ e_{y,01}(\overline{z}_\ell) \\ e_{y,10}(\overline{z}_\ell) \\ e_{y,11}(\overline{z}_\ell) \end{bmatrix} - \begin{bmatrix} n_{y0} & 0 & 0 & 0 \\ 0 & n_{y1} & 0 & 0 \\ 0 & 0 & n_{y0} & 0 \\ 0 & 0 & 0 & n_{y1} \end{bmatrix} \begin{bmatrix} e_{x,00}(\overline{z}_\ell) \\ e_{x,01}(\overline{z}_\ell) \\ e_{x,10}(\overline{z}_\ell) \\ e_{x,11}(\overline{z}_\ell) \end{bmatrix} \tag{67}$$

$$\begin{aligned} &\begin{bmatrix} \overline{\varepsilon}_{zz,00} & \overline{\varepsilon}_{zz,0-1} & \overline{\varepsilon}_{zz,-10} & \overline{\varepsilon}_{zz,-1-1} \\ \overline{\varepsilon}_{zz,01} & \overline{\varepsilon}_{zz,00} & \overline{\varepsilon}_{zz,-11} & \overline{\varepsilon}_{zz,-10} \\ \overline{\varepsilon}_{zz,10} & \overline{\varepsilon}_{zz,1-1} & \overline{\varepsilon}_{zz,00} & \overline{\varepsilon}_{zz,0-1} \\ \overline{\varepsilon}_{zz,11} & \overline{\varepsilon}_{zz,10} & \overline{\varepsilon}_{zz,01} & \overline{\varepsilon}_{zz,00} \end{bmatrix} \begin{bmatrix} e_{z,00}(\overline{z}_\ell) \\ e_{z,01}(\overline{z}_\ell) \\ e_{z,10}(\overline{z}_\ell) \\ e_{z,11}(\overline{z}_\ell) \end{bmatrix} = \begin{bmatrix} n_{y0} & 0 & 0 & 0 \\ 0 & n_{y1} & 0 & 0 \\ 0 & 0 & n_{y0} & 0 \\ 0 & 0 & 0 & n_{y1} \end{bmatrix} \begin{bmatrix} h_{x,00}(\overline{z}_\ell) \\ h_{x,01}(\overline{z}_\ell) \\ h_{x,10}(\overline{z}_\ell) \\ h_{x,11}(\overline{z}_\ell) \end{bmatrix} \\ &- \begin{bmatrix} n_{x0} & 0 & 0 & 0 \\ 0 & n_{x0} & 0 & 0 \\ 0 & 0 & n_{x1} & 0 \\ 0 & 0 & 0 & n_{x1} \end{bmatrix} \begin{bmatrix} h_{y,00}(\overline{z}_\ell) \\ h_{y,01}(\overline{z}_\ell) \\ h_{y,10}(\overline{z}_\ell) \\ h_{y,11}(\overline{z}_\ell) \end{bmatrix} - \begin{bmatrix} \overline{\varepsilon}_{zx,00} & \overline{\varepsilon}_{zx,0-1} & \overline{\varepsilon}_{zx,-10} & \overline{\varepsilon}_{zx,-1-1} \\ \overline{\varepsilon}_{zx,01} & \overline{\varepsilon}_{zx,00} & \overline{\varepsilon}_{zx,-11} & \overline{\varepsilon}_{zx,-10} \\ \overline{\varepsilon}_{zx,10} & \overline{\varepsilon}_{zx,1-1} & \overline{\varepsilon}_{zx,00} & \overline{\varepsilon}_{zx,0-1} \\ \overline{\varepsilon}_{zx,11} & \overline{\varepsilon}_{zx,10} & \overline{\varepsilon}_{zx,01} & \overline{\varepsilon}_{zx,00} \end{bmatrix} \begin{bmatrix} e_{x,00}(\overline{z}_\ell) \\ e_{x,01}(\overline{z}_\ell) \\ e_{x,10}(\overline{z}_\ell) \\ e_{x,11}(\overline{z}_\ell) \end{bmatrix} \\ &- \begin{bmatrix} \overline{\varepsilon}_{zy,00} & \overline{\varepsilon}_{zy,0-1} & \overline{\varepsilon}_{zy,-10} & \overline{\varepsilon}_{zy,-1-1} \\ \overline{\varepsilon}_{zy,01} & \overline{\varepsilon}_{zy,00} & \overline{\varepsilon}_{zy,-11} & \overline{\varepsilon}_{zy,-10} \\ \overline{\varepsilon}_{zy,10} & \overline{\varepsilon}_{zy,1-1} & \overline{\varepsilon}_{zy,00} & \overline{\varepsilon}_{zy,0-1} \\ \overline{\varepsilon}_{zy,11} & \overline{\varepsilon}_{zy,10} & \overline{\varepsilon}_{zy,01} & \overline{\varepsilon}_{zy,00} \end{bmatrix} \begin{bmatrix} e_{y,00}(\overline{z}_\ell) \\ e_{y,01}(\overline{z}_\ell) \\ e_{y,10}(\overline{z}_\ell) \\ e_{y,11}(\overline{z}_\ell) \end{bmatrix} \end{aligned} \tag{68}$$

The full-component coupled-wave equation for the normal field $\mathbf{f}_{\hat{n}}$ then can be extended as:

$$\mathbf{f}_{\hat{n}} = \begin{bmatrix} \vec{e}_z \\ \vec{h}_z \end{bmatrix} = \begin{bmatrix} -\tilde{\varepsilon}_{zz}^{-1}\tilde{\varepsilon}_{zx} & -\tilde{\varepsilon}_{zz}^{-1}\tilde{n}_x & -\tilde{\varepsilon}_{zz}^{-1}\tilde{\varepsilon}_{zy} & \tilde{\varepsilon}_{zz}^{-1}\tilde{n}_y \\ -\tilde{n}_y & 0 & \tilde{n}_x & 0 \end{bmatrix} \cdot \begin{bmatrix} \vec{e}_x \\ \vec{h}_y \\ \vec{e}_y \\ \vec{h}_x \end{bmatrix} \equiv \mathbf{D} \cdot \mathbf{f}_{\hat{t}} \tag{69}$$

Here, the symbol $\vec{(.)}$ represents a $N_g N_h \times 1$ vector, and the symbol $\tilde{(.)}$ represents a $N_g N_h \times N_g N_h$ matrix, indicating the considered $g(h)$ ranged from $g_{min}(h_{min})$ to $g_{max}(h_{max})$ with $N_g = g_{min} + g_{max} + 1 (N_h = h_{min} + h_{max} + 1)$. As indicated in Equation (69), the normal field $\mathbf{f}_{\hat{n}}$ can be obtained straightforwardly if the tangential field $\mathbf{f}_{\hat{t}}$ is given.

A.4 Derive the coupled-wave equation of the tangential field $\mathbf{f}_{\hat{t}}$

Further, we derive the coupled-wave equation of the tangential field $\mathbf{f}_{\hat{t}}$, in which the component fields of $\mathbf{f}_{\hat{n}}$ are replaced by those of $\mathbf{f}_{\hat{t}}$ via equation (69). Similarly, we consider the associated formulas of $\frac{\partial e_{y,gh}(\bar{z}_\ell)}{\partial \bar{z}}$, $\frac{\partial e_{x,gh}(\bar{z}_\ell)}{\partial \bar{z}}$, $\frac{\partial h_{y,gh}(\bar{z}_\ell)}{\partial \bar{z}}$, and $\frac{\partial h_{x,gh}(\bar{z}_\ell)}{\partial \bar{z}}$ in Equations (58), (60), (64), and (66), respectively, i.e.:

$$\frac{\partial e_{x,gh}(\bar{z}_\ell)}{\partial \bar{z}} = -in_{xg} e_{z,gh}(\bar{z}_\ell) - ih_{y,gh}(\bar{z}_\ell) \tag{60}$$

$$\frac{\partial e_{y,gh}(\bar{z}_\ell)}{\partial \bar{z}} = -in_{yh} e_{z,gh}(\bar{z}_\ell) + ih_{x,gh}(\bar{z}_\ell) \tag{58}$$

$$\begin{aligned}\frac{\partial h_{y,gh}(\bar{z}_\ell)}{\partial \bar{z}} &= -in_{yh} h_{z,gh}(\bar{z}_\ell) - i\sum_{u'v'} \bar{\varepsilon}_{xx,(g-u')(h-v')} e_{x,u'v'}(\bar{z}_\ell) \\ &\quad -i\sum_{u'v'} \bar{\varepsilon}_{xy,(g-u')(h-v')} e_{y,u'v'}(\bar{z}_\ell) - i\sum_{u'v'} \bar{\varepsilon}_{xz,(g-u')(h-v')} e_{z,u'v'}(\bar{z}_\ell)\end{aligned} \tag{64}$$

$$\begin{aligned}\frac{\partial h_{x,gh}(\bar{z}_\ell)}{\partial \bar{z}} &= -in_{xh} h_{z,gh}(\bar{z}_\ell) + i\sum_{u'v'} \bar{\varepsilon}_{yx,(g-u')(h-v')} e_{x,u'v'}(\bar{z}_\ell) \\ &\quad +i\sum_{u'v'} \bar{\varepsilon}_{yy,(g-u')(h-v')} e_{y,u'v'}(\bar{z}_\ell) + i\sum_{u'v'} \bar{\varepsilon}_{yz,(g-u')(h-v')} e_{z,u'v'}(\bar{z}_\ell)\end{aligned} \tag{66}$$

With equation (69), these equations can matrixize the coupled-wave equation of $\mathbf{f}_{\hat{t}}$ as:

$$\begin{aligned}\frac{\partial \mathbf{f}_{\hat{t}}}{\partial \bar{z}} &= i \begin{bmatrix} 0 & -1 & 0 & 0 \\ -\tilde{\varepsilon}_{xx} & 0 & -\tilde{\varepsilon}_{xy} & 0 \\ 0 & 0 & 0 & 1 \\ \tilde{\varepsilon}_{yx} & 0 & \tilde{\varepsilon}_{yy} & 0 \end{bmatrix} \begin{bmatrix} \vec{e}_x \\ \vec{h}_y \\ \vec{e}_y \\ \vec{h}_x \end{bmatrix} + i \begin{bmatrix} -\tilde{n}_x \vec{e}_z \\ -\tilde{n}_y \vec{h}_z - \tilde{\varepsilon}_{xz} \vec{e}_z \\ -\tilde{n}_y \vec{e}_z \\ -\tilde{n}_x \vec{h}_z + \tilde{\varepsilon}_{yz} \vec{e}_z \end{bmatrix} \\ &= \begin{bmatrix} \tilde{n}_x \tilde{\varepsilon}_{zz}^{-1} \tilde{\varepsilon}_{zx} & \tilde{n}_x \tilde{\varepsilon}_{zz}^{-1} \tilde{n}_x - 1 & \tilde{n}_x \tilde{\varepsilon}_{zz}^{-1} \tilde{\varepsilon}_{zy} & -\tilde{n}_x \tilde{\varepsilon}_{zz}^{-1} \tilde{n}_y \\ \tilde{\varepsilon}_{xz} \tilde{\varepsilon}_{zz}^{-1} \tilde{\varepsilon}_{zx} - \tilde{\varepsilon}_{xx} + \tilde{n}_y \tilde{n}_y & \tilde{\varepsilon}_{xz} \tilde{\varepsilon}_{zz}^{-1} \tilde{n}_x & \tilde{\varepsilon}_{xz} \tilde{\varepsilon}_{zz}^{-1} \tilde{\varepsilon}_{zy} - \tilde{\varepsilon}_{xy} - \tilde{n}_y \tilde{n}_x & -\tilde{\varepsilon}_{xz} \tilde{\varepsilon}_{zz}^{-1} \tilde{n}_y \\ \tilde{n}_y \tilde{\varepsilon}_{zz}^{-1} \tilde{\varepsilon}_{zx} & \tilde{n}_y \tilde{\varepsilon}_{zz}^{-1} \tilde{n}_x & \tilde{n}_y \tilde{\varepsilon}_{zz}^{-1} \tilde{\varepsilon}_{zy} & -\tilde{n}_y \tilde{\varepsilon}_{zz}^{-1} \tilde{n}_y + 1 \\ -\tilde{\varepsilon}_{yz} \tilde{\varepsilon}_{zz}^{-1} \tilde{\varepsilon}_{zx} + \tilde{\varepsilon}_{yx} + \tilde{n}_x \tilde{n}_y & -\tilde{\varepsilon}_{yz} \tilde{\varepsilon}_{zz}^{-1} \tilde{n}_x & -\tilde{\varepsilon}_{yz} \tilde{\varepsilon}_{zz}^{-1} \tilde{\varepsilon}_{zy} + \tilde{\varepsilon}_{yy} - \tilde{n}_x \tilde{n}_x & \tilde{\varepsilon}_{yz} \tilde{\varepsilon}_{zz}^{-1} \tilde{n}_y \end{bmatrix} \\ &\quad \cdot i\mathbf{f}_{\hat{t}} \equiv i\mathbf{G} \cdot \mathbf{f}_{\hat{t}}\end{aligned} \tag{70}$$

Definitely, the equation (70) turns the Maxwell's curl equations into a eigen-system problems. Up to now, with the known structured layers for equations (44)-(45) and the known incidence related to equations (46)-(47), the transition behaviors of the tangential field $\mathbf{f}_{\hat{t}}$ can be formulated layer by layer via equation (70), and the corresponding normal field $\mathbf{f}_{\hat{n}}$ can be evaluated sequentially via equation (69).

In the following contexts, we continue to describe (a) the solutions of the transition fields within stack layers via equation (70), especially for these uniform layers with isotropic materials which bring in the degenerate eigen-states, and (b) the continuum of fields conditioned at interfaces between stack layers. Consequently, a complete analysis for fields through all stacks can be fulfilled, and the associated near/far field optics can be evaluated.

A.5 Eigen-system solutions

As indicated in equation (70), the tangential fields $\mathbf{f}_{\hat{t}}$ within the layers proceed an eigen-system process, in which the eigen-states are independent to each other and allow individual/straightforward analyses to evaluate the transition behaviors through the layers. At the interfaces among the layers, the tangential fields $\mathbf{f}_{\hat{t}}$ associated with the composite phases/amplitudes of the eigen-states follow the physical continuous conditions in the laboratory framework. These characteristics lead to the necessary transform between the laboratory and eigen-system frameworks as described below. Besides, for these uniform layers with isotropic materials, especially for the incident and emitted regions, the eigen-system shows the degenerate status, and a reasonable choice of the eigen-states corresponding to the physical conditions is emphasized below. Implemented with all these, the behaviors of the tangential fields $\mathbf{f}_{\hat{t}}$ through all stacks layers including the in-between interfaces can be decided. The normal fields $\mathbf{f}_{\hat{n}}$ are then obtained by equation (69), and thereby the complete light waves are understood.

A.5.1 Uniform layers with isotropic materials

For the uniform layers with isotropic materials, i.e. $\varepsilon(\overline{r})$ is a scalar constant, the coupled-wave equation of the tangential fields $\mathbf{f}_{\hat{t}}$ in equation (70) can be simplified as:

$$\frac{\partial}{\partial \overline{z}}\begin{bmatrix}\vec{e}_x\\ \vec{h}_y\\ \vec{e}_y\\ \vec{h}_x\end{bmatrix} = i\mathbf{C}\cdot\begin{bmatrix}\vec{e}_x\\ \vec{h}_y\\ \vec{e}_y\\ \vec{h}_x\end{bmatrix}$$

$$= i\begin{bmatrix}0 & \tilde{n}_x\tilde{\varepsilon}^{-1}\tilde{n}_x-1 & 0 & -\tilde{n}_x\tilde{\varepsilon}^{-1}\tilde{n}_y\\ -\tilde{\varepsilon}+\tilde{n}_y\tilde{n}_y & 0 & -\tilde{n}_y\tilde{n}_x & 0\\ 0 & \tilde{n}_y\tilde{\varepsilon}^{-1}\tilde{n}_x & 0 & -\tilde{n}_y\tilde{\varepsilon}^{-1}\tilde{n}_y+1\\ \tilde{n}_x\tilde{n}_y & 0 & \tilde{\varepsilon}-\tilde{n}_x\tilde{n}_x & 0\end{bmatrix}\cdot\begin{bmatrix}\vec{e}_x\\ \vec{h}_y\\ \vec{e}_y\\ \vec{h}_x\end{bmatrix} \tag{71}$$

Here, all the submatrices in $\mathbf{C}$ are diagonal and thereby the component states are independent. By straightforward calculation, its eigen-values as well as the corresponding eigen-vectors for (g,h)-order component can be obtained:

$$eigval \equiv \kappa_{gh} = \begin{bmatrix}-\zeta_{gh} & 0 & 0 & 0\\ 0 & -\zeta_{gh} & 0 & 0\\ 0 & 0 & \zeta_{gh} & 0\\ 0 & 0 & 0 & \zeta_{gh}\end{bmatrix} \tag{72}$$

with $\xi_{gh} = \sqrt{\varepsilon - n_{yh}n_{yh} - n_{xg}n_{xg}}$ while the corresponding eigen-vector matrix are:

$$eigvec = \left[\mathbf{v}'_{gh1} \; \mathbf{v}'_{gh2} \; \mathbf{v}'_{gh3} \; \mathbf{v}'_{gh4} \right]$$

$$= \begin{bmatrix} \frac{-\xi_{gh}}{n_{xg}n_{yh}} & \frac{n_{xg}n_{xg}-\varepsilon_{gh}}{n_{xg}n_{yh}} & \frac{\xi_{gh}}{n_{xg}n_{yh}} & \frac{n_{xg}n_{xg}-\varepsilon_{gh}}{n_{xg}n_{yh}} \\ \frac{n_{yh}n_{yh}-\varepsilon_{gh}}{n_{xg}n_{yh}} & \frac{-\varepsilon_{gh}\xi_{gh}}{n_{xg}n_{yh}} & \frac{n_{yh}n_{yh}-\varepsilon_{gh}}{n_{xg}n_{yh}} & \frac{\varepsilon_{gh}\xi_{gh}}{n_{xg}n_{yh}} \\ 0 & 1 & 0 & 1 \\ 1 & 0 & 1 & 0 \end{bmatrix} \tag{73}$$

Due to the degeneracy in $(\kappa_{gh,1}, \kappa_{gh,2})$ and $(\kappa_{gh,3}, \kappa_{gh,4})$, the eigenvector $(\mathbf{v}'_{gh1}, \mathbf{v}'_{gh2})$ as well as $(\mathbf{v}'_{gh3}, \mathbf{v}'_{gh4})$ can be remixed by arbitrary linear combinations. Choosing

$$m_{gh} = \sqrt{n_{yh}n_{yh} + n_{xg}n_{xg}} \tag{74}$$

$$\mathbf{v}_{gh1} = \left(n_{xg}\xi_{gh}\mathbf{v}'_{gh1} - n_{xg}\mathbf{v}'_{gh2} \right) / m_{gh} \tag{75}$$

$$\mathbf{v}_{gh2} = \left(-\frac{\varepsilon_{gh}n_{yh}}{\xi_{gh}}\mathbf{v}'_{gh1} + n_{yh}\mathbf{v}'_{gh2} \right) / m_{gh} \tag{76}$$

$$\mathbf{v}_{gh3} = \left(-n_{xg}\xi_{gh}\mathbf{v}'_{gh3} - n_{xg}\mathbf{v}'_{gh4} \right) / m_{gh} \tag{77}$$

$$\mathbf{v}_{gh4} = \left(\frac{\varepsilon_{gh}n_{yh}}{\xi_{gh}}\mathbf{v}'_{gh3} + n_{yh}\mathbf{v}'_{gh4} \right) / m_{gh} \tag{78}$$

the equation (73) is then shown as:

$$eigvec = \mathbf{T}_{gh} = \left[\mathbf{v}_{gh1} \; \mathbf{v}_{gh2} \; \mathbf{v}_{gh3} \; \mathbf{v}_{gh4} \right]$$

$$= \begin{bmatrix} \frac{n_{yh}}{m_{gh}} & \frac{n_{xg}}{m_{gh}} & \frac{n_{yh}}{m_{gh}} & \frac{n_{xg}}{m_{gh}} \\ \frac{n_{yh}\xi_{gh}}{m_{gh}} & \frac{\varepsilon_{gh}n_{xg}\xi_{gh}^{-1}}{m_{gh}} & \frac{-n_{yh}\xi_{gh}}{m_{gh}} & \frac{-\varepsilon_{gh}n_{xg}\xi_{gh}^{-1}}{m_{gh}} \\ \frac{-n_{xg}}{m_{gh}} & \frac{n_{yh}}{m_{gh}} & \frac{-n_{xg}}{m_{gh}} & \frac{n_{yh}}{m_{gh}} \\ \frac{n_{xg}\xi_{gh}}{m_{gh}} & \frac{-\varepsilon_{gh}n_{yh}\xi_{gh}^{-1}}{m_{gh}} & \frac{-n_{xg}\xi_{gh}}{m_{gh}} & \frac{\varepsilon_{gh}n_{yh}\xi_{gh}^{-1}}{m_{gh}} \end{bmatrix} \tag{79}$$

Hence, $\mathbf{v}_{gh1}$ and $\mathbf{v}_{gh2}$ correspond to the (g, h)-order forward TE and TM (transverse electric and transverse magnetic) representations (with respect to the plane of the diffraction wave), respectively. $\mathbf{v}_{gh3}$ and $\mathbf{v}_{gh4}$ then correspond to the backward TE and TM ones. For example, with equation (69) and (79), $\mathbf{v}_{gh1}$ denotes the component fields:

$$\mathbf{e}_{gh} = \frac{n_{yh}}{m_{gh}}\hat{\imath} - \frac{n_{xg}}{m_{gh}}\hat{\jmath} \tag{80}$$

$$\mathbf{h}_{gh} = \frac{n_{xg}\xi_{gh}}{m_{gh}}\hat{\imath} + \frac{n_{yh}\xi_{gh}}{m_{gh}}\hat{\jmath} - m_{gh}\hat{k} \tag{81}$$

along the direction $\mathbf{n_{gh}} = n_{xg}\hat{\imath} + n_{yh}\hat{\jmath} + \xi_{gh}\hat{k}$. It can be seen that the characteristic fields in equations (80)-(81) associated with the eigen-solution $\propto exp(-i\xi_{gh}z)$ and constitutes the forwards TE wave. It is noted that the field amplitudes are normalized to $|\mathbf{e}_{gh}| = 1$, $|\mathbf{h}_{gh}| = \sqrt{\varepsilon}$, and $\mathbf{e}_{gh} \cdot \mathbf{n}_{gh} = \mathbf{h}_{gh} \cdot \mathbf{n}_{gh} = \mathbf{e}_{gh} \cdot \mathbf{h}_{gh} = 0$ - that is, $\mathbf{n}_{gh}$, $\mathbf{e}_{gh}$, and $\mathbf{h}_{gh}$ are mutually perpendicular. Similarly, the remaining eigen-vectors can characterize the forwards and backwards TE/TM waves and are omitted here. In this way, a unit-amplitude incident wave then can be given as $\vec{E}_q^+ = [0...1...0]^t, \vec{M}_q^+ = 0$ for TE wave, and $\vec{M}_q^+ = [0...1...0]^t, \vec{E}_q^+ = 0$ for TM wave as defined below.
Considering the full components $g_{min} \leq g \leq g_{max}$ and $h_{min} \leq h \leq h_{max}$, the coupled-wave equation (71) can be straightforwardly written as:

$$\frac{\partial}{\partial_{\overline{z}}}\mathbf{f}_{\hat{t}} = i\mathbf{C}f_{\hat{t}}$$

$$\Rightarrow \frac{\partial}{\partial_{\overline{z}}}\mathbf{T}^{-1}\mathbf{f}_{\hat{t}} = i\mathbf{T}^{-1}\mathbf{C}\mathbf{T}\mathbf{T}^{-1}f_{\hat{t}}$$

$$\Rightarrow \frac{\partial}{\partial_{\overline{z}}}\mathbf{q}_{\hat{t}} = i\kappa\mathbf{q}_{\hat{t}} \quad with \;\; \mathbf{f}_{\hat{t}} = \mathbf{T}\mathbf{q}_{\hat{t}} \tag{82}$$

where

$$\mathbf{T} = \begin{bmatrix} \dot{\mathbf{n}}_y & \dot{\mathbf{n}}_x & \dot{\mathbf{n}}_y & \dot{\mathbf{n}}_x \\ \dot{\mathbf{n}}_y\xi & \varepsilon\dot{\mathbf{n}}_x\xi^{-1} & -\dot{\mathbf{n}}_y\xi & -\varepsilon\dot{\mathbf{n}}_x\xi^{-1} \\ -\dot{\mathbf{n}}_x & \dot{\mathbf{n}}_y & -\dot{\mathbf{n}}_x & \dot{\mathbf{n}}_y \\ \dot{\mathbf{n}}_x\xi & -\varepsilon\dot{\mathbf{n}}_y\xi^{-1} & -\dot{\mathbf{n}}_x\xi & \varepsilon\dot{\mathbf{n}}_y\xi^{-1} \end{bmatrix}, \quad \mathbf{q}_{\hat{t}} = \begin{bmatrix} \vec{E}_q^+ \\ \vec{M}_q^+ \\ \vec{E}_q^- \\ \vec{M}_q^- \end{bmatrix} \tag{83}$$

Note that $\dot{\mathbf{n}}_y$ and $\dot{\mathbf{n}}_x$ are the $N_gN_h \times N_gN_h$ diagonal matrices with diagonal elements $\frac{n_{yh}}{m_{gh}}$ and $\frac{n_{xg}}{m_{gh}}$ respectively. ξ^{-1} is the matrix with elements $1/\xi_{gh}$, not the inverse of the matrix ξ. Moreover, $\vec{E}_q^+$ and $\vec{M}_q^+$ ($\vec{E}_q^-$ and $\vec{M}_q^-$) correspond to the physical forward (backward) TE and TM waves, respectively. The transition of fields $\mathbf{q}_{\hat{t}}$ within the considered layer are now solved as:

$$\mathbf{q}_{\hat{t}}(\overline{z}) = exp\left[i\kappa\left(\overline{z} - \overline{z}_0\right)\right]\mathbf{q}_{\hat{t}}(\overline{z}_0) \tag{84}$$

A.5.2 Periodic-structured layers with isotropic/birefringent materials

For the in-between periodic layers, the transition equations of tangential fields $\mathbf{f}_{\hat{t}}$ in equation (70) can be generally expressed as:

$$\frac{\partial}{\partial_{\overline{z}}}\mathbf{q}_{\hat{t}} = i\kappa^{(a)}\mathbf{q}_{\hat{t}} \quad with \;\; \mathbf{f}_{\hat{t}} = \mathbf{T}^{(a)}\mathbf{q}_{\hat{t}} \tag{85}$$

with the transition of $\mathbf{q}_{\hat{t}}$

$$\mathbf{q}_{\hat{t}}(\overline{z}) = exp\left[i\kappa^{(a)}\left(\overline{z} - \overline{z}_0\right)\right]\mathbf{q}_{\hat{t}}(\overline{z}_0) \tag{86}$$

Here, $\mathbf{T}^{(a)}$ is the eigen-vector matrix of $\mathbf{G}$ of equation (70) with column eigen-vectors, and $\kappa^{(a)}$ is the corresponding diagonal eigen-value matrix.

A.6 Boundary conditions

Now for each layer, we have been able to independently solve the transition of electromagnetic fields in the individual layers, but the continuum of fields on interfaces has still not been included. Considering the tangential components in $\mathbf{f}_{\hat{t}}$ are continuous across i_{th} interface at $\overline{z}_i$, the constriction equations can be shown as

$$\mathbf{T}_i^{(a)}\mathbf{q}_{\hat{t},i}(\overline{z}_i) = \mathbf{T}_{i+1}^{(a)}\mathbf{q}_{\hat{t},i+1}(\overline{z}_i) \tag{87}$$

Grouping this condition into the fields $\mathbf{q}_{\hat{t}}$ in equation (86), and introducing two virtual layers to consider the Fresnel refraction and reflection at surfaces of the media as described in the texts, a general expression for $N-$multilayered periodic structures can be obtained as in equation (13). This argument ignores the effects of multiple reflections as applied by (extended) Jones method, and similarly supplies as a easy-manipulated method. Further, an alternative process to consider the multiple reflections is described as below for references. Similarly, group the equation (87) with (86), the consecutive matrix equation with undecided diffraction/reflection waves can be written as:

$$\begin{aligned}
\mathbf{q}_{\hat{t},N+1}(\overline{z}_n) &= \mathbf{T}_{N+1}^{-1}\mathbf{T}_N^{(a)}\mathbf{q}_{\hat{t},N}(\overline{z}_N) \\
&= \mathbf{T}_{N+1}^{-1}\mathbf{T}_N^{(a)}\exp\left[i\kappa_N^{(a)}\left(\overline{z}_N-\overline{z}_{N-1}\right)\right]\mathbf{q}_{\hat{t},N}(\overline{z}_{N-1}) \\
&= \mathbf{T}_{N+1}^{-1}\mathbf{T}_N^{(a)}\exp\left[i\kappa_N^{(a)}\left(\overline{z}_N-\overline{z}_{N-1}\right)\right] \\
&\quad\times(\mathbf{T}_N^{(a)})^{-1}\mathbf{T}_{N-1}^{(a)}\exp\left[i\kappa_{N-1}^{(a)}\left(\overline{z}_{N-1}-\overline{z}_{N-2}\right)\right] \\
&\quad\times\ldots \\
&\quad\times(\mathbf{T}_1^{(a)})^{-1}\mathbf{T}_0\mathbf{q}_{\hat{t},0}(\overline{z}_0)
\end{aligned} \tag{88}$$

where the first boundary is indexed as 0. Consequently, the relation between fields in the incident region 0 and in the emitted region $N+1$ can be obtained as:

$$\mathbf{q}_{\hat{t},N+1} = \begin{bmatrix} \vec{E}_{q,N+1}^{+} \\ \vec{M}_{q,N+1}^{+} \\ \vec{E}_{q,N+1}^{-} \\ \vec{M}_{q,N+1}^{-} \end{bmatrix} = \mathbf{T}_{N+1}^{-1}\mathbf{T}_N^{a}\exp\left[i\kappa_N^{a}\left(\overline{z}_N-\overline{z}_{N-1}\right)\right]\ldots\left(\mathbf{T}_1^{a}\right)^{-1}\mathbf{T}_0\begin{bmatrix} \vec{E}_{q,0}^{+} \\ \vec{M}_{q,0}^{+} \\ \vec{E}_{q,0}^{-} \\ \vec{M}_{q,0}^{-} \end{bmatrix} \equiv \mathbf{W}^{-1}\begin{bmatrix} \vec{E}_{q,0}^{+} \\ \vec{M}_{q,0}^{+} \\ \vec{E}_{q,0}^{-} \\ \vec{M}_{q,0}^{-} \end{bmatrix} = \mathbf{W}^{-1}\mathbf{q}_{\hat{t},0} \tag{89}$$

or alternatively:

$$\mathbf{q}_{\hat{t},0} \equiv \begin{bmatrix} \mathbf{q}_{\hat{t},0}^{+} \\ \mathbf{q}_{\hat{t},0}^{-} \end{bmatrix} = \begin{bmatrix} \mathbf{W}_1 & \mathbf{W}_2 \\ \mathbf{W}_3 & \mathbf{W}_4 \end{bmatrix}\begin{bmatrix} \mathbf{q}_{\hat{t},N+1}^{+} \\ \mathbf{q}_{\hat{t},N+1}^{-} \end{bmatrix} = \mathbf{W}\mathbf{q}_{\hat{t},N+1} \tag{90}$$

Consider that the reflective field in the emitted region is zero, i.e. $\mathbf{q}_{\hat{t},N+1}^{-} = \left[\vec{E}_{q,N+1}^{-}\ \vec{M}_{q,N+1}^{-}\right]^T = 0$. The transmittance field in the emitted region can be obtained as:

$$\mathbf{q}_{\hat{t},N+1}^{+} = \mathbf{W}_1^{-1}\mathbf{q}_{\hat{t},0}^{+} \tag{91}$$

and the reflective field in the incident region is:

$$\mathbf{q}_{\hat{f},0}^{-} = \mathbf{W}_3\mathbf{W}_1^{-1}\mathbf{q}_{\hat{f},0}^{+} \tag{92}$$

A.7 Diffraction efficiency

To evaluate the diffraction efficiency with the obtained $\mathbf{q}_{\hat{f}}^{\pm}$, the x, y, and z components of the transmittance/reflection fields of the diffraction order (g, h) can be calculated by equations 19 (or 82, 83) and 69 for emitted/incident regions, and thereby the standard definitions of diffraction efficiency can be followed. Note that the incident fields should be excluded when calculating the reflection fields in the incident region.

B. Program codes of Wolfram Mathematica for Coupling Matrix Method

In this appendix, the program codes of Wolfram Mathematica for the (numerical) study case in the previous section are added as follows. It could be able to do the simulations by copy and paste the codes, while few characters may need to be adjusted, e.g., the superscript of W' (W'') and the power symbol on no^2 (ne^2).

```
(*Initialize one − period LC profiles (θo, φo) for single LC layer*)
dx = 0.1; dy = dx; (*um/grid; grid interval *)
GridNx = 100; GridNy = 100; (*grid num. in x and y *)
Λx = GridNx*dx; Λy = GridNx*dy; (* unit cell *)
θo = Table[π*i/GridNx, {i, GridNx}, {j, GridNy}];
φo = Table[π/2.0, {i, GridNx}, {j, GridNy}];
dz = 2.0; (*um; the thickness of the LC layer *)

(*Define optical − related parameters*)
nI = 1.0; nE = 1.0; θ = 0.001; φ = 0.0; λ = 0.55;
ne = 1.5; no = 1.6;
grng = 1; hrng = 1; (* − grng ≤ g ≤ grng; −hrng ≤ h ≤ hrng*)
Ng = 2*grng + 1; Nh = 2*hrng + 1; (*Note Ng < GridNx, Nh < GridNy*)

(*Initialize relevant wave − vector matrixes related to nxg, nyh, respectively*)
gindx = Table[Floor[(i − 1.0)/Nh] − grng, {i, Ng*Nh}]; (* g sequence in ei or hi fields *)
hindx = Table[Mod[(i − 1), Nh] − hrng, {i, Ng*Nh}]; (* h sequence in ei or hi fields *)
nx = DiagonalMatrix[Table[nI*Sin[θ]*Cos[φ] − gindx[[i]]*λ/Λx, {i, Ng*Nh}]];
ny = DiagonalMatrix[Table[nI*Sin[θ]*Sin[φ] − hindx[[i]]*λ/Λy, {i, Ng*Nh}]];
m = DiagonalMatrix[Table[Sqrt[nx[[i, i]]^2 + ny[[i, i]]^2], {i, Ng*Nh}]];
ξ = DiagonalMatrix[Table[Sqrt[nI^2 − nx[[i, i]]^2 − ny[[i, i]]^2], {i, Ng*Nh}]];
ξinv = DiagonalMatrix[Table[1.0/Sqrt[nI^2 − nx[[i, i]]^2 − ny[[i, i]]^2], {i, Ng*Nh}]];

(*Calculate εijgh by Fourier transform of εij(x, y; z) for the single LC layer*)
εxxgh=InverseFourier[no^2+(ne^2−no^2)*Sin[θo]^2*Cos[φo]^2]/Sqrt[GridNx]/Sqrt[GridNy];
εxygh=InverseFourier[(ne^2−no^2)*Sin[θo]^2*Sin[φo]Cos[φo]]/Sqrt[GridNx]/Sqrt[GridNy];
εxzgh=InverseFourier[(ne^2−no^2)*Sin[θo]Cos[θo]Cos[φo]]/Sqrt[GridNx]/Sqrt[GridNy];
εyygh=InverseFourier[no^2+(ne^2−no^2)*Sin[θo]^2*Sin[φo]^2]/Sqrt[GridNx]/Sqrt[GridNy];
```

```
εyzgh=InverseFourier[(ne^2 − no^2)*Sin[θo]Cos[θo]Sin[ϕo]]/Sqrt[GridNx]/Sqrt[GridNy];
εzzgh=InverseFourier[no^2 + (ne^2 − no^2)*Cos[θo]^2]/Sqrt[GridNx]/Sqrt[GridNy];

(* Define the matrix εij with element εijgh *)
εxx = Table[0, {i, Ng*Nh}, {j, Ng*Nh}]; εxy = Table[0, {i, Ng*Nh}, {j, Ng*Nh}];
εxz = Table[0, {i, Ng*Nh}, {j, Ng*Nh}]; εyy = Table[0, {i, Ng*Nh}, {j, Ng*Nh}];
εyz = Table[0, {i, Ng*Nh}, {j, Ng*Nh}]; εzz = Table[0, {i, Ng*Nh}, {j, Ng*Nh}];
εzzinv = Table[0, {i, Ng*Nh}, {j, Ng*Nh}];
For[i = 1, i ≤ Ng*Nh, For[j = 1, j ≤ Ng*Nh,
g = gindx[[i]] − gindx[[j]]; h = hindx[[i]] − hindx[[j]];
gp = If[g ≥ 0, g = g + 1, g = g + GridNx + 1];
(* follow arrangements of components in εijgh *)
hp = If[h ≥ 0, h = h + 1, h = h + GridNy + 1];
εxx[[i, j]] = εxxgh[[gp, hp]]; εxy[[i, j]] = εxygh[[gp, hp]]; εxz[[i, j]] = εxzgh[[gp, hp]];
εyy[[i, j]] = εyygh[[gp, hp]]; εyz[[i, j]] = εyzgh[[gp, hp]]; εzz[[i, j]] = εzzgh[[gp, hp]];
j++; ]; i++; ];
εzzinv = Inverse[εzz];

(* Calculate matrix G for the single LC layer*)
G11 = Dot[nx, εzzinv, εxz]; G12 = Dot[nx, εzzinv, nx] − IdentityMatrix[Ng*Nh];
G13 = Dot[nx, εzzinv, εyz]; G14 = −Dot[nx, εzzinv, ny];
G21 = Dot[εxz, εzzinv, εxz] − εxx + Dot[ny, ny]; G22 = Dot[εxz, εzzinv, nx];
G23 = Dot[εxz, εzzinv, εyz] − εxy − Dot[ny, nx]; G24 = −Dot[εxz, εzzinv, ny];
G31 = Dot[ny, εzzinv, εxz]; G32 = Dot[ny, εzzinv, nx];
G33 = Dot[ny, εzzinv, εyz]; G34 = −Dot[ny, εzzinv, ny] + IdentityMatrix[Ng*Nh];
G41 = −Dot[εyz, εzzinv, εxz] + εxy + Dot[nx, ny]; G42 = −Dot[εyz, εzzinv, nx];
G43 = −Dot[εyz, εzzinv, εyz] + εyy − Dot[nx, nx]; G44 = Dot[εyz, εzzinv, ny];
G1i = Join[G11, G12, G13, G14, 2]; G2i = Join[G21, G22, G23, G24, 2];
G3i = Join[G31, G32, G33, G34, 2]; G4i = Join[G41, G42, G43, G44, 2];
G = Join[G1i, G2i, G3i, G4i];
Ta = Transpose[Eigenvectors[G]]; (*eigen − vecotr matrix*)
Tainv = Inverse[Ta]; (*inverse of the eigen − vecotr matrix *)
κa = Dot[Tainv, G, Ta]; (*eigen − value matrix corresponding to the arrangement of Ta*)

(*Calculate the matrixes related to incidnet and emitted air regions, i.e. nI = nE = 1*)
nxd = DiagonalMatrix[Table[nx[[i, i]]/m[[i, i]], {i, Ng*Nh}]];
nyd = DiagonalMatrix[Table[ny[[i, i]]/m[[i, i]], {i, Ng*Nh}]];
T11 = nyd; T12 = nxd; T13 = nyd; T14 = nxd;
T21 = Dot[nyd, ζ]; T22 = nI^2*Dot[nxd, ζinv]; T23 = −Dot[nyd, ζ];
T24 = −nI^2*Dot[nxd, ζinv];
T31 = −nxd; T32 = nyd; T33 = −nxd; T34 = nyd;
T41 = Dot[nxd, ζ]; T42 = −nI^2*Dot[nyd, ζinv]; T43 = −Dot[nxd, ζ];
T44 = nI^2*Dot[nyd, ζinv];
T1i = Join[T11, T12, T13, T14, 2]; T2i = Join[T21, T22, T23, T24, 2];
T3i = Join[T31, T32, T33, T34, 2]; T4i = Join[T41, T42, T43, T44, 2];
Ti = Join[T1i, T2i, T3i, T4i]; (* transform matrix Ti*)
Tiinv = Inverse[Ti]; (* inverse of the transform matrix Ti *)
```

```
(*Solution(1) : solve diffractions and reflections with multi − reflections*)
(*set incident plane wave indfd, e.g. set the value of the component withg = h = 0 to be 1*)
indfd = Table[0, {i, 2*Ng*Nh}]; indfd[[Round[(Ng*Nh + 1)/2]]] = 1.0; (* forward incidence*)

(* Calculate the matrix Winv*)
expκ = DiagonalMatrix[Table[Exp[I*κa[[i, i]]*dz*2π/λ], {i, 4*Ng*Nh}]];
Winv = Dot[Tiinv, Ta, expκ, Tainv, Ti]; (* the total transfer matrix *)
W = Inverse[Winv];

(* Calculate the diffraction and reflection fields *)
diff1 = Table[0, {i, 2*Ng*Nh}]; ref1 = Table[0, {i, 2*Ng*Nh}]; (*initialize*)
W1 = W[[1;;2*Ng*Nh, 1;;2*Ng*Nh]];
W3 = W[[(2*Ng*Nh + 1);;4*Ng*Nh, 1;;2*Ng*Nh]];
diff1 = Dot[Inverse[W1], indfd]; (* diffraction fields *)
ref1 = Dot[W3, Inverse[W1], indfd]; (* reflection fields *)

(*Print diffraction and reflection fields as well as the corresponding g, h orders*)
Print["TE mode with multi-reflections"];
For[i = 1, i ≤ Ng*Nh,
Print[gindx[[i]], ", ", hindx[[i]], ", ", Abs[diff1[[i]]], ", ", Abs[ref1[[i]]]]; i++; ];

(*Solution(2) : solve diffractions and reflections without multi − reflections *)
(*Calculate the matrixes related to virtual layer with n = (ne + no)/2*)
ξavg = DiagonalMatrix[Table[Sqrt[((ne + no)/2.0)^2
− nx[[i, i]]^2 − ny[[i, i]]^2], {i, Ng*Nh}]];
ξavginv = DiagonalMatrix[Table[1.0/ξavg[[i, i]], {i, Ng*Nh}]];
T11 = nyd; T12 = nxd; T13 = nyd; T14 = nxd; T21 = Dot[nyd, ξavg];
T22 = nI^2*Dot[nxd, ξavginv]; T23 = −Dot[nyd, ξavg]; T24 = −nI^2*Dot[nxd, ξavginv];
T31 = −nxd; T32 = nyd; T33 = −nxd; T34 = nyd; T41 = Dot[nxd, ξavg];
T42 = −nI^2*Dot[nyd, ξavginv]; T43 = −Dot[nxd, ξavg]; T44 = nI^2*Dot[nyd, ξavginv];
T1i = Join[T11, T12, T13, T14, 2]; T2i = Join[T21, T22, T23, T24, 2];
T3i = Join[T31, T32, T33, T34, 2]; T4i = Join[T41, T42, T43, T44, 2];
Tavg = Join[T1i, T2i, T3i, T4i]; (* transform matrix Ti*)
Tavginv = Inverse[Tavg]; (* inverse of the transform matrix Ti *)
ClearAll[T11, T12, T13, T14, T21, T22, T23, T24, T31, T32, T33, T34, T41, T42, T43, T44];
ClearAll[T1i, T2i, T3i, T4i];

(* Calculate the transfer matrixes *)
S1 = Dot[Ta, expκ, Tainv];
Sent = Table[0, {i, 4*Ng*Nh}, {j, 4*Ng*Nh}];
Sext = Table[0, {i, 4*Ng*Nh}, {j, 4*Ng*Nh}];
W′ = Inverse[Dot[Tavginv, Ti]]; W1′ = W′[[1;;2*Ng*Nh, 1;;2*Ng*Nh]];
W″ = Inverse[Dot[Tiinv, Tavg]]; W1″ = W″[[1;;2*Ng*Nh, 1;;2*Ng*Nh]];
Sent[[1;;2*Ng*Nh, 1;;2*Ng*Nh]] = Inverse[W1′]; Sent = Dot[Tavg, Sent];
Sext[[1;;2*Ng*Nh, 1;;2*Ng*Nh]] = Inverse[W1″]; Sext = Dot[Sext, Tavginv];
```

```
(* Calculate the diffraction and reflection fields *)
indfd2 = Table[0, {i, 4*Ng*Nh}]; indfd2[[Round[(Ng*Nh + 1)/2]]]
= 1.0; (* ignore backward *)
diff2 = Dot[Sext, S1, Sent, indfd2];

(*Print diffraction and reflection fields as well as the corresponding g, h orders*)
Print["TE mode without multi-reflections"];
For[i = 1, i ≤ Ng*Nh,
Print[gindx[[i]], ", ", hindx[[i]], ", ", Abs[diff2[[i]]]]; i++; ];
```

5. References

Berreman, D. W. (1972). Optics in Stratified and Anisotropic Media: 4-by-4-Matrix Formulation, *Journal of the Optical Society of America*, Vol. 62, Iss. 4, April 1972, pp. 502-510.

Blinov, L. M.; Cipparrone, G.; Pagliusi, P.; Lazarev, V. V. & Palto, S. P. (2006). Mirrorless lasing from nematic liquid crystals in the plane waveguide geometry without refractive index or gain modulation, *Applied Physics Letters*, Vol. 89, Iss. 3, July 2006, pp. 031114-3.

Blinov, L. M.; Lazarev, V. V.; Palto, S. P.; Cipparrone, G.; Mazzulla, A. & Pagliusi, P. (2007). Electric field tuning a spectrum of nematic liquid crystal lasing with the use of a periodic shadow mask, *Journal of Nonlinear Optical Physics & Materials*, Vol. 16, Iss. 1, March 2007, pp. 75-90.

Galatola, P; Oldano, C & Kumar, P. B. S. (1994). Symmetry properties of anisotropic dielectric gratings, *Journal of the Optical Society of America A*, Vol. 11, Iss. 4, April 1994, pp. 1332-1341.

Glytsis, E. N. & Gaylord, T. K. (1987), Rigorous three-dimensional coupled-wave diffraction analysis of single and cascaded anisotropic gratings, *Journal of the Optical Society of America A*, Vol. 4, Iss. 11, November 1987, pp. 2061-2080.

Ho, I. L.; Chang, Y. C.; Huang, C. H.& Li, W. Y. (2011), A detailed derivation of rigorous coupled wave algorithms for three-dimensional periodic liquid-crystal microstructures, *Liquid Crystals*, Vol. 38, No. 2, February 2011, 241ąV252.

Jones, R. C. (1941). A new calculus for the treatment of optical systems, *Journal of the Optical Society of America*, Vol. 31, Iss. 7, July 1941, pp. 488ąV493.

Kriezisa, E. E. & Elston, S. J. (1999). A wide angle beam propagation method for the analysis of tilted nematic liquid crystal structures, *Journal of Modern Optics*, Vol. 46, Iss. 8, 1999, pp. 1201-1212.

Kriezis, E. E.; Filippov, S. K. & Elston, S. J. (2000). Light propagation in domain walls in ferroelectric liquid crystal devices by the finite-difference time-domain method, *Journal of Optics A: Pure and Applied Optics*, Vol. 2, No. 1, January 2000, pp. 27-33.

Kriezis, E. E. & Elston, S. J. (2000). Wide-angle beam propagation method for liquid-crystal device calculations, *Applied Optics*, Vol. 39, Iss. 31, November 2000, pp.5707-5714.

Kriezis, E. E.; Newton, C. J. P.; Spiller, T. P. & Elston, S. J. (2002). Three-dimensional simulations of light propagation in periodic liquid-crystal microstructures, *Applied Optics*, Vol. 41, Issue 25, September 2002 , pp. 5346-5356.

Lien, A.(1997). A detailed derivation of extended Jones matrix representation for twisted nematic liquid crystal displays, *Liquid Crystals*, Vol. 22, No. 2, February 1997, pp. 171-175.

Olivero, D & Oldano, C. (2003). Numerical methods for light propagation in large LC cells: a new approach, *Liquid Crystals*, Vol. 30, Iss. 3, 2003, pp. 345-353.

Rokushima K. & Yamakita, J. (1983). Analysis of anisotropic dielectric gratings, *Journal of the Optical Society of America*, Vol. 73, Iss. 7, July 1983, pp. 901-908.

Sutkowski M.; Grudniewski T.; Zmijan R.; Parka J. & Nowinowski K. E. (2006). Optical data storage in LC cells, *Opto-Electronics Review*, Vol. 14, No. 4, December 2006, pp. 335-337.

Witzigmann, B; Regli, P & Fichtner, W. (1998). Rigorous electromagnetic simulation of liquid crystal displays, *Journal of the Optical Society of America A*, Vol. 15, Iss. 3, March 1998, pp.753-757.

Zhang, B. & Sheng, P. (2003). Optical measurement of azimuthal anchoring strength in nematic liquid crystals, *Physical Review E*, Vol. 67, Iss. 4, April 2003, pp. 041713-9.

5

Wavelet Network Implementation on an Inexpensive Eight Bit Microcontroller

Lyes Saad Saoud, Fayçal Rahmoune,
Victor Tourtchine and Kamel Baddari
*Laboratory of Computer Science, Modeling, Optimization,
Simulation and Electronic Systems (L.I.M.O.S.E),
Department of Physics, Faculty of Sciences,
University M'hamed Bougara Boumerdes
Algeria*

1. Introduction

The approximation of general continuous functions by nonlinear networks is very useful for system modeling and identification. Such approximation methods can be used, for example, in black-box identification of nonlinear systems, signal processing, control, statistical data analysis, speech recognition, and artificial intelligence. Recently neural networks have been established as a general approximation tool for fitting nonlinear models from input/output data due to their ability of learning rather than complicated process functions (Gao, 2002). Their attractive property is the self-learning ability. A neural network can extract the system features from historical training data using the learning algorithm, requiring little or no a priori knowledge about the process (Patan, 2008). This is why during the past few years the nonlinear dynamic modelling of processes by neural networks has been extensively studied (Narendra & Parthasarathy, 1990; Nerrand et al., 1993; Levin, 1992; Rivals & Personnaz, 1996). In standard neural networks, the nonlinearities are approximated by superposition of sigmoidal functions (Cybenko, 1989).

In the other hand, the wavelet theory has found many applications in function approximation, numerical analysis and signal processing. Though this attractive theory has offered efficient algorithms for various purposes, their implementations are usually limited to wavelets of small dimension. The reason is that constructing and storing wavelet basis of large dimension are of prohibitive cost. In order to handle problems of larger dimension, it is necessary to develop algorithms whose implementation is less sensitive to the dimension. And it is known that neural networks are powerful tools for handling problems of large dimension.

Due to the similarity between wavelet decomposition and one-hidden-layer neural networks, the idea of combining both wavelets and neural networks has been proposed in various works (Zhang & Benveniste, 1992; Pati & Krishnaprasad, 1993; Hong, 1992; Bakshi & Stephanopoulos, 1993; Tsatsanis & Giannakis, 1993;et al., 1994; Delyon et al., 1995; Saad Saoud & Khellaf, 2009). For example, in (Zhang & Benveniste, 1992) wavelet network is introduced as a class of feedforward networks composed of wavelets, in (Pati & Krishnaprasad, 1993) the discrete wavelet transform is used for analyzing and synthesizing feedforward neural networks, in (Hong, 1992) orthogonal wavelet bases are used for constructing wavelet-based neural network, and in (Saad Saoud & Khellaf, 2009) the dynamic wavelet networks is proposed and used to control the chemical reactor. Combining wavelets and neural networks can hopefully remedy the weakness of each other, resulting in networks with efficient constructive methods and capable of handling problems of moderately larger dimensions.

Hence, we can say that the neural network and the wavelet network are capable of modeling non-linear systems. On the basis of supplied training data the neural or the wavelet networks learn the relationship between the process input and output. The data have to be examined carefully before they can be used as a training set for network methods. The training sets consist of one or more input data and one or more output data (Roffel & Betlem, 2006). After the training of the network, a test-set of data should be used to verify whether the desired relationship was learned. These two operations (train and validate the network) are achieved generally by using the computer, the finding weights and bias are implemented either in the computer itself or through the implementation of the optimal network's parameters in the microcontroller (Gulbag et al., 2009; Cotton et al., 2008; Liung et al., 2003; Neelamegamand & Rajendran , 2005). One common drawback is that in both cases, in order to find the network's parameters precise calculations that are very processor intensive are required. This robust processing equipment can be expensive and rather large.

In several applications such as adaptive control (Plett, 2003), or predictive control (Liu et al., 1998), we need to adapt the network's parameters in real time and in this case the computer is very important to adapt the parameters. These problems were overcome with the implementation of the whole neural network with its backpropagation algorithm in the microcontroller, which is proposed in our previous work (Saad Saoud & Khellaf, 2011). In this later work a multilayer neural network is trained and validated using a very inexpensive and low end microcontroller, but the problems of larger dimensions still exist. For this case the real implementation of the whole wavelet network into an inexpensive microcontroller is proposed in this study.

The low end and inexpensive microcontroller PIC16F877A of Microchip trains and validates the wavelet network, and the well-known backpropagation algorithm is implemented to obtain the optimal network parameters. All the operations done by the microcontroller are shown through an alphanumeric liquid crystal display and several buttons are added in the embedded system which produces an ergonomic communication interface human/machine. The wavelet network takes more program memory place, for this reason the assembly language is preferred. The Continuous Stirred Tank Reactor (CSTR) system is chosen as a realistic nonlinear system to demonstrate the feasibility and the performance of the results found using the microcontroller. Several results will be presented in this chapter to give the reader more information about this field. A comparative study is made between the microcontroller and the computer.

The chapter is organized as follows: After the description of the nonlinear dynamic system identification in general and by using the wavelet network in particular, the implementation of the backpropagation algorithm for the wavelet network in the microcontroller. A comparison between wavelet network based on the eight bit microcontroller and those based on the computer is presented. To illustrate how effectively the eight bit microcontroller can learn nonlinear dynamic models, results for a Continuous Stirred Tank Reactor are given. All the electronic tools, electrical schemes and the implemented algorithm are discussed. The chapter concludes with few final remarks.

2. Nonlinear dynamic system identification

The key problem in system identification is to find a suitable model structure within which a good model is to be found (Sjöberg et al. 1995). In general, the nonlinear dynamic system identification is the operation to determine a transformation operator T_i, for some desired $\varepsilon>0$, so that (Efe & Kaynak, 1997):

$$\left\| T_i(u)-T_p(u) \right\| \leq \varepsilon \ , u \in U \tag{1}$$

With $T_p(u)$ and $T_i(u)$ denote the system to be identified, which maps the compact set $U \in R^n$ and $Y \in R^m$, and the identification model outputs respectively for the same input u.
The purpose is to find a class T_i such that T_p is represented by T_i adequately well. The operator T_p is defined by specific input-output pairs that are obtained from the inputs and the outputs of the system to be identified. Fig. 1 summarizes a system identification structure.
Several nonlinear identification models can be found in the literature. The wavelet networks prove their capabilities (Zhang & Benveniste, 1992). The most important aspect of a wavelet network based identification scheme is the determination of an adaptive algorithm that minimizes the difference between the actual plant and the outputs of the identified model by using a set of training pairs which represent the approximate behavior of the actual plant.

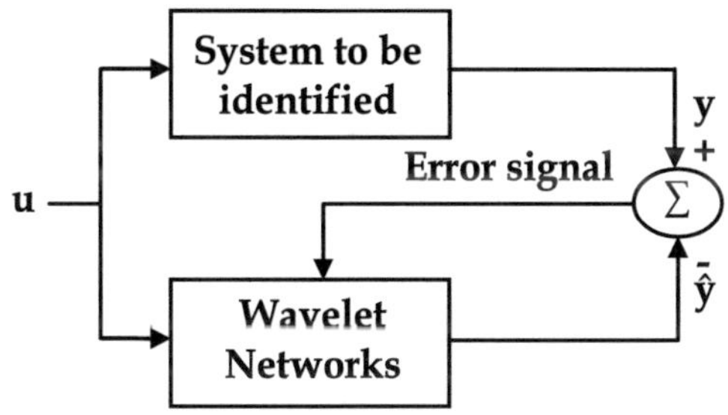

Fig. 1. Wavelet networks identification

As mentioned in the introduction, it exists a great similarity between the wavelet network and the one hidden layer neural networks which gives it the opportunity to be an attractive alternative to the feed forward neural network or the Radial Basis Function (RBF) networks (Khalaf, 2006). The use of the wavelet as an activate function gives power to the network because it reduces naturally the noise and hence the resulting network becomes more simple due to the linear output. In this work, another aspect is given to this powerful architecture, which is the implementation of the whole wavelet network with its backpropagation algorithm in the PIC microcontroller.
Wavelet function is a waveform that has a limited duration and an average value of zero. There exist several types of wavelet functions. In this work the Mexican Hat wavelet (Abiyev, 2003) given by the equation 2 is used as a transfer function.

$$\psi(\tau)=\left(2/\sqrt{3}\right)\ \pi^{-1/4}(1-\tau^2)\exp\left(-\tau^2/2\right) \tag{2}$$

With:

$$\tau = \frac{u_i(k) - t_i}{d_i}$$

The network output is simply the sum of the neuron's outputs of the hidden layer and it is given by the equation (3) as follows:

$$\hat{y}(k) = \sum_{i=1}^{n} w_i \psi(\tau) \qquad (3)$$

Where :
$u(k)$ is the neuron input vector of the dimension $[nx1]$.
$y(k)$ is the neuron output at time instant k.
ψ is wavelet activation function of the neuron with a translation t and a dilation d.
Figures 2 and 3 are the representation of the Mexican Hat wavelet function for different translations and dilations:

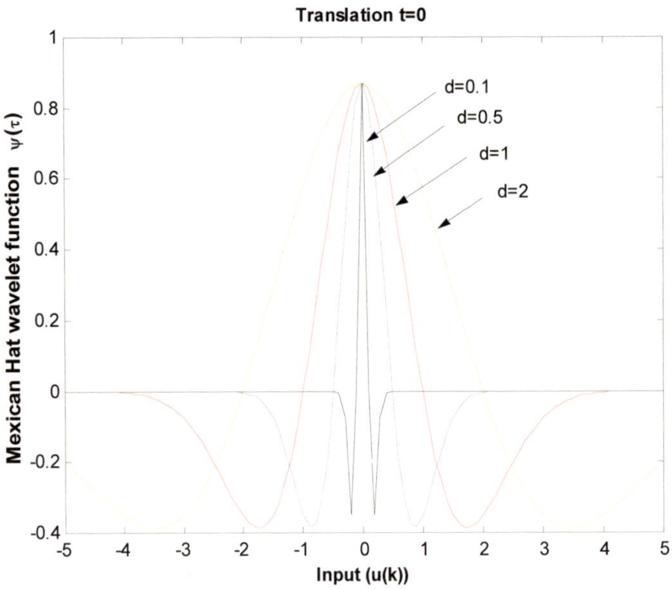

Fig. 2. Mexican Hat wavelet function for different dilations and a translation equals to zero

The backpropagation algorithm is the most popular training method which is widely used in the neural network applications (Efe, 1996). The backpropagation algorithm is, however, very general and not limited to one-hidden-layer sigmoid neural network models. Instead, it could be applied to all network models (Sjöberg et al., 1995).
In this chapter we use this elegant technique to train the wavelet network. The method minimizes the performance or the so-called the cost function defined on the actual and desired outputs of the network by the equation (4). When updating each individual network parameter (weight, translation and dilation), the gradient information obtained from the differentiation of the cost function is used. As a matter of fact, we are looking for least mean squares. This can be attained by moving the network parameters vectors in a direction such that the performance function decreases. It is obvious that this direction is the negative gradient direction of the performance function.

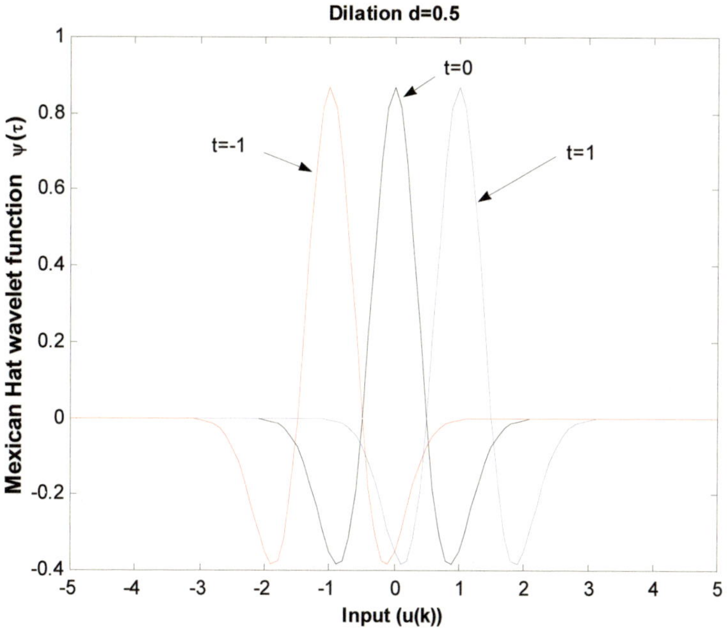

Fig. 3. Mexican Hat wavelet function for different translations and a dilation equals to 0.5

$$E = \frac{1}{2}\sum_{k=1}^{m} e_k^2 \tag{4}$$

With

$$e_k = \left(y_k - \hat{y}_k\right) \tag{5}$$

Assuming that the objective is to minimize this kind of performance function, the network parameters updating rule is given by the equations (6) to (9):

$$w_{new} = w_{old} + \eta\psi(\tau)e_k \tag{6}$$

$$t_{new} = t_{old} + \eta w^T h e_k \tag{7}$$

$$d_{new} = d_{old} + \eta \tau w^T h e_k \tag{8}$$

With:

$$h = \left(2/\sqrt{3}\right)\pi^{-1/4}\tau(3-\tau^2)\exp\left(-\tau^2/2\right)/d_{old} \tag{9}$$

As we can see from the above equations, the updating network parameters are performed by evaluating the gradient of the performance function with respect to each individual network parameter in the network.

3. PIC Implementation of the full wavelet network

In our previous paper (Saad Saoud & Khellaf, 2011) the neural network is implemented and trained using Microchip microcontroller (PIC16F876A). In this work, the extension of this card is used to train and validate another more powerful model which is the wavelet network. In this work, the microcontroller type is replaced by the PIC16F877A to give more power and utility to the final realized card.
First of all, why did we choose this type of microcontroller?
The PIC microcontrollers not only are the most widely used and well known microcontrollers, they are also the best supported. In fact, PIC system design and programming has become a powerful specialization with a large number of professional and amateur specialists. There are hundreds of websites devoted to PIC-related topics. An entire cottage industry of PIC software and hardware has flourished around this technology (Sanchez & Canton, 2007).
In this chapter and like our previous work we keep the mid-range PIC microcontroller, but we move to an upgraded version: the PIC16F877A, to give the new designed card better extensions, such as, more inputs outputs giving the user more flexibility when using the card in the control or other real practical situations.
The wavelet network implementation is written in assembly language based on the MPLAB tool distributed freely by Microchip Company. A sample of the 9162 assembly lines taken as the principal program is shown in the figure 4. The wavelet network hexadecimal source (Wavelet.hex) can be found in Intech publisher. Like the old realized card, the microcontroller has to find the optimal wavelet network parameters by using the well known backpropagation algorithm based on the memorized input output pairs in the EEPROM memory. The photo of the realized card based on the PIC16F877A microcontroller is shown in figure 5.
As shown in the program sample, the loops (Loop 1 and Loop 2) are used to read the 500 pairs input-output. The "Loop 3" in Fig. 4 is simply used to train the network. It can be noticed that the program is very flexible, and it gives the designers the opportunity to change the iteration number, the number of the pairs of modeling data and also the type of the data itself, just they have to respect the input regressor.

4. Practical results

To test the performance of the realized card (Fig.5), the Continuous Stirred Tank Reactor (CSTR) chemical reactor is chosen. The secrets of this system are the high nonlinearity and simplicity of the mathematical model. For this reason several researchers use the CSTR as a simulated or practical plant to validate their results such as the references (Lightbody & Irwin , 1997; Morningred et al.,1990; Henson & Seborg , 1990; Espinosa et al., 2005; Saad Saoud & Khellaf, 2009; Saad Saoud et al., 2011), and it consists simply of an irreversible, exothermic reaction. A→B, in a constant volume rate cooled by a single coolant stream which can be modeled by the following equations :

$$\dot{C}_a(t) = \frac{q}{v}(C_{ao} - C_a(t)) - k_0 C_a(t) e^{-\frac{E}{RT(t)}} \tag{10}$$

$$\dot{T}(t) = \frac{q}{v}(T_0 - T(t)) - k_1 C_a(t) e^{-\frac{E}{RT(t)}} + k_2 q_c(t)(1 - e^{-\frac{k_3}{q_c(t)}})(T_{c0} - T(t)) \tag{11}$$

```
;*********************************************************************************************;
; This is a sample program of the wavelet network approach written in assembly language. ;
; © Lyes Saad Saoud, Fayçal Rahmoune, Victor Tourtchine & Kamel Baddari, 2011             ;
;*********************************************************************************************;
START
        :
        :
        MOVLW     .100                    ; 100 iterations for the
                                          ; backpropagation
                                          ; procedure
        MOVWF     Counter3                ; In Counter3
Loop3
;---------------------------------------------------------------------------------------------;
; For read 500 values from the extern EEPROM memory                                           ;
;---------------------------------------------------------------------------------------------;
        MOVLW     .5                      ; For 5 times
        MOVWF     Counter2                ; In Counter2
Loop2
        MOVLW     .100                    ; For 100 times
        MOVWF     Counter1                ; In Counter1
Loop1
        CALL      LOAD_DATA               ; Load one input, output regressor
        CALL      WAVENET_OUTPUT          ; Compute the wavelet network output
        CALL      ERROR                   ; Compute the error between the
                                          ; modeled and estimated outputs
        CALL      ADAPTATION              ; Call the backpropagation subprogram
        CALL      MSE                     ; Compute the last mean squared error
        DECFSZ    Counter1,f              ; Decrease  the counter 1
                                          ; If counter 1 equal zero jump
        GOTO      Loop1                   ; Else, loop to counter 1
        DECFSZ    Counter2,f              ; Decrease counter 2
                                          ; If counter 2 equal zero jumps
        GOTO      Loop2                   ; Else, loop to counter 2
;-------------------------------------------------------------------------------------------;
        BCF       PORTC,1                 ; Orient LCD for control
        SENDCAR   B'11000000'             ; SENDCAR is a macro to send data
                                          ; to the display, this data tell the
                                          ; display to start writing in the line 2
                                          ; This macro always validate the
                                          ; instruction at the end
        BSF       PORTC, 1                ; Orient LCD for data
        SENDCAR   'M'                     ; Send the letters "M, S, E, ="
        SENDCAR   'S'                     ; for display
        SENDCAR   'E'                     ; MSE=
        SENDCAR   '='                     ;
```

```
CALL      Display_MSE        ; Calculate and display the mean
                             ; squared error MSE
CLRF      MSEEXP             ; Clear register MSEEXP
CLRF      MSEAARGB0          ; Clear register MSEAARGB0
CLRF      MSEAARGB1          ; Clear register MSEAARGB1
CLRF      eepm_output        ; Point to the first output address
CLRF      eepm_output+1      ; Point to the next output address
                             ; The initial output address equal to
                             ; zero
MOVLW     0x0E               ; Point to the first output address
MOVWF     eepm_input         ;
MOVLW     0x14               ; Point to the next output address
MOVWF     eepm_input+1       ;
                             ; The initial input address
                             ; equal to 0x0E14
DECFSZ    Counter3,f         ; Decrease  the counter 3
                             ; If counter 3 equal zero jumps
GOTO      Loop3              ; Else, loop to counter 3
  .
  .
  .
;--------------------------------------------------------------------------------;
; Validate with new input-output pairs and read the trained and validated data from ;
; the EEPROM memory and display them through the alphanumerical LCD                 ;
;--------------------------------------------------------------------------------;
CALL      Validation_Part    ; Validate the founding model
CALL      Read_data          ; Read the trained and validated
                             ; data from the EEPROM memory
                             ; and display them through the
                             ; alphanumerical LCD
GOTO      START              ; Start from the beginning
  .
  .
  .
```

Fig. 4. A sample program of the assembly wavelet network implementation

The process describes the reaction of two products, which are mixed and react to generate a compound A having a concentration $C_a(t)$, with the temperature of the mixture $T(t)$. This reaction is exothermic. The generated heat acts to slow the reaction. The reaction is controlled by introducing a coolant flow rate $q_c(t)$, which helps to change the temperature and thereby the concentration. C_{a0} is the inlet feed concentration, q is the process flow rate, T_0 and T_{c0} are the inlet feed and coolant temperatures. All these values are assumed constant at nominal values. In the same way, k_0, E/R, υ, k_1, k_2 and k_3 are thermodynamic and chemical constants.

$$k_1 = -\frac{\Delta H k_0}{\rho C_p} \qquad k_2 = -\frac{\rho_c C_{pc}}{\rho C_p \upsilon} \qquad k_3 = -\frac{h_a}{\rho_c C_{pc}}$$

The nominal conditions for a product concentration $C_a = 0.1\ mol/l$ are:
$T = 438.54K$, $q_c = 103.41\ l/min$.

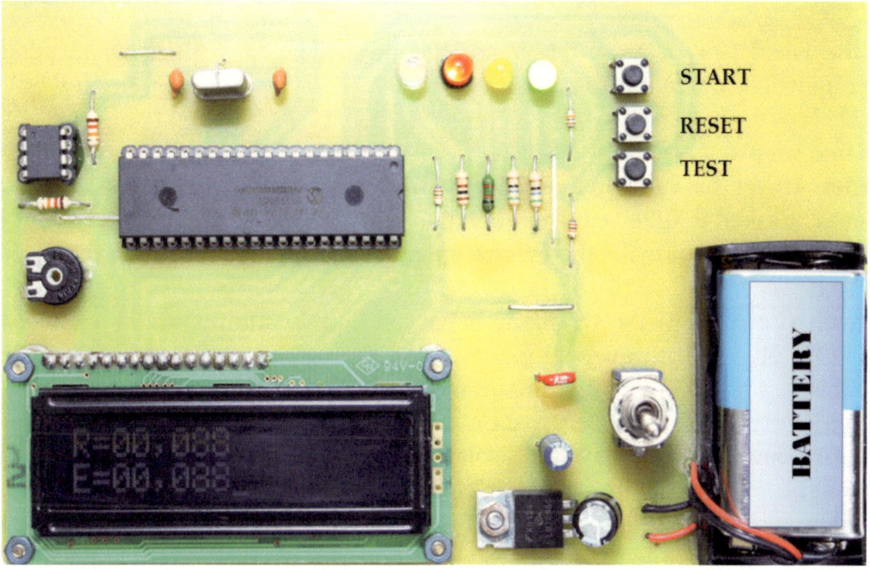

Fig. 5. The realized card based on the PIC microcontroller using the wavelet network

The data used for modeling can be found in (De Moor, 1998) and it is shown in figure 7. We choose randomly a sequence of 750 samples of the input-output from the whole given data. As we have made in our previous works (Saad Saoud & Khellaf, 2011; Saad Saoud et al., 2011), before using the data, we have to make several changes to be accepted by the network. The used data saved in the EEPROM passed through two necessary operations, first it has been normalized to 0.1 and 0.9 and second it is multiplied by 36408 to cover the maximum range and not exceeding the positive signed range of 16 bit (7FFF hex) of data. It should be mentioned that in very few cases, error values could be displayed on the LCD and this is due the floating point format difference between the computer and the PIC microcontroller (PIC 16F877A, 24 bit floating point format). In the microcontroller, all these operations will be reversed to have real values. Also it can be seen in the previous sample assembly language program that 500 samples were used to train the network and the rest of data is used for the network validation. At the beginning, the microcontroller trains the network to find the optimal parameters. All the parameters except the dilations were initialized randomly, hence the network dilations were initialized ones to avoid zeros in the denominators.

These operations and the others are summarized in the following algorithm:

1. Initialize the weights and translations randomly small and the dilations by ones.
2. Load the data
3. Calculate the wavelet network output using the equation (3)
4. Calculate the error using the equation (5)
5. Update the weights, the translations and the dilations with the equations (6) to (9).
6. If the $E < \varepsilon$ or the number of the iteration is achieved, End the operation Else goto the step 2.

To illustrate this algorithm and enter deeply into the microcontroller program memory we give the following example :

Let's take for example the concentration $C_a=0.11944\ mol/l$.

We have to normalize this value between 0.1 and 0.9, for this we make this operation:

$C_{a\_Normalized} = a\_N x C_a + b\_N$

When the coefficients a_N and b_N are the normalization coefficients for the whole concentration data and in this case are :

$a\_N=9.2937$ *and* $b\_N=-0.4638$

We can find simply $C_{a\_Normalized}= 0.64625$.

And before saving this value in the EEPROM we have to multiply it by 36408. So the value that should be saved is 5BE8 hex. This operation will be done with all the taken samples.

The microcontroller will make the inverse operations at each time it reads the values from the EEPROM.

In the realized card shown in the figure 5, we can find, in addition to the microcontroller, the EEPROM 24C512, the power supply part and the alphanumeric Liquid Cristal Display (LCD). This last one has a capital importance, it is the bridge between the users and the embedded card and without it the user cannot manipulate the card. Its use gives the realized card power and simplicity. It displays all the necessary operations. For example, the user can start and see the training part of the network, the mean squared error will be displayed at each time the program finishes the three loops (Loop 1 to 3) in the sample program of the figure 4. The users can read the validated data memorized in the EEPROM at the end of the training part and at each time the system can be interrupted by pushing on the rest button.

As shown in the electronic circuit (Fig. 6) it contains few components, and we can say that the deserved work can be done with three principal components: the microcontroller with its basic circuits, the memory and the LCD.

The comparison between the obtained results using two instruments (the computer and the microcontroller) and based on the extensive practical results carried out within the course of this study is shown by figures 8 and 9 and it is given in the table 1. As remarked, the obtained results using the realized card is very close to the real results, and sometimes better than the results obtained using the computer especially in the validation part (Fig. 9). This is due to the well known problem of the wavelet network, which is the choice of the initial wavelet network parameters. These results give the microcontroller applications an expansion in the artificial intelligence field.

On the other hand, when we want to compare the card based on the neural network which has been realized in our previous work and the card presented in this chapter, we can find several points. First, the assembly language line number is smaller in the wavelet network based card than the program based on the neural network. This difference gives the card an important consequence which is the time reduction 2.77 second which is greater in the card based on the neural network (Saad Saoud & Khellaf, 2011). These two important advantages are the result of the network architecture itself, because the wavelet network has a linear output which makes the calculations easier.

Second, the new card gives the designer more simplicity of use, because the assembly program is written in the way that we can make a call of the standard subroutines, and we can also change at any time the network parameters such as the number of samples for training and validation or the number of iterations. We can say also that there are several disadvantages of the new card but not so important. The card cost is a little more expensive that the old card (Saad Saoud & Khellaf, 2011), and the wavelet network needs a special care during parameters initialization.

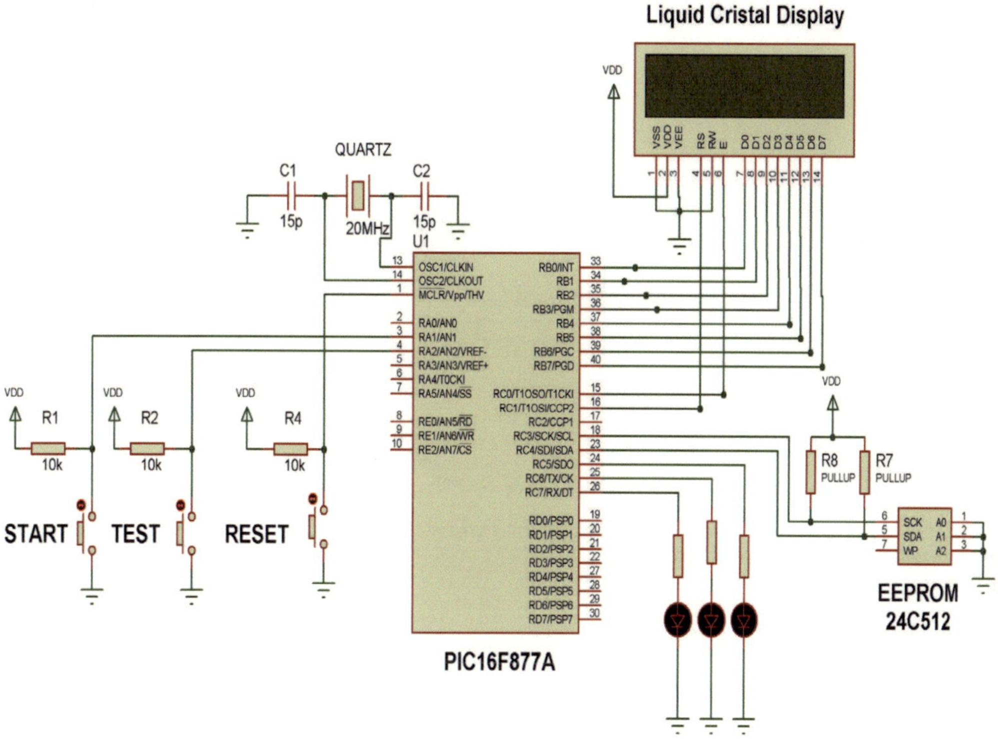

Fig. 6. The electronic circuit of the realized card

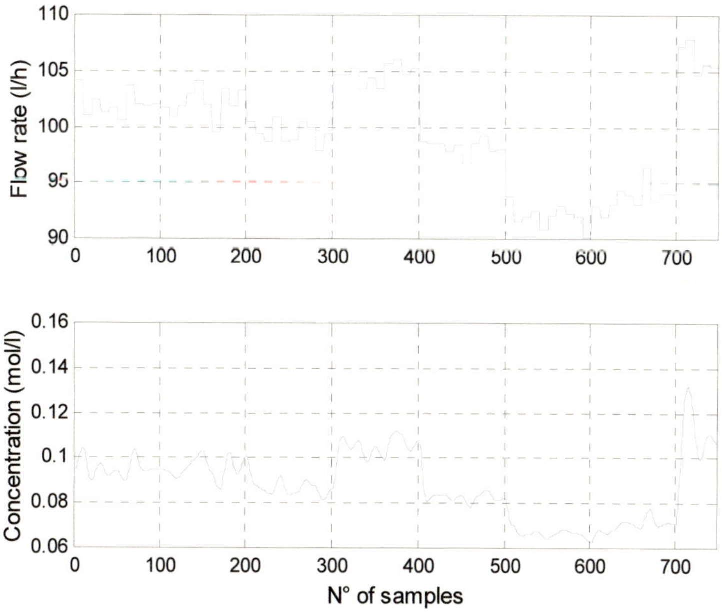

Fig. 7. Data used to train and validate the wavelet network using the PIC microcontroller

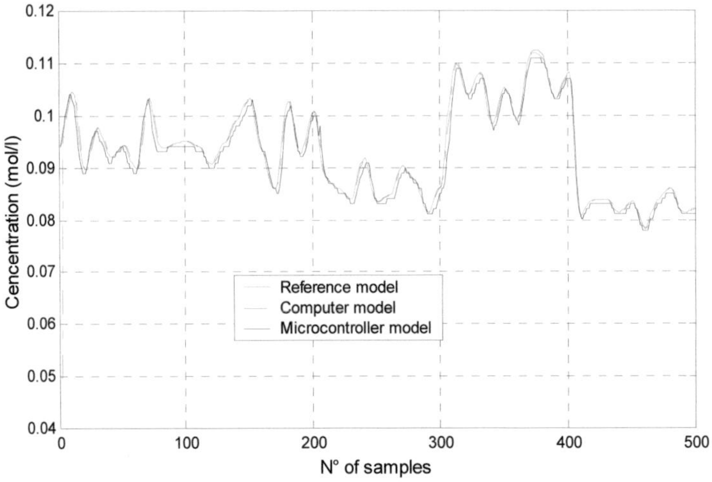

Fig. 8. Comparison results between computer and microcontroller for the training part of the CSTR using wavelet network based.

5. Moving to other LCD technologies

In this chapter, a direct Liquid Crystal Display (LCD) application is illustrated. Even though it is used as a slave of the PIC16F877A, it could be noted that without this tool the electronic card can not be used. This proves the usefulness of the liquid crystal display in the industrial life.

This card is normally operated with any LCD type, which is actually in our work the Twisted Nematic (TN)-Display. But it can be improved with other display types such us the In-Plane Switching (IPS)- and Multidomain-Vertical-Alignment (MVA)-Displays. In this work, we take the reference (Willem den Boer, 2005), as an example to describe these modes.

	Wavelet network trained through	
	Computer Pentium Dual-core inside speed 1.73 GHz	**Microcontroller PIC16F877A speed 20 MHz**
One iteration execution time (sec)	0.183 for 500 samples	2.77 for 500 samples
Instrument's cost	Expensive	Inexpensive
Space occupation	Big space consumed	Embedded

Table 1. Comparison between the two strategies (computer and microcontroller)

An often-quoted drawback of conventional TN LCDs has been their poor viewing angle behavior. Several dramatic improvements in viewing angle have been developed over the past few years. The most important ones are (Willem den Boer, 2005): The In-Plane-Switching (IPS) LC mode and the Multidomain-Vertical-Alignment (MVA) LC mode.

In both cases, the aim is to obtain a good viewing angle. The use of one of them gives the realized card another aspect, and improves the quality of the electronic system. Like all systems, moving to another type has advantages and disadvantages. We can cite here the principal disadvantages:

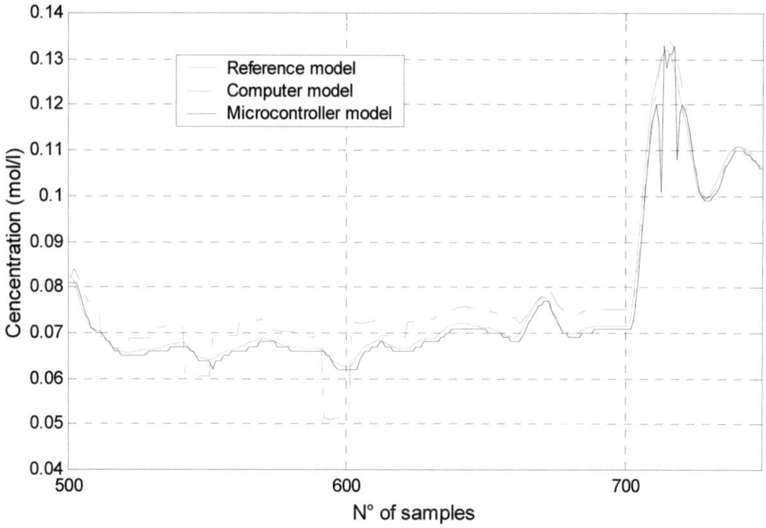

Fig. 9. Validation results of the simulated CSTR reactor using neural network based on the microcontroller and the computer.

- The cost of the card: the IPS and MVA types are more expensive than the simple TN display type.
- The occupied space: on one hand and because the LCD is controlled by the microcontroller, and the IPS-, MVA-Display types are more sophisticated, they require more assembly instructions which make the program more complicated. On the other hand, bigger LCD occupies more place which means large card dimension.

Hence, we can say, the choice of the LCD type is based on the practical situation in where the realized card will be implemented.

6. Conclusion

In this chapter, a direct practical application of the Liquid Cristal Display (LCD) is presented and proved by the real implementation of the full wavelet network with its backpropagation in a very inexpensive microcontroller. The realized cad is embedded and very simple to use by people having a small knowledge in the electronic and the assembly language programming. The embedded card is tested using the celebrated nonlinear system which is the CSTR chemical reactor. And in all cases, the programmer can at any time change the system to be modeled using the wavelet network respecting the input regressor and the data manipulation before saving it. In our future work, we try to use the realized card in real control situations and giving the architecture a dynamic aspect.

7. Appendix

In this appendix we try to give the reader of this chapter the tools to realize and test by himself the proposed card in this chapter.

First, it is very important to realize the card, for this reason the card's PCB is in pdf format that can be found within this book (wavelet mask.pdf and wavelet card.pdf). Second, the users should implement the components like it is shown in figures 5 and 6.

By a simple PIC and EEPROM programmer, we can charge the hexadecimal file (Wavelet.hex) and the binary file (CSTR.bin) which contain the hexadecimal file of the main assembly language program and data used for modeling and validation respectively. Verify the existence of the 9 V battery and that is all.
All necessary files can be found with the publisher of this book Intech.

8. References

Abiyev, R.H. (2003). Fuzzy Wavelet Neural Network for Control of Dynamic Plants, *International Journal of Computational Intelligence*, 1 (2), pp.139-143.

Bakshi, B. R. & Stephanopoulos, G. (1993). Wave-net: A multi resolution hierarchical neural network with localized learning, *American Institute of Chemical Engineering Journal*, vol. 39, pp. 57–81.

Cotton, N. J.; Wilamowski, B. M. & Dündar, G. (2008). A Neural Network Implementation on an Inexpensive Eight Bit Microcontroller, *12th International Conference on Intelligent Engineering Systems*, February 25–29, 2008, Miami, Florida.

Cybenko, G. (1989) Approximation by Superpositions of a Sigmoidal Function, *Mathematics of control, signals and systems*, Vol.2, pp. 303-314.

Delyon, B.; Juditsky, A. & Benveniste, A. (1995). Accuracy analysis for wavelet approximations, *IEEE Transaction on Neural Networks*, vol. 6, pp. 332–348.

De Moor, B. (1998). Daisy: Database for the identification of systems. Department of Electrical Engineering -ESAT- K.U. Leuven, Belgium. http://www.esat.kuleuven.ac.be/sista/daisy/, Used data set: Continuous Stirred Tank Reactor, Section: Process Industry Systems, code: 98-002.

Den Boer, W. (2005). *Active Matrix Liquid Crystal Displays: Fundamentals and Applications*. Elsevier.

Efe, M. Ö. (1996). Identification and Control of Nonlinear Dynamical Systems Using Neural Networks. M.S. Thesis, Boðaziçi University.

Efe, M. Ö. & Kaynak, O. (1997). Identification and Control of a Nonlinear Bioreactor Plant Using Classical and Dynamical Neural Networks.' Proceeding of the International Symposium on Industrial Electronics (ISIE'97), Guimaraes, Portugal, Vol.3, 7-11 July 1997, pp. 1211-1215.

Espinosa, J.; Vandewalle J. & Wertz V. (2005) *Fuzzy Logic, Identification and Predictive Control*, Springer-Verlag, UK.

Gao, X. (2002). A comparative research on wavelet neural networks. *Proceedings of the 9th International Conference on Neural Information Processing*, vol.4, pp. 1699 – 1703, 18-22 Nov. 2002.

Gulbag, A.; Temurtas, F.; Tasaltin C. & Ozturk, Z. Z. (2009). A neural network implemented microcontroller system for quantitative classification of hazardous organic gases in the ambient air, *International Journal of Environment and Pollution*, Vol. 36-3, pp. 151 – 165.

Khalaf, N. M. A. (2006). Wavelet Network Identifier for Nonlinear Functions, thesis, University of Technology, Iraq.

Henson, M. & Seborg D. (1990). Input-output linearization of general nonlinear processes, *AIChE Journal*, pp. 1753–1757.

Hong, J. (1992). Identification of stable systems by wavelet transform and artificial neural networks, Ph.D. dissertation, Univ. Pittsburgh, PA.

Kreinovich, V.; Sirisaengtaksin, O. & Cabrera, S. (1994). Wavelet neural networks are asymptotically optimal approximators for functions of one variable, *in Proceeding IEEE International Conference on Neural Networks*, Orlando, FL, June 1994, pp. 299-304.

Levin, A. U. (1992). Neural networks in dynamical systems: a system theoretic approach, PhD Thesis, Yale University, New Haven, CT.

Lightbody, G. & Irwin, G. (1997). Nonlinear control structures based on embedded neural system models. *IEEE Transaction on Neural Networks*, 8, pp. 553–567.

Liu, G. P.; Kadirkamanathan, V. & Billings, S. A. (1998). Predictive control for non-linear systems using neural networks," *International Journal of Control*, Vol. 71, No. 6, pp.1119- 1132.

Liung, T. K.; Mashor, M. Y.; Isa, N. A. M.; Ali, A. N. & Othman, N. H. (2003). Design of a neural network based cervical cancer diagnosis system: a microcontroller approach, *ICAST* 2003.

Morningred, J. D.; Paden, B. E.; Seborg D. E. & Mellichamp, D. A.. An Adaptive Nonlinear Predictive Controller, *American Control Conference,* 23-25 May 1990, San Diego, CA, USA, pp. 1614 - 1619.

Narendra, K. S. & Parthasarathy, K. (1990). Identification and Control of Dynamical Systems Using Neural Networks, *IEEE Transactions on Neural Networks*, Vol.1 (1), pp. 4-27.

Neelamegamand, P. & Rajendran, A. (2005). Neural Network Based Density Measurement, *Bulgarian Journal of Physics*. Vol. 31, pp. 163–169.

Nerrand, O.; Roussel-Ragot, P.; Personnaz, L. & Dreyfus G. (1993). Neural Networks and Nonlinear Adaptive Filtering: Unifying Concepts and New Algorithms, *Neural Computation*, Vol.5 (2), pp. 165-199.

Patan, K. (2008). *Artificial Neural Networks for the Modelling and Fault Diagnosis of Technical Processes*, Springer-Verlag Berlin Heidelberg.

Pati, Y. C. & Krishnaprasad, P. S. (1993). Analysis and synthesis of feedforward neural networks using discrete affine wavelet transformations, *IEEE Transactions on Neural Networks*, vol. 4, pp. 73–85.

Plett, G. L. (2003). Adaptive Inverse Control of Linear and Nonlinear Systems Using Dynamic Neural Networks," IEEE *Transactions on Neural Networks*, Vol. 14, No. 2, pp. 360-372.

Rivals, I. & Personnaz, L. (1996). Black Box Modeling With State-Space Neural Networks, in: *Neural Adaptive Control Technology*. Zbikowski, R. & Hunt, K. J., pp. 237-264 World Scientific, Singapore.

Roffel, B. & Betlem, B. (2006). *Process Dynamics and Control, Modeling for Control and Prediction*, John Wiley & Sons, 2006.

Saad Saoud, L. & Khellaf, A. (2009) Identification and Control of a Nonlinear Chemical process Plant Using Dynamical Neural Units, *Third International Conference on Electrical Engineering Design and technologies*, Oct. 31- Nov. 2, 2009, Tunisia.

Saad Saoud, L. & Khellaf, A. (2011) A Neural Network Based on an Inexpensive Eight Bit Microcontroller, *Neural computing and application*, 20(3), pp. 329-334.

Saad Saoud, L.; Rahmoune, F.; Tourtchine, V. & Baddari, K. (2011). An Inexpensive Embedded Electronic Continuous Stirred Tank Reactor (CSTR) Based on Neural Networks, *The 2nd International Conference on Multimedia Technology (ICMT2011)*, Accepted, July 26-28, 2011, Hangzhou , China.

Sanchez, J. & Canton, M. P. (2007) *Microcontroller Programming: The Microchip PIC.* Taylor & Francis, Boca Raton.

Sjöberg, J.; Zhang, Q.; Ljung, L.; Benveniste, A.; Delyon, B.; Glorennec, P.Y.; Hjalmarssont, H. & Juditskys, A. (1995). Nonlinear Black-box Modeling in System Identification: a Unified Overview, *Automarica*, Vol. 31, No. 12, pp. 1691-1724.

Zhang, Q. & Benveniste, A. (1992). Wavelet Networks, *IEEE Transactions on Neural Networks*, Vol. 3 (6), pp. 889-898.

Tsatsanis, M. K. & Giannakis, G. B. (1993). Time-varying system identification and model validation using wavelets, *IEEE Transaction on Signal Processing*, vol. 41, pp. 3512–3523.

Part 3

Liquid Crystal Displays - Future Developments

6

Active Matrix Driving and Circuit Simulation

Makoto Watanabe
Sony Mobile Display Corporation
Japan

1. Introduction

This chapter explains the principle of active matrix driving which is the most popular driving method used in current liquid crystal displays (LCDs). It then discusses issues that designers must overcome to avoid the malfunctioning and introduces a liquid crystal model for conducting circuit simulations to optimize the circuit parameters efficiently.

1.1 Equivalent circuit of a pixel in LCDs

The equivalent circuit of a pixel operated by active matrix driving is shown in Fig. 1.
Data lines are connected to a data driver for generating the signal pulses for the picture data. Scan lines are connected to a scan driver for generating the scan pulses which enable the addressing driving. Vsig and Vg are applied to the data lines and scan lines, respectively. The thin film transistor (TFT) has three terminals of MOS transistors, and each terminal gate, drain, and source is connected to a scan line, a data line, and a pixel electrode, respectively. Cgs means the parasitic capacitance between the gate and source terminal in the TFT. Liquid crystal is injected into the gap between the pixel electrode and the counter backplane electrode, and it forms a liquid crystal cell capacitance (Clc). Clc is a variable capacitor that changes value according to the applied voltage between the pixel electrode and a counter backplane electrode. The voltage of the pixel electrode and counter backplane electrode are denoted by Vpix and Vcom, respectively. The storage capacitor is denoted Csc, and it is connected in parallel to Clc. Its function is to hold charges on a pixel electrode while the TFT is switched off.

1.2 Timing chart for each signal pulse

Fig. 2 shows the wave forms applied to each bus line and electrode. The period during which the pixel electrode voltage (Vpix) is higher than the counter electrode voltage (Vcom) is called the "plus frame" (Fig.2 (a)), whereas the period during which Vpix is lower than Vcom is called the "minus frame" (Fig.2 (b)). The plus and minus frames are switched every frame period (Tf).
When the voltage of the gate terminal connecting to the scan line rise to a high level, the resistance between the drain and source terminals becomes very low (Ron). As a result, electrical charges flow into the pixel electrode from the data line till the voltage of the pixel electrode achieves the voltage of the data line during the writing time (Tw). This process is called the "Charge Process". When the voltage of the scan line starts to drop, the pixel electrode voltage shows a negative shift ΔVfd because of the coupling with the gate terminal

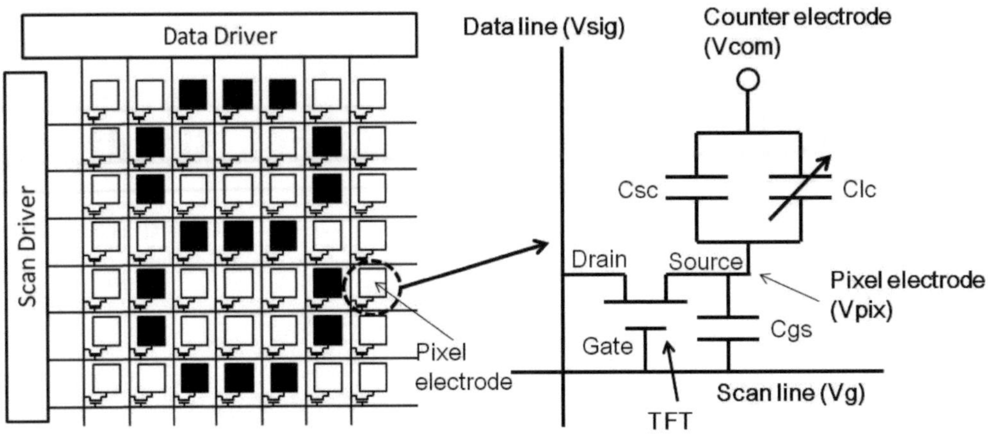

Fig. 1. Equivalent circuit of LCD panel operated with active matrix driving

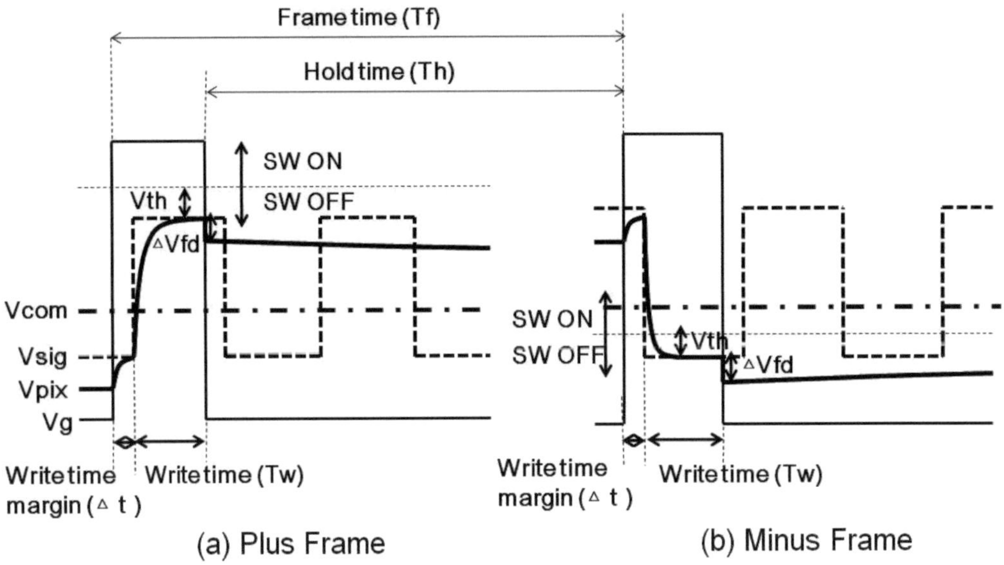

Fig. 2. Timing chart for each signal pulse applied to bus lines and electrodes

via the gate-source capacitance of TFT (Cgs) . This process is called the "Coupling Process". ΔVfd is generally called the feed-through voltage. When the voltage of the scan line becomes low, the resistance between the drain and source terminal becomes very high (Roff) . Ideally charges on the pixel electrode are kept for a hold time (Th) until TFT is switched on at the beginning of the next frame and the pixel electrode voltage keeps constant value. However some amount of current leaks out between the drain and source terminal, and in turn, the charges on the pixel electrode decreases gradually during the hold time (Th). This process is called the "Hold Process". The counter electrode voltage (Vcom) is just set to be the central value between the pixel electrode voltages in the plus and minus frames. By setting Vcom in this manner, we can obtain a constant transmittance that does not depend upon the frame; that is, we can get a flicker-free image since the applied voltages in both frames are the same. As stated above, there are three stages (a) Charge Process (b) Hold Process (c) Coupling Process. In the following, the operation and key design points of the circuit are explained in detail.

Let us begin by explaining the relationships among the frame time (Tf), hold time (Th) and write time (Tw). The frame time (Tf) is generally taken to be 1/60 (sec) for historical reasons. For an application which cannot be allowed to have a flicker malfunction, however, the frame time is usually set to less than 1/60 (sec). The reason why flicker malfunction occurs will be explained later. In Fig. 2, Δt is called the "write time margin" and it means the offset time between the scan line pulse and the data line pulse. (Tw+Δt)=Tf/N is satisfied if there are N scan lines. For instance, Tw+Δt=7.3μ(sec) in the case of high-definition TVs which have 1125 scan lines with Tf=1/120(sec)≅8.3m(sec). The write time margin (Δt) is usually designed to be around 2μ(sec), although it depends on the expected amount of pulse decay. Consequently, Tw is about 5μ(sec). Strictly speaking, the hold time (Th) should be of the difference between the frame time (Tf) and (Tw+Δt). However, considering that Tw+Δt is on the order of microseconds, the hold time (Th) can be approximated to be the frame time (Tf).

1.2.1 Charge process

When TFT behaves just as an electrical switch, it operates in the linear region of the MOS transistor (Vds<Vgs-Vth). Vds is the voltage between the drain and source terminal, Vgs is the voltage between the gate and source terminal, and Vth is the threshold voltage of the TFT. In the linear region, the current between the drain and source terminal (Ids) can be described as (Sze, 1981)

$$I_{ds} = \mu C_{ox}\left(\frac{W}{L}\right)\left(V_{gs}-V_{th}\right)V_{ds} \tag{1}$$

Here μ is mobility, Cox is the gate insulator capacitance per unit, and L and W are the channel length and width. Therefore, the resistance while the TFT is switched on (Ron) can be expressed as

$$R_{on} = \frac{V_{ds}}{I_{ds}} = \frac{1}{\mu C_{ox}\left(\frac{W}{L}\right)\left(V_{gs}-V_{th}\right)} \tag{2}$$

As shown in Fig. 2, the source voltage of the TFT during a plus frame is higher than that during a minus frame. In other words, Vgs is smaller in a plus frame than in a minus frame. Therefore, according to Eq.(2), Ron becomes higher in a plus frame than in a minus frame. Referring to the equivalent circuit in Fig. 1, the time constant (τ_{on}) for charging to the pixel electrode can be described as

$$\tau_{on} = R_{on}(C_{gs} + C_{lc} + C_{sc}) \tag{3}$$

The charge time (Tw) should be sufficiently long compared with the time constant τ_{on}. In general, the TFT and the voltage applied to bus lines are designed so as to satisfy Tw=3 τ_{on} ~6 τ_{on}.

1.2.2 Hold process

The TFT switching-off resistance (Roff) cannot be expressed with a simple equation like Ron since it has complicated physical mechanisms (Jacunski, 1999). We note that ambient light dramatically decreases Roff. Referring to the equivalent circuit in Fig.1, the time constant (τ_{OFF}) for holding charges on the pixel electrode can be described as

$$\tau_{off} = R_{off}(C_{gs} + C_{lc} + C_{sc}) \tag{4}$$

The time constant τ_{OFF} should be sufficiently long compared with the hold time (Th).

1.2.3 Coupling process

When the voltage for the scan line falls from the high level (Vgon) to the low-level (Vgoff), the pixel electrode voltage shifts by the coupling with the parasitic capacitance (Cgs) between the gate and the source terminal of the TFT. This voltage shift of the pixel electrode is called the feed-through voltage (ΔVfd), and it can be expressed as

$$\Delta V_{fd} = \frac{C_{gs}}{C_{gs} + C_{lc} + C_{sc}}\left(V_{gon} - V_{goff}\right) \tag{5}$$

As discussed earlier, ΔVfd should be a constant independent of any conditions since it uniquely determines the voltage for the counter backplane electrode. This is important in that we can get high-quality flicker free images and get high reliability without residual DC voltages in the liquid crystal cell. ΔVfd, however, depends on various factors, such as Clc, Cgs, and pulse wave distortions.

The first factor is the liquid crystal capacitance (Clc) dependences on the voltage between the pixel and counter electrode. As this voltage increases, the electric field across the liquid crystal cell gets reinforced and the liquid crystal molecule tends to reorient itself along the electric field direction. Consequently, the capacitance of the liquid crystal cell (Clc) increases as shown in Fig. 3. From Eq.(5), ΔVfd changes in accordance with the value of Clc.

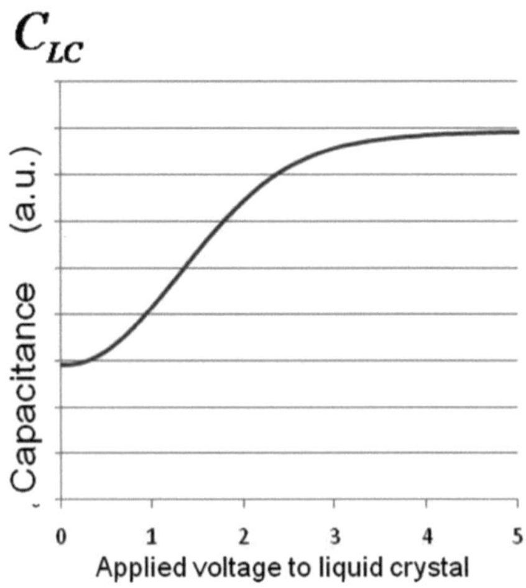

Fig. 3. Clc dependences on the applied voltage to liquid crystal

The second factor is the capacitance between the gate and source terminal (Cgs), which depends on the state (on or off) of the TFT. Fig. 4. shows a cross-sectional view of a general a-Si TFT. While TFT is the on-state (Fig.4 (a)), a conducting channel forms at the bottom of the a-Si layer and a capacitance between the channel and gate electrode (Con) has been generated. Half of Con can be regarded as Cgs in the on-state of TFT. A conducting channel does not form while TFT is the off-state (Fig.4 (b)). Therefore, Cgs during the off-state is the same as the capacitance between the source and gate electrode (Coff). Fig. 5 shows the dependence of Cgs on Vgs.

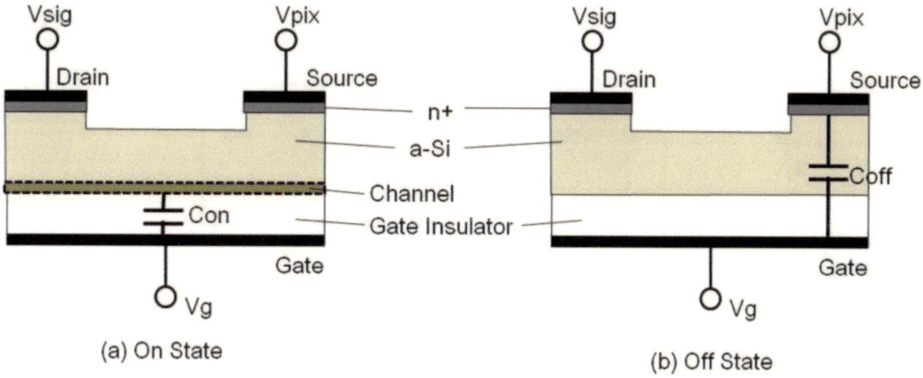

Fig. 4. Gate–source capacitance (On state / Off state)

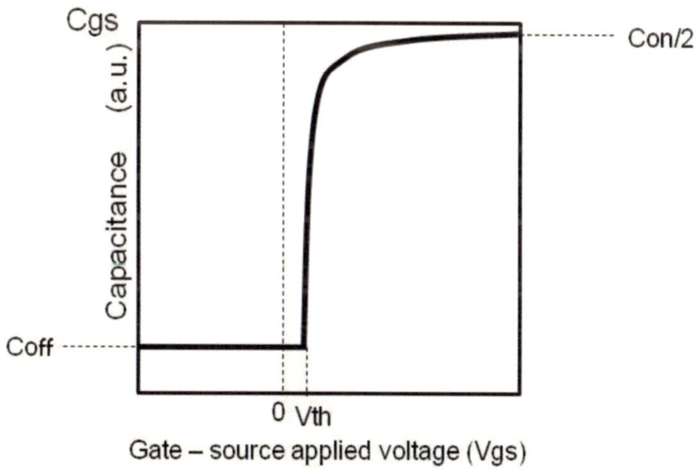

Fig. 5. Dependence of Cgs on the applied voltage

The cut-off voltage which TFT changes form on-state to off-state is Vgs=Vth.
The equation for ΔVfd is updated by taking this behavior into account. In Eq. (6), ΔVfd in a plus frame is smaller than it is in a minus frame since Con/2 is bigger than Coff.

$$\Delta V_{fd} = \frac{(C_{on}/2)(V_{gon} - (V_{PIX} + V_{th}))}{C_{on}/2 + C_{lc} + C_{sc}} + \frac{C_{off}((V_{PIX} + V_{th}) - V_{goff})}{C_{off} + C_{lc} + C_{sc}} \tag{6}$$

The third factor is the influence of a scan pulse distortion on ΔVfd. The scan pulse significantly decays at pixels farther away from the scan driver. Such a decay could cause problems especially in larger displays (Watanabe, 1996). Fig. 6(a) shows the feed through voltage (ΔVfd) under the condition that the scan pulse does not decay. Fig.6 (b) shows the feed through voltage (ΔVfd) under the condition that the scan pulse decays.

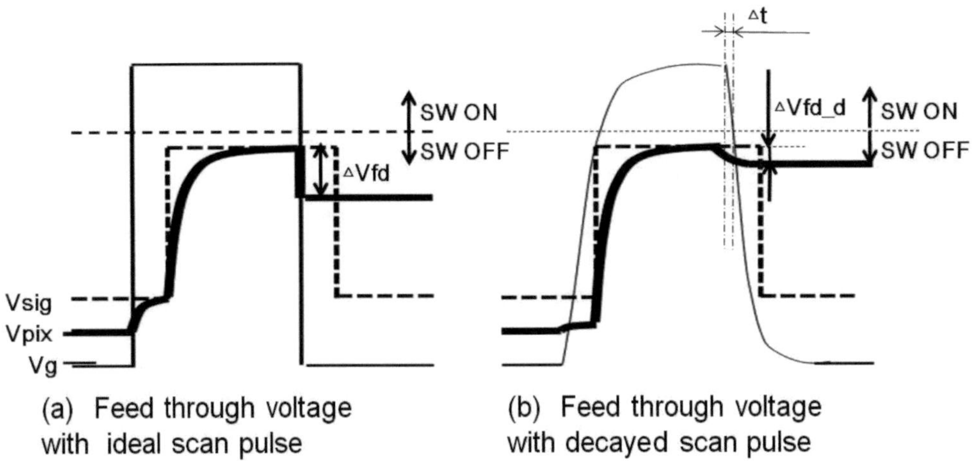

Fig. 6. Influence of a scan pulse distortion on the feed through voltage

When the scan pulse decays, it takes a finite amount of time Δt for the TFT to switch off. During Δt, current Ids(t) continues to flow from a data line to the pixel electrode. Therefore, ΔVfd is modified to Eq. (7).

$$\Delta V_{fd\_d} = \Delta V_{fd} - \frac{1}{C_{gs} + C_{lc} + C_{sc}} \int_0^{\Delta t} I_{ds}(t)dt \tag{7}$$

According to Eq.(7), the effective feed through voltage ΔVfd_d is smaller in a minus frame than in a plus frame because the cut-off voltage becomes lower and Δt becomes longer in a minus frame than in a plus frame.

2. Display quality problems

This section discusses some issues LCD designers must overcome in order to avoid such as shading, cross-talk (vertical / horizontal), flicker malfunctions, low response time and charge leakage in liquid crystal cells.
Even though active matrix driving is a dramatic improvement upon passive matrix driving (Pochi, 1999), the above malfunctions remains in the specific displayed patterns, so-called "killer pattern".
Here we explain what these modes of malfunction are, why they appear, and how to suppress them. This information would be very useful for designers not only in optimizing the design parameters but also in analyzing problems they may encounter. These problems tend to be apparent especially in large and high-resolution LCDs. In the example illustrated in the following explanation, we assume that the treated LCDs are of the dot inversion type

which same pixel polarities are aligned in a checker pattern (Fig.7 (a)) and normally white mode which has the maximum transmittance with no applied voltage (Fig.7 (b)). It will, however, be very easy for readers to apply them to other types of active matrix LCDs.

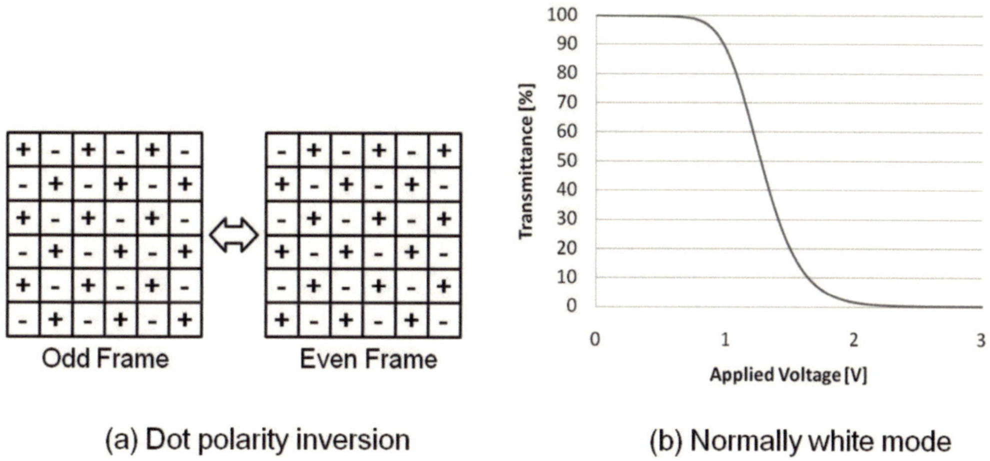

Fig. 7. Driving method assumed in the example

2.1 Shading

When a checker pattern of gray and black is displayed in background, as shown in Fig.8 (a), we see an unexpected gradational change vertically (Fig.8 (b)).

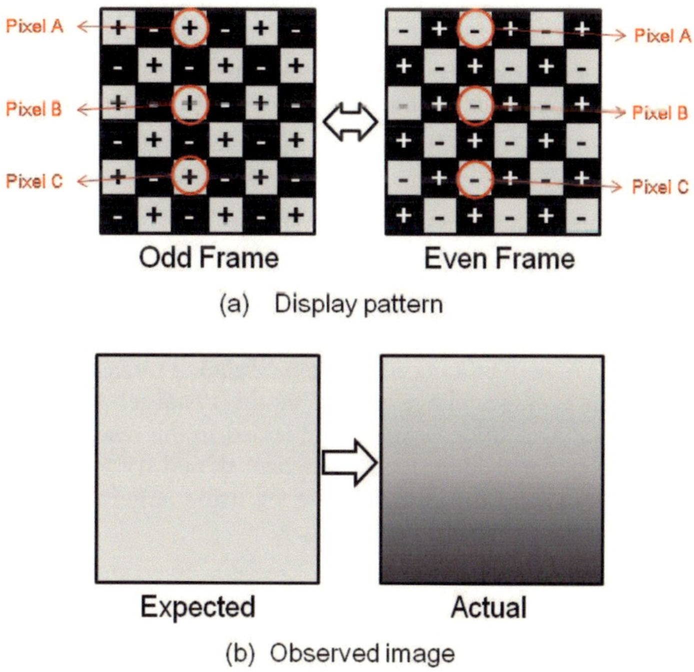

Fig. 8. Shading

The mechanism causing this phenomenon arises from the parasitic capacitance Cdp between the data line, or adjacent data line, and the pixel electrode (see Fig. 9). Here, we shall discuss the voltage modulation of a pixel electrode by the data line voltage (Vsig) and adjacent data line voltage (Vsiga) via the parasitic capacitance Cdp.

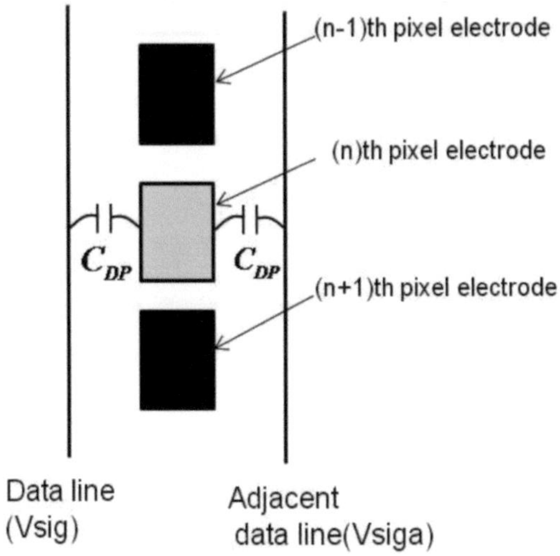

Fig. 9. Circuit model for the shading phenomenon

Fig.10 (a) shows the fluctuation in the data line voltage and Fig. 10 (b) shows the voltages of pixel electrodes located at A, B and C in Fig.8. At the moment when the frame changes from odd(even) to even(odd), the symmetry between the fluctuation in a data line and that of the adjacent data line is offset by the value of ΔVsig. Consequently, the voltage of the pixel electrode is modulated via Cdp. The modulated voltage of a pixel electrode δV can be expressed as

$$\delta V = \frac{2C_{dp}}{C_{tot}} \Delta V_{sig} \tag{8}$$

Here Ctot means the total of the capacitances connected to the pixel electrode. The voltages of the pixels in the upper area of display (such as Pixel A), is updated immediately after modulation at the moment of a frame switching. In contrast, pixels in the lower area of display (such as Pixel C) remain influenced by the modulation remains for a long time up to one frame. Therefore, the average voltage across a liquid crystal cell is higher in Pixel C than in Pixels A and B and a lower transmittance is obtained in the case of the normally white mode of LCDs. To reduce the shading level, designers should try to decrease the parasitic capacitance Cdp or to increase Ctot , for example, by using larger Csc in the pixel design process.

2.2 Vertical crosstalk

There is a variety of modes for a vertical crosstalk (Watanabe, 1997). Here we will describe the coupling mode between a pixel electrode and a data line: this sort of crosstalk will be a critical issue when very high-resolution LCDs become available.

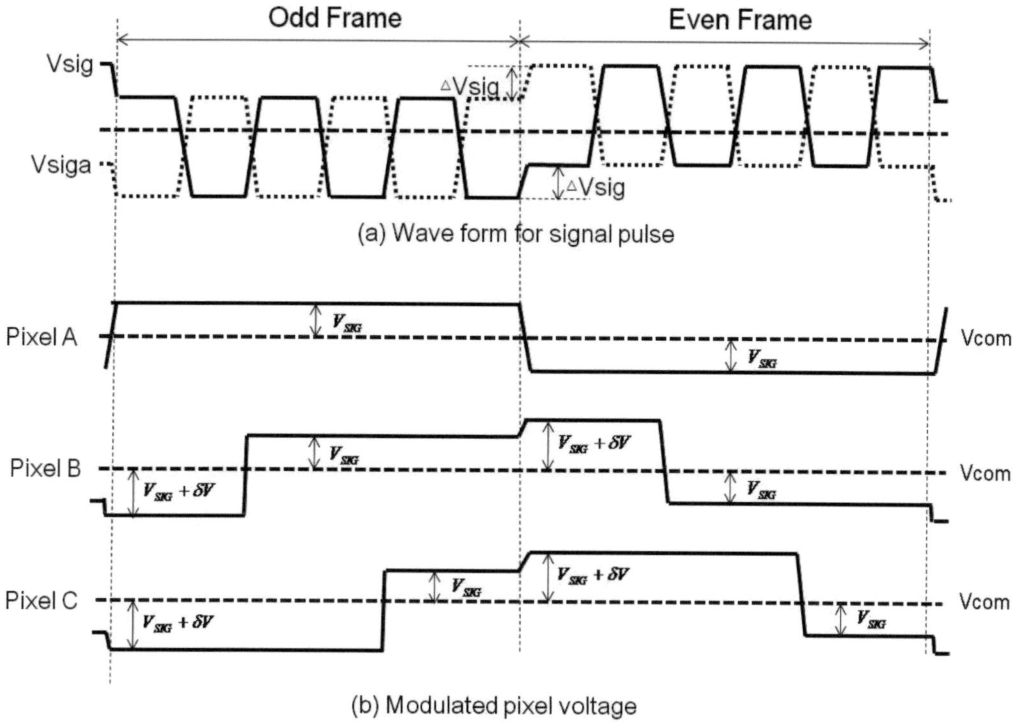

Fig. 10. Pixel voltage modulation by data lines via parasitic capacitances

When a white window is displayed in the middle of the display area on a background composes of a gray tone and black checker pattern as shown in Fig.11 (a), we see an unexpected brightness in the area above and below the window (Fig.11 (b)). The area above the window becomes darker and the area below the window becomes lighter than the tone of the background.

To explain the mechanism causing such vertical crosstalk, we will assume that the shading phenomenon discussed earlier can be ignored. However, readers should be aware that shading and vertical crosstalk can coexist on the same screen. The circuit model to explain vertical crosstalk is the same as that for the shading (Fig.9).

Fig. 12 (a) shows the voltage fluctuation in the data line and the pixel electrode voltage at point A in Fig. 11. Fig. 12 (b) shows the voltage fluctuation in the data line and the pixel electrode at points B and C in Fig. 11. Note that there is a white window below B and above C. In a pixel at B and C, the symmetrical relationship between the fluctuation of the data line and that of the adjacent data line disappears, when the period for drawing the widow starts or ends. Consequently, the pixel electrode voltage is modulated via the parasitic capacitance Cdp. The modulated voltage δV can be expressed by using the same equation as in the case of shading (Eq. (8)).

The brightness difference between pixels B and C can be explained as follows. The voltage of pixel B is affected by drawing the window after the charge process is completed in this frame. Meanwhile the voltage of the pixel C is affected before the charge process is completed in this frame. Therefore the voltage across the liquid crystal cell increases by δV in the pixel B and decreases by δV in the pixel C.

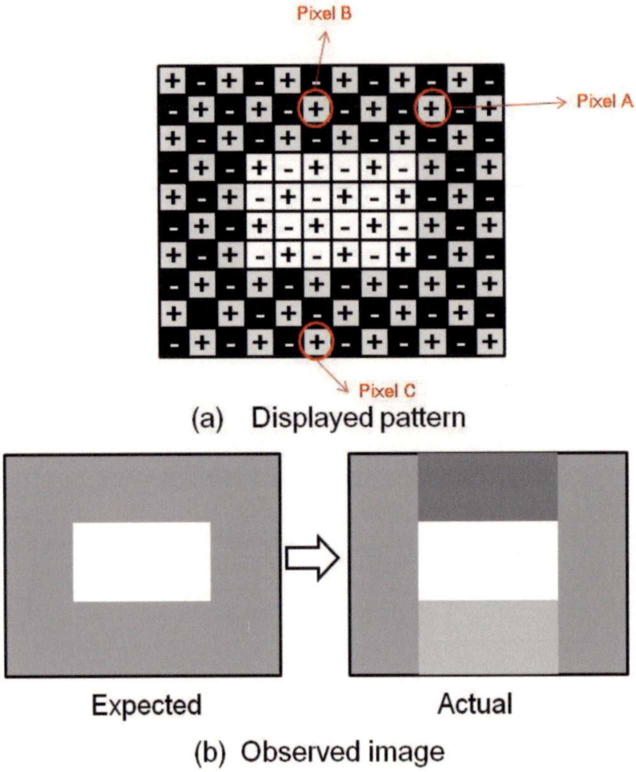

Fig. 11. Vertical crosstalk

To reduce vertical crosstalk, as is done in shading, designers should try to decrease the parasitic capacitance Cdp or increase Ctot, for example, by using a larger Csc in a pixel design process.

2.3 Horizontal crosstalk

When a white window is displayed in the middle of a display area with a background consisting of a striped pattern with gray tones and black as shown in Fig. 13(a), we see the brightness difference in the left and the right areas of the window compared with other area (Fig.13 (b)).

Here we should consider that the tones in the right and left area of the window are rather closed to the expected value compared with the tone in the other area.

Now let us explain the mechanism causing the horizontal crosstalk. Several models of the horizontal crosstalk have been proposed (Kimura, 1994). One of the circuit models is shown in Fig.14. The DC voltage Vcom is provided to a backplane counter electrode at the edge of the display area, which means there are no supply points for Vcom in the display area. Each node of this resistance network in the counter electrode is unintentionally connected to the data line via the parasitic capacitance Cd-com. The voltage of the counter electrode is thus modulated by the fluctuation of the data line voltage (Vsig).

Fig. 15 (a) shows the fluctuation in the voltage of all data lines in Fig. 13. The numbers on the left of the waves correspond to those assigned to the data lines in Fig. 13. The combination of the voltage fluctuations for all data lines modulates the counter electrode

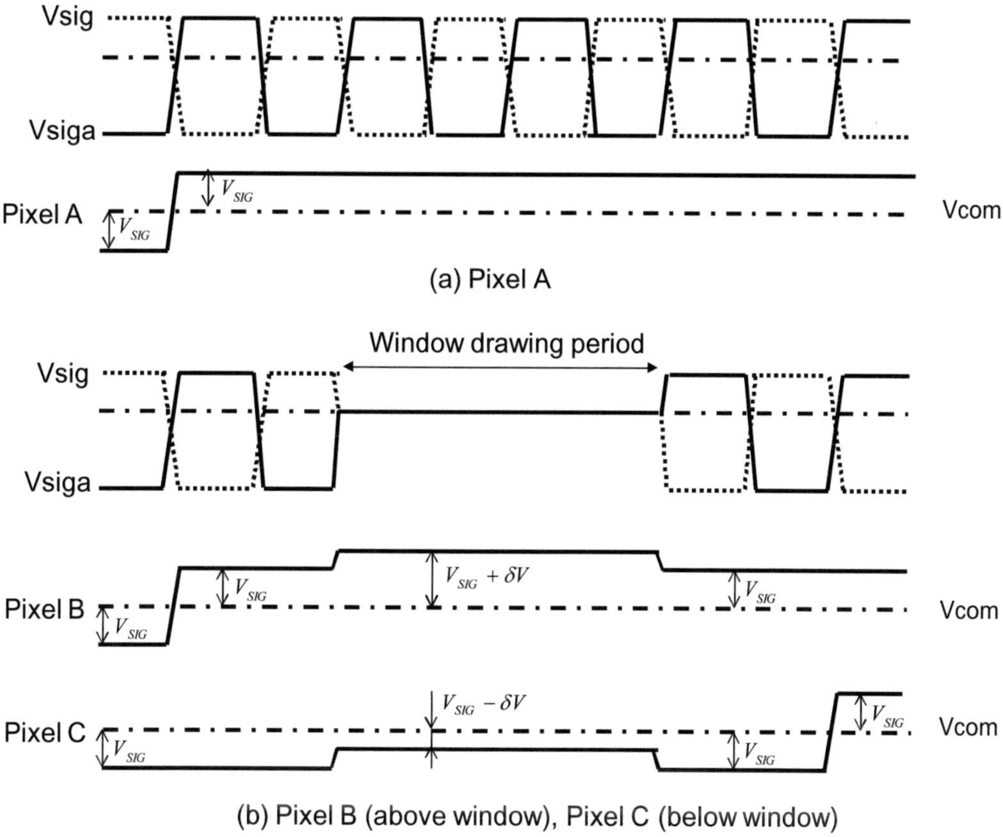

Fig. 12. Pixel voltage modulation by data lines via parasitic capacitances

voltage, as shown in Fig.15 (b). We can see the level of the modulation is lower in the period for drawing the window than in the other periods. ΔVcom and τ_{com} in Fig. 15(b) can be approximately expressed as

$$\Delta V_{com} = \frac{1}{C_{comtot}} \sum_i C_{d(i)-com} \Delta V_{sig(i)} \tag{9}$$

$$\tau_{com} = R_{com} C_{comtot} \tag{10}$$

Here Rcom means the total resistance of the counter electrode, and Ccomtot means the total capacitance connected to the counter electrode.

Referring to Fig. 16, if the wave for the counter electrode has not been restored to the DC level (Vcom) during the write time (Tw), an unexpected voltage will be applied to a liquid crystal cell Vlc' (>Vlc). The situation in pixel A is similar to Fig.16 (a) since the counter electrode voltage is not so modulated during pixel A's write time. Meanwhile the situation in a pixel B is similar to Fig.16 (b) since the counter electrode voltage is too modulated during pixel B's write time. As a result, we can see horizontal crosstalk since the effective voltage across the liquid crystal cell is bigger in pixel B than in pixel A. The best way to suppress the horizontal crosstalk is to decrease the counter electrode resistance. It is, however, difficult since optical properties such as transparency are often sacrificed.

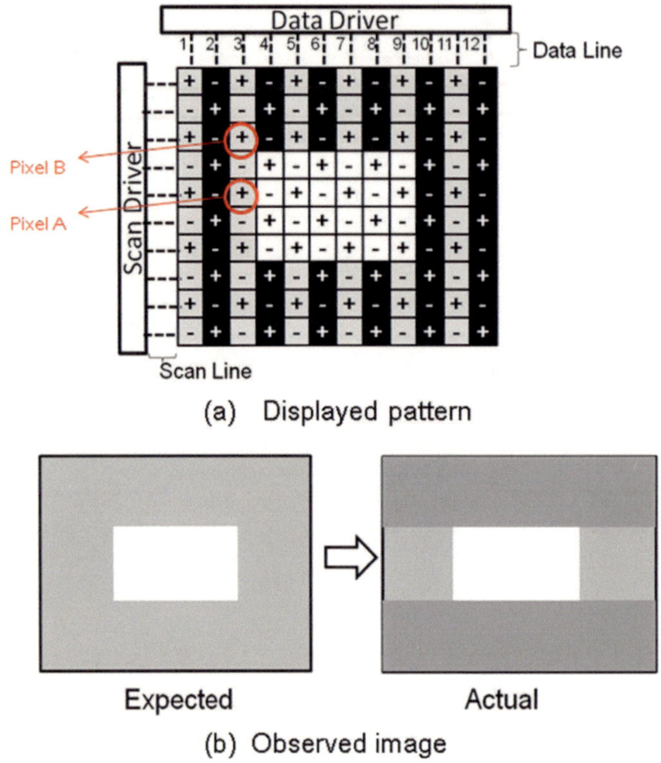

Fig. 13. Horizontal crosstalk

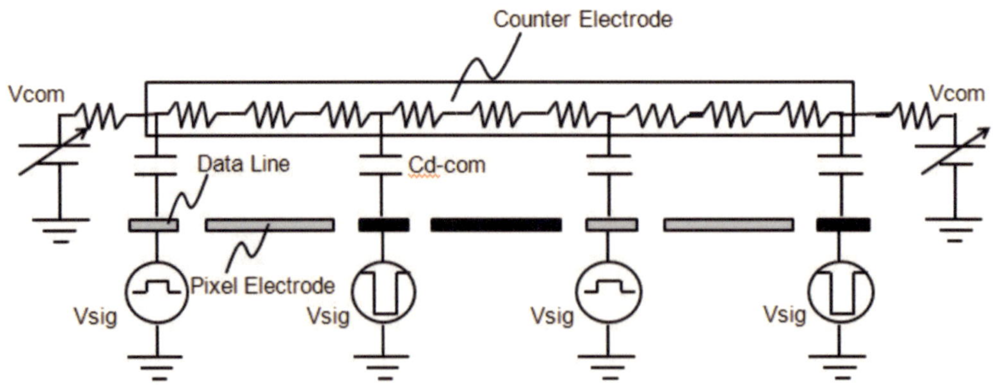

Fig. 14. Circuit model for horizontal crosstalk

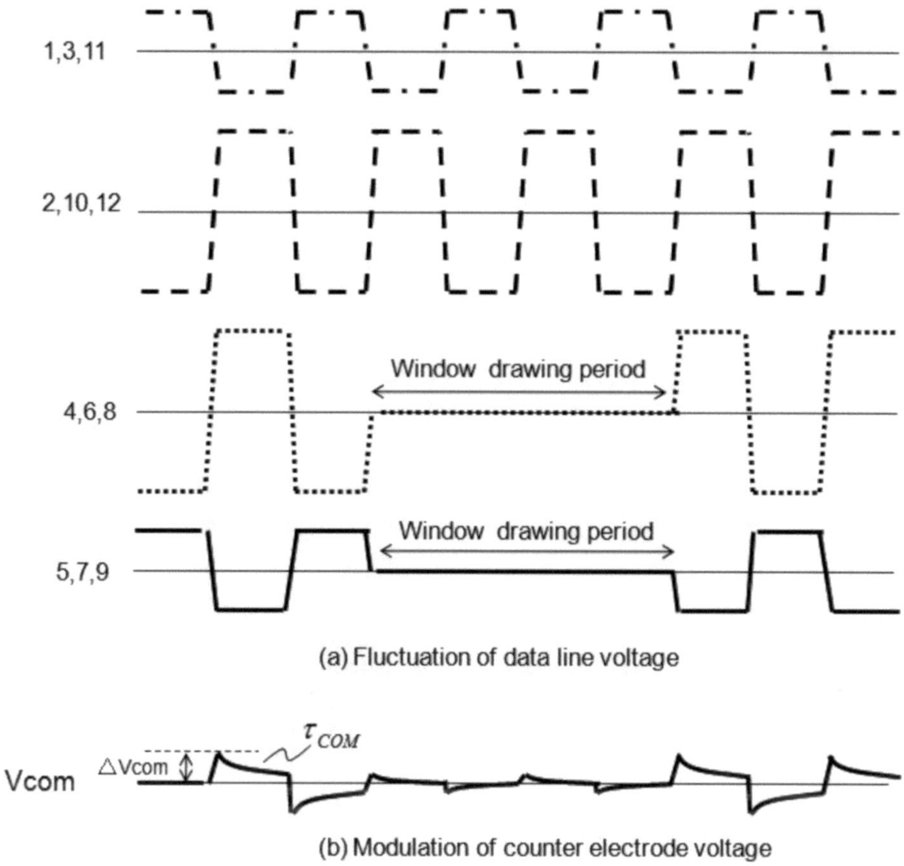

Fig. 15. Counter electrode voltage modulation by data lines via parasitic capacitances

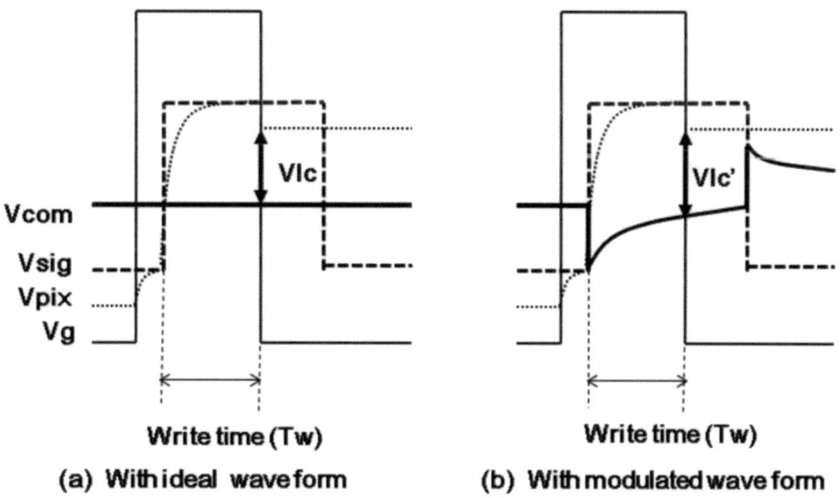

Fig. 16. Change in effective voltages across a liquid crystal cell resulting from counter electrode voltage modulation.

2.4 Flicker

When a gray tone and black checker pattern is displayed in the background as shown in Fig. 17 (a), we sometimes can see the flicker caused by the brightness differences between odd frames and even frames (Fig17 (b)). Note that all the gray tone dots with gray tone have the same polarity during one frame.

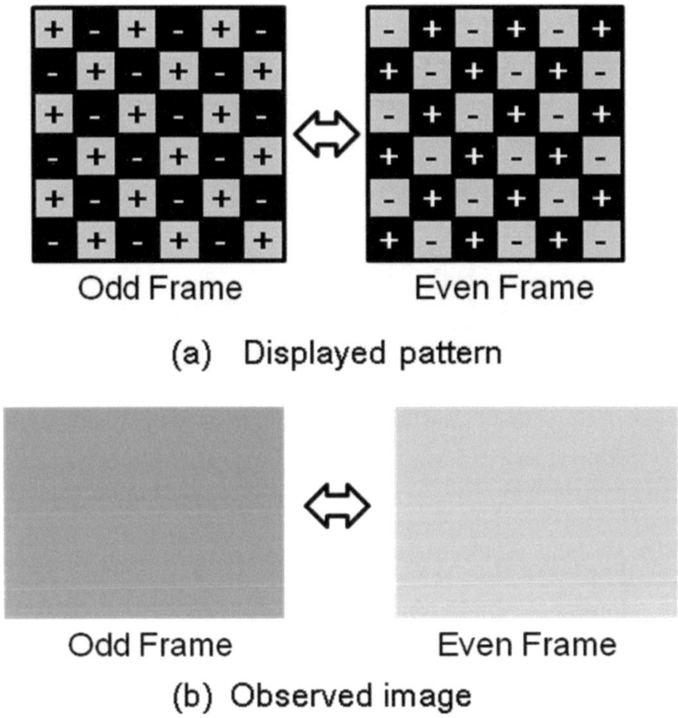

Fig. 17. Flicker

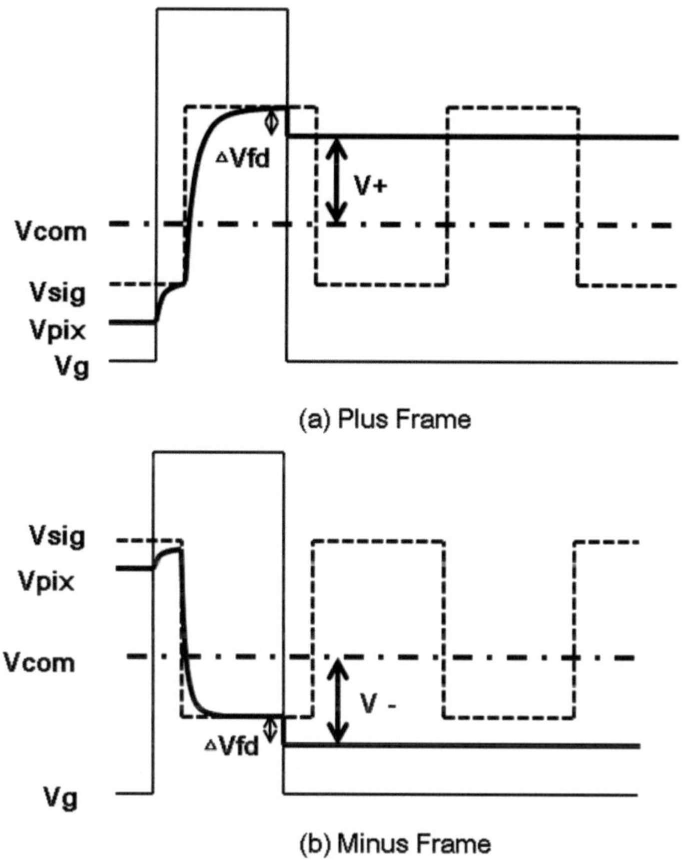

Fig. 18. Adjustment of the counter electrode voltage (Vcom) to suppress flicker (V+ = V-)

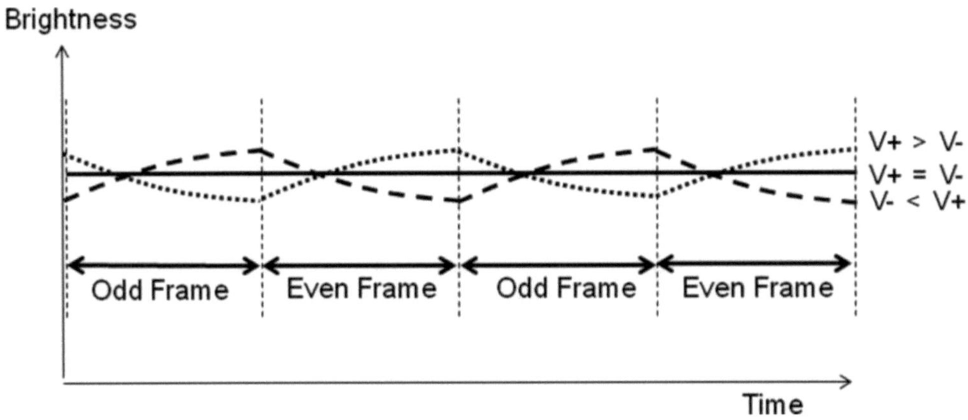

Fig. 19. Brightness fluctuation due to mis-adjustment of the counter electrode voltage(Vcom)

Ideally, the counter electrode voltage (Vcom) should be set so that the voltage between the pixel electrode and the counter electrode $|V_{pix} - V_{com}|$ is the same from one frame to the next (V+=V- in Fig.18). In so doing, we cannot see any flicker since the voltage across a liquid crystal cell stays constant value. However, as discussed earlier, it is too difficult to adjust Vcom such that V+=V- under any condition since the feed-through voltage ΔVfd depends on several factors: variation of Clc, Cgs, pulse distortion, and so on. If the voltage in each frame is not same (V+≠V-), a brightness fluctuation can be seen, as shown in Fig. 19. It is often said that the human eyes cannot perceive flicker of more than 50Hz. Therefore, making a frame rate much higher than 100Hz is one of the effective means to suppress flicker.

2.5 Low response time

If the response time of LCDs is not sufficiently short, the outline of the moving object on the screen blurs, as shown in Fig. 20. Although the properties of the liquid crystal material and cell structures are the dominant factors determining the response time (Wittek, 2008) , the influence of the circuit parameters is not negligible.

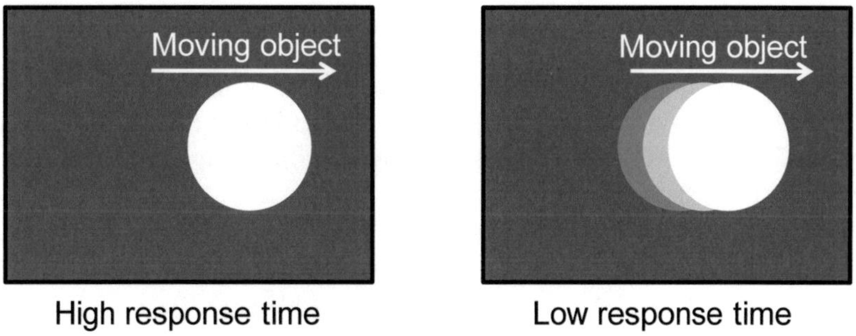

Fig. 20. Influence of response time on displayed image

Fig. 21 shows the effect of a storage capacitor (Csc) reducing the response time of LCDs.
If Csc is not sufficiently large (solid line), the response time is much longer than the case of the ideal operation (dashed line). In the worst case, the transmittance cannot achieve the expected level even in the saturated state.

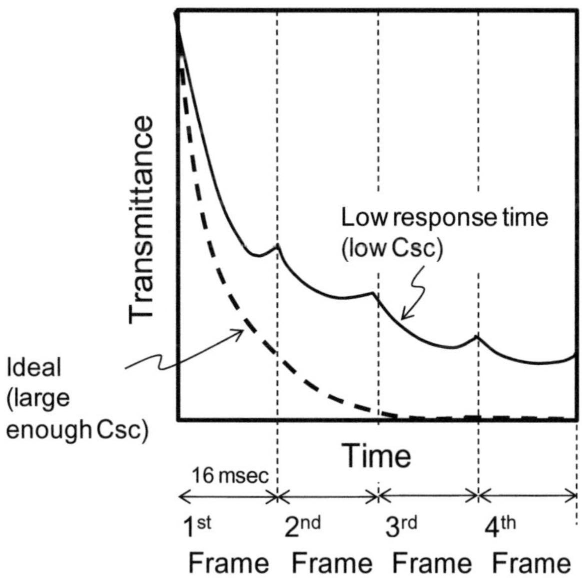

Fig. 21. Dependence of the response time on storage capacitance (Csc)

Now let us explain the mechanism of the low response time, as stated above using Fig. 22. In order to display the black pattern, V0 is applied to the liquid crystal capacitor and the storage capacitor (Csc). Here we should note that the liquid crystal capacitance (Clc0) is smaller just after the period of the charging process explained in section 1.2.1 because a white pattern was displayed in the previous frame and a finite amount of time is necessary to rotate the liquid crystal. The total accumulated charge (q0) on Csc and Clc0 can be described as

$$q_0 = (C_{SC} + C_{LC0})V_0 \tag{11}$$

After that, during the hold process, the liquid crystal tries to align itself parallel to the electric field and the liquid crystal capacitance becomes larger (Clc1), as shown in Fig. 3. Because the total charge on the electrodes is constant (q0), the applied voltage between the electrodes (V1) can be described as

$$V_1 = \frac{q_0}{C_{SC} + C_{LC1}} V_0 = \frac{C_{SC} + C_{LC0}}{C_{SC} + C_{LC1}} V_0 = \frac{1 + (C_{LC0} / C_{SC})}{1 + (C_{LC1} / C_{SC})} V_0 \tag{12}$$

In Eq. (12), we can see that V1 is smaller (larger) than V0 when Clc1 is larger (smaller) than Clc0. This means that the voltage applied to the liquid crystal cell drops (rises) until the pixel electrode is recharged in the next frame when the liquid crystal capacitance for the previous pattern is smaller (larger) than that for the latter pattern. Eq. (12) also shows that Csc should be sufficiently large compared with Clc0 and Clc1 in order to reduce the response time. This problem is especially evident in the ferroelectric liquid crystal, which has a large spontaneous polarization on its own.

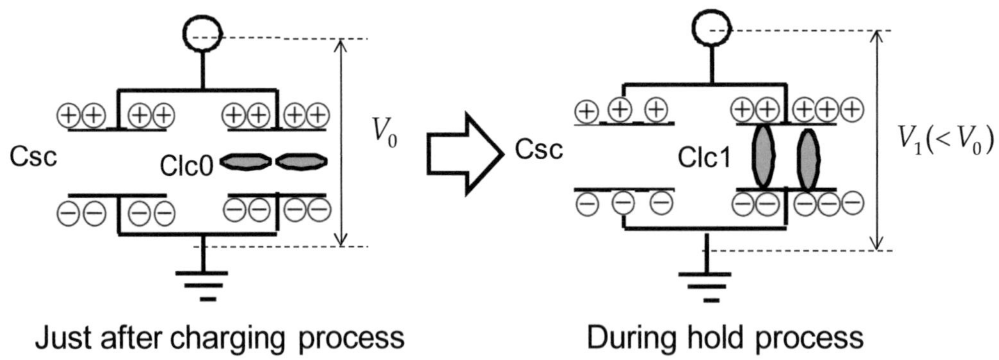

Fig. 22. Applied voltage drop caused by the liquid crystal response

2.6 Charge leakage in liquid crystal cells

An actual liquid crystal cell has a parasitic resistance because it includes many impurity ions. The resistivity of liquid crystal is normally on the order of 10^{10}~10^{13}(Ω • cm).

Therefore, even if the TFT works as an ideal switch, that is, the switching-off resistance (Roff) is infinite, the charge on the electrode cannot be kept during the hold process because of the parasitic resistance in the liquid crystal cell.

Here, we discuss the requirements of the circuit parameters in order to reduce the above influence. Fig. 23(a) shows the equivalent circuit for a pixel during the hold process with the assumption that the TFT is an ideal switch with an infinite resistance. The liquid crystal cell is composed of two electrodes with an area S and with a gap d, as shown in Fig. 23(b). The liquid crystal is injected into the gap between two electrodes. Clc, Rlc and Csc denote liquid crystal capacitance, parasitic resistance, and storage capacitance, respectively. Referring to Fig. 23(b), Rlc and Clc can be described as $R_{LC} = \rho(d / S)$, $C_{LC} = \varepsilon_0\varepsilon_r(S / d)$. Here, ρ and, εr mean the resistivity and, permittivity of the liquid crystal material, and ε0 means the vacuum permittivity.

The decay time constant (τ) for the voltage between two electrodes (Vcell) in Fig. 23(a) can be described as

$$\tau = R_{LC}(C_{LC} + C_{SC}) = R_{LC}C_{LC}\left(1 + \frac{C_{SC}}{C_{LC}}\right) = \varepsilon_0\varepsilon_r\rho\left(1 + \frac{C_{SC}}{C_{LC}}\right) \tag{13}$$

To keep the charge during the hold process, τ should be sufficiently long compared with the hold time (Th) in Fig. 2.

$$\tau = \varepsilon_0\varepsilon_r\rho\left(1 + \frac{C_{SC}}{C_{LC}}\right) \gg Th \tag{14}$$

Eq. (14) can be transformed into

$$\frac{C_{SC}}{C_{LC}} \gg \frac{Th}{\varepsilon_0\varepsilon_r\rho} - 1 \tag{15}$$

Given Th = 16 (msec), **ε0** = 8.854x10^{12}(F/m), **εr**=10, **ρ**=10^{10}(Ω • cm), we find that Csc/Clc should be designed to be more than 0.2.

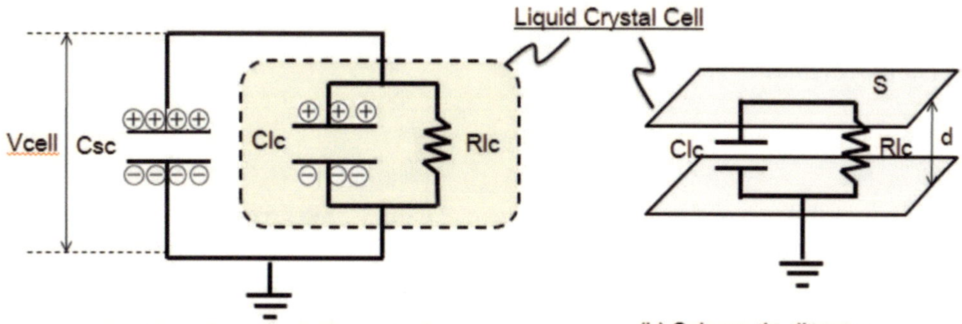

(a) Equivalent circuit for a pixel during hold process

(b) Schematic diagram for a liquid crystal cell

Fig. 23. Charge leakage model with a parasitic resistance in the liquid crystal cell

3. Design optimization by circuit simulation

A circuit simulation is a great tool for efficiently optimizing the design parameters given the complicated trade-off relationship among the electrical characteristics of displays. Although commercial circuit simulators can also be used for some design work, their device models are not detailed enough to have sufficient accuracy for liquid crystal capacitor and TFT since there are some important properties which are not considered in the implemented models into commercial simulator. To solve this problem, we have developed our own device models for a liquid crystal capacitor (Watanabe, 2007) and TFT (Ishihara, 2008) including the photo-leak effect. This section introduces the liquid crystal capacitor model as a representative. Moreover the Appendix provides codes written in VerilogA for implementing our device model into a circuit simulator.

3.1 Liquid crystal capacitor model

To make competitive products, the optical and electrical characteristics of LCDs should simultaneously be optimized to the highest level. Liquid crystal cells behave as a non-liner history-dependent capacitors from the electrical point of view, and as light valves, the transmittance of which can be varied with an applied voltage from the optical point of view. The following presents a macro-model for liquid crystal cells that includes these electrical and optical behaviors. We have enhanced Smet's approach (Smet, 2004) to improve the accuracy.

3.1.1 Modeling of a liquid crystal cell

The average orientation of liquid crystal molecule is assumed to be represented by one-dimensional variable 'x' as shown in Fig.24. In practice, three kinds of torques are in play (See Fig. 24).

1. The elastic torque $F_{elas} = Kx$ (K: constant)
   This torque pulls the molecules back to their resting position x=0 (parallel to the alignment layer). This torque is assumed to follow Hooke's law.
2. The electrical torque $F_{elec} = cE^2$ (c: constant)
   This torque aligns the molecules parallel to the field E. It is proportional to E^2.
3. The viscosity torque $F_{vis} = \gamma \frac{dx}{dt}$ (γ: constant)
   This torque hinders any movement and is proportional to the velocity at which the molecule move.

Since the moment of inertia of a liquid crystal molecule is small, it can be neglected. The equilibrium of torques in Fig. 24 states that:

$$cE^2 = Kx + \gamma \frac{dx}{dt} \tag{16}$$

This is well-known first-order system with time constant $\tau = \gamma / K$ and $x(t) \to \frac{c}{K}\left[\frac{V_{ext}}{d}\right]^2 (t \to \infty)$. Eq. (16) can be solved with the low pass filter circuit shown in Fig. 25. The resistance Rd and the capacitance Cd should be determined so that their product corresponds to the time constant $\tau = \gamma / K$. Note that the magnitude of the electric field E in the cell is described as Vext/d. Here, Vext and d are respectively the applied voltage between two electrodes and the cell gap in Fig.24. After solving for x(t) from this circuit, the effective voltage Vi at time t is calculated as $V_i = \sqrt{\frac{Kd^2}{c} x(t)}$.

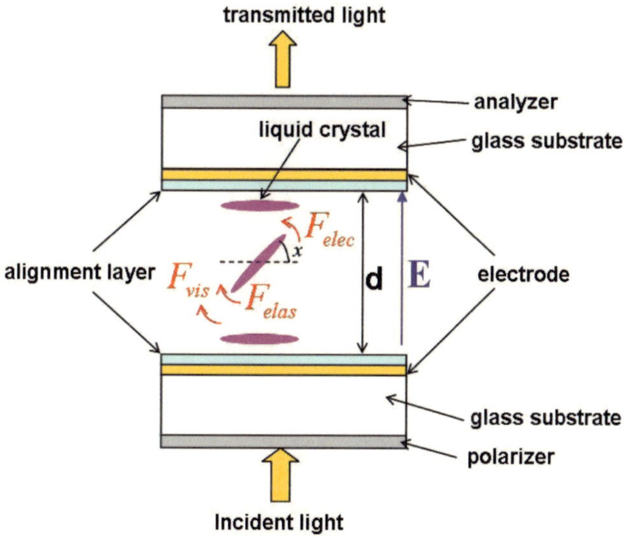

Fig. 24. Schematic diagram of liquid crystal cell

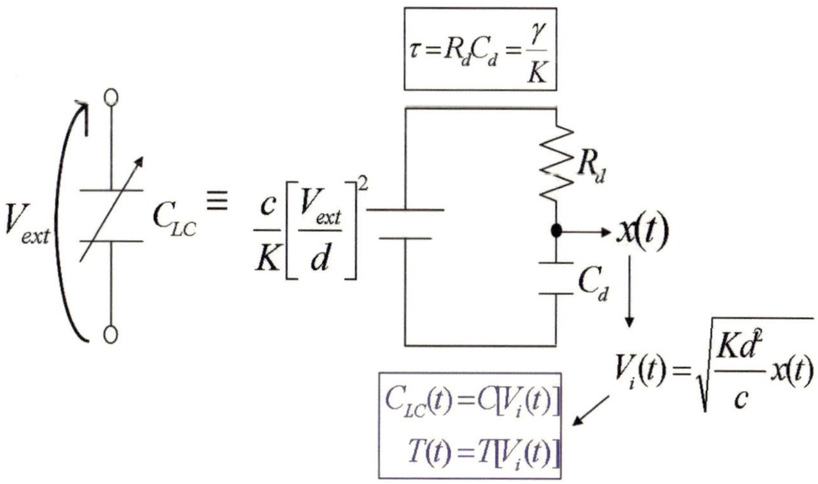

Fig. 25. Macro-model for liquid crystal cell

The static behaviors for the capacitance and transmittance of liquid crystal cells can be expressed empirically with Eq. (17) and Eq. (19) as function of the effective applied voltage Vi.

$$C(V_i) = C_{\perp} + \frac{2}{\pi}\left(C_{\parallel} - C_{\perp}\right) arctan\left[\frac{\alpha + \left(a^2 + \delta^2\right)^{1/2}}{2}\right] \tag{17}$$

$$\alpha = \frac{V_i - V_{tc}}{V_{mc}} \tag{18}$$

$$T(V_i) = T_{min} + (1 - T_{min}) tanh\left[\frac{\beta + \left(\beta^2 + \eta^2\right)^{1/2}}{2}\right] \tag{19}$$

$$\beta = \frac{V_i - V_{to}}{V_{mo}} \tag{20}$$

C||, C⊥, δ, Vtc, Vmc, in Eq.(17) and Tmin, η, Vto, Vmo in Eq.(19) are treated as model parameters extracted from experimental results.

To improve accuracy, we have modified the form of the time constant (τ) in the equation proposed by Smet by considering the following points.

First, we should improve the accuracy for the external applied voltage dependency. Generally, the response time of liquid crystal depends on the external applied voltage (Vext). We got the following expression for time constant (τ) after further investigating of the relationship between the electric torque(Felec) and the applied voltage(Vext) (Watanabe, 2007).

$$\tau = \frac{1}{a_1 + a_2 V_{ext}^m} \tag{21}$$

Here a_1, a_2 and m are model parameters. Basically, m takes a value around 2.

The second point is to give the parameter to rise and fall process individualy. A rise (fall) process is defined as the case that the orientation angle x of the liquid crystal molecule is increasing (decreasing). The time constant for each process does not generally coincide. To implement this factor into our model, the parameters in Eq. (21) are selected after the voltage for both terminals of Rd are compared:

If

$$\frac{c}{K}\left(\frac{V_{ext}}{d}\right)^2 \geq x(t)$$

Then

$$a_1 = a1\_r, a_2 = a2\_r \quad \text{(for rise process)}$$

Else

$$a_1 = a1\_f, a_2 = a2\_f \quad \text{(for fall process)}$$

This additional routine enables individual control of the rise and fall behaviors. The VerilogA code for the above macro model is in Appendix.

3.1.2 Model evaluation

We verified the macro-model by comparing its results with experimental data. Fig. 26 compares modeled and experimental data for the static behavior. The experimental data is for a twisted-nematic liquid crystal cell with a 3.5μm gap. The model exhibits good agreement with the experimental data in both its transmittance and dielectric constant characteristics.

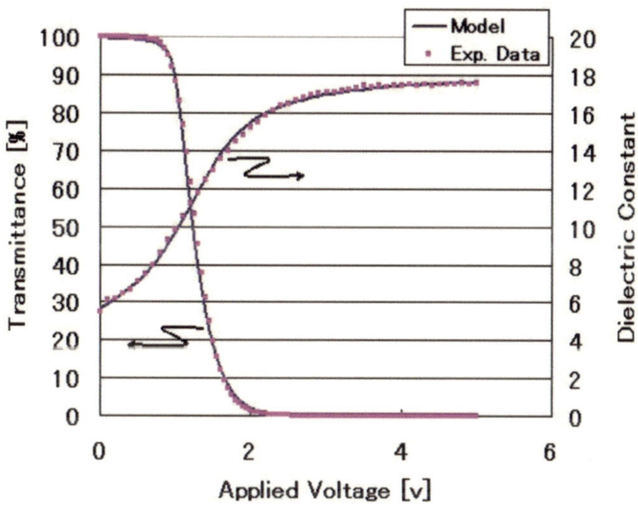

Fig. 26. Comparison of static behaviors of experimental data and model

Fig. 27 compares modelled and experimental data for the dynamic behavior for externally applied voltages. The transient behavior from low voltage to high voltage (rising process) is shown in Fig. 27 (a), and from high voltage to low voltage (falling process) in Fig. 27 (b). The transient behavior of the dielectric constant is not shown here because no measuring procedure has been established for it. We assumed that the dynamic parameters for the transmittance and dielectric constant are the same when this model is used in the actual design for the liquid crystal cell, though this assumption cannot be verified directly.

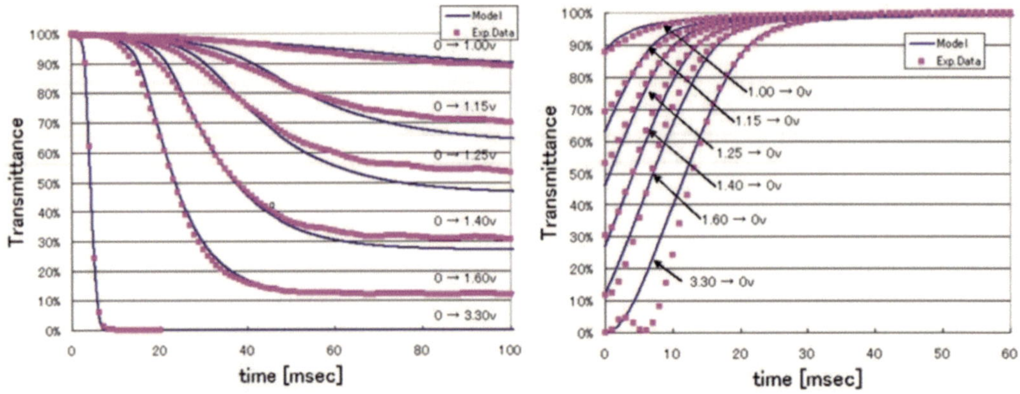

(a) For a rising process (b) For a falling process

Fig. 27. Comparison of dynamic behaviors of experimental data and modelled

In Fig.27, the model exhibits good agreement with the experimental data except for the following points.

1. The bounce around 5msec in 3.30v → 0v in Fig.27 (b)
   This phenomenon is well known as the back-flow of liquid crystal molecule (Doom, 1975). This shape of the curve could not be expressed in principle since Eq.(19) is monotonous function. A new model supporting this phenomenon should be developed.
2. The disagreement between model and experiment after ample time elapses in Fig. 27(a) and in the initial state in Fig.27 (b)
   The origin of this disagreement is considered to be measurement error, since the modelled transmittances after an ample time has passed in Fig. 27(a) and at the initial time in Fig. 27(b) do not coincide with the static measurements ones in Fig. 26. It is hard to improve the measurement accuracy because the transmittance is very sensitive to the applied voltage around 1v, as can be seen in Fig.26.

4. Conclusion

Thanks to its individual control of each pixel, active matrix driving is definitely superior to passive driving in its capability to display higher quality pictures. Even in an active matrix driving, however, some malfunctions such as shading, crosstalk and flicker become apparent on screen when a specific picture pattern is displayed. In this chapter, we made the mechanism behind such malfunctions and some of the key design parameters. Readers may find the complicated trade-off relationship among key design parameters and a difficulty in optimizing the parameters simultaneously. A circuit simulator would be very useful for designers to optimize the design parameters more efficiently. However, general commercial circuit simulators often do not have enough device models such as liquid crystal cell capacitors, TFTs supporting the ambient light effects, and so on. Therefore, for the sake of accurate circuit simulations, LCD designers must aggressively develop device models and continuously improve them. In this chapter, we described a macro model for a liquid crystal cell capacitor as a representative model for making accurate designs. The current trend of developing larger and higher resolution of LCDs will spur a need for accurate circuit simulation technologies since there will be less margin for each design parameter.

5. Appendix

List 1. VerilogA code for Liquid Crystal Macro Model

```
// Liquid Crystal Cell Macro Model ver1.0
// M.Watanabe
//
`include "discipline.h"
`include "constants.h"
//
`define m_pi      (3.14159265358979323846)
//
module lccap(a,b);
inout a,b;
electrical a,b;
//
// Internal nodes
electrical vc, vt;
//
// Parameters for Static behavior
//
// For Capacitance
parameter real  cv = 8.82;
parameter real  cp = 27.1;
parameter real vtc = 0.88;
parameter real vmc = 0.88;
parameter real delta = 0.1;
//
// For Transmittance
parameter real tmin= 8.7e-4;
parameter real vto = 1.1;
parameter real vmo = 0.54;
parameter real eta = 0.24;
//
// Parameters for Dynamic behavior
//
// For Capacitance
parameter real  a1c_r = 0.014;
parameter real  a2c_r = 0.05;
parameter real  a1c_f = 0.014;
parameter real  a2c_f = 0.05;
parameter real  cdc   = 1e-3;
parameter real  mc_r  = 2.4;
parameter real  mc_f  = 2.05;
//
// For Transmittance
parameter real  a1t_r = 0.020;
parameter real  a2t_r = 0.02;
parameter real  a1t_f = 0.0205;
parameter real  a2t_f = 0.025;
parameter real  cdt   = 1e-3;
parameter real  mt_r  = 2.4;
parameter real  mt_f  = 2.05;
//
// Geometrical Parameters
//
parameter real area = 1;
parameter real vini = 0;
parameter real scale =1;
//
real a1c, a2c, a1t, a2t, cap, rdc, rdt;
real trans,alpha,beta;
real vi_c, vrms_c, vv, qi_c, vin, vi_t, vrms_t, qi_t, mc, mt ;
real qlc, cdiff;
//
```

```
analog
 begin
@(initial_step)
begin
cdiff = (2/`m_pi) * (cp - cv);
qlc = 0;
vi_c = vini * vini;
vi_t = vini * vini;
end

//
vin = V(a,b);
vv  = vin * vin;
//
// For Capacitance
//
if (vin >= vi_c)
 begin //rising
  a1c = a1c_r;
  a2c = a2c_r;
  mc  = mc_r;
 end
else
 begin //falling
  a1c = a1c_f;
  a2c = a2c_f;
  mc  = mc_f;
 end

rdc = 1/(a1c + a2c * pow(vi_c, mc) );
I(vc) <+ ddt(cdc * V(vc));
I(vc) <+ (V(vc)-vv)/rdc;

vi_c = V(vc);
vrms_c = sqrt(vi_c + 0.01);
alpha = (vrms_c - vtc)/vmc;
cap = scale * area * (cv + cdiff * atan((alpha+sqrt(alpha * alpha + delta * delta ))/2));
//
// For Transmittance
//
if (vin >= vi_t)
 begin //rising
  a1t = a1t_r;
  a2t = a2t_r;
  mt  = mt_r;
 end
else
 begin //falling
  a1t = a1t_f;
  a2t = a2t_f;
  mt  = mt_f;
 end

rdt = 1/(a1t + a2t * pow(vi_t, mt) );
I(vt) <+ ddt(cdt * V(vt));
I(vt) <+ (V(vt)-vv)/rdt;
vi_t = V(vt);
vrms_t = sqrt(vi_t + 0.01);
beta  = (vrms_t - vto)/vmo;

trans = 100*(1-(1-tmin)*tanh((beta+sqrt(beta * beta + eta * eta))/2));
//
// Description for device behavior
//
qlc = cap * vin;
I(a,b) <+ ddt(qlc);

end
endmodule
```

6. References

Doom, Z., Dynamic behavior of twisted nematic liquid-crystal layers in switched field, Journal of Applied Physics, Vol.46, No.9, September 1975, pp.3783-3745

Ishihara, K.(2008), Implementation of optical response of thin film transistor, *Proceedings of IEEE BMAS 2008 Conference*, pp.39-44, 2008.

Jacunski, M. D. (1999). A Short-Channel DC SPICE Model for Polysilicon Thin-Film Transistors Including Temperature Effects, *IEEE TRANSACTIONS ON ELECTRON DEVICES*, Vol.46, No.6, June 1999, pp.1146-1158, ISSN 0018-9383

Kimura, M. (1994). Simulation Techniques for Horizontal Crosstalk in TFT-LCDs, *Proceedings of AM-LCD'94*, pp.228-231, ISBN 4-930813-62-10, Shinjuku, Tokyo, Japan, November 30 - December 1, 1994

Pochi, Y. (1999). *Optics of Liquid Crystal Displays*, pp.248-252, John Wiley & Sons, ISBN 0-471-18201-X, the United States of America

Smet, H.D., Electrical model of a liquid crystal pixel with dynamic, voltage history-dependent capacitance value, *Liquid Crystal*, Vol.31, No.5, May 2004, pp.705-711

Sze, S. M. (1981). *Physics of Semiconductor Devices Second Edition*, pp.440, John Wiley & Sons, ISBN 0-471-05661-8, the United States of America

Watanabe, M.(1996), Development of a 20.1 Diagonal Super Fine TFT LCD, *Proceedings of Euro Display'96*, pp.587-590, 1996.

Watanabe, M.(1997), Novel pixel structre for IPS TFT-LCD with color-shift free,*Proceedings of International Display Research Conference'97*, pp.L-9, 1997.

Watanabe, M.(2007), Macro modeling of liquid crystal cell with VerilogA, *Proceedings of IEEE BMAS 2007 Conference*, pp.132-137, 2007.

Wittek,M. (2008), Advanced LC Materials for Ultra-Fast Switching for Active-Matrix-Device (AMD) Applications, Proceedings of IDRC '08 Conference, pp.253-255, 2008

7

Intelligent and Green Energy LED Backlighting Techniques of Stereo Liquid Crystal Displays

Jian-Chiun Liou
Industrial Technology Research Institute
Taiwan

1. Introduction

As modem society has been reshaped rapidly into the multimedia information society, large size of display devices is required. Thin-film-transistor liquid- crystal-displays (TFT-LCDs) are one of the most popular display devices from small to large size. TFT-LCDs need backlight source, because they are not a self luminance device. Backlight for LCDs is becoming more important with growth of the LCD size. Up to now, multiple fluorescent lamps such as cold cathode fluorescent lamp (CCFL), external electrode fluorescent lamp (EEFL) and flat fluorescent lamp (FFL) are generally used as LCD backlight source. Since, the conventional backlight of LCDs illuminates at the full luminance regardless of the images to be displayed, it wasted power and contrast ratio is row due to the light leakage in dark state.

Recently, as the luminance efficiency of light emitting diode (LED) has been improved and the cost of LED is going down, the LED is the substitutive solution for the backlight source. Moreover, since LED has many advantages such as long lifetime, wide color gamut, fast response, and so on, LEDs are expected to replace the conventional fluorescent lamps for backlight source of LCD in near future. Although, LED backlight driving systems have been developed and introduced to the market, further reduction of power consumption and cost reduction are still demanded to be widely used as backlight source. In segmented dimming and local dimming methods which have usually adapted to CCFL and LED backlight source, the whole backlight is divided into several segmented and block, backlight scaling is adapted to each segmented and block, respectively. The more division of backlight, the more power saving can be achieved. Therefore, local dimming method can save power consumption more effectively than segmented dimming one. Especially, segmented dimming method has limitation of power saving when the image is bright vertically. Therefore, local dimming is the best way to have local dimming effects. However, local dimming method results in huge increase in the number of drivers needed for large number of division. To compromise between segmented dimming and local dimming methods, a new LED backlight system for LCD TVs which involves the time division X-Y segmented driving method that utilizes row and column switches to control the individual division screen is proposed.

Even though the refresh time of an LCD panel is as fast as a CRT display, there are still some problems to be solved before this LCD panel can be implemented to be a time-multiplexed stereo display, vision and panel seem to be redundant. First of all, LCD is a hold-type display, which is different from the impulse operation type of CRT. This has been a very good property of an LCD while it works at 2D display mode because it can avoid blinking or flickering at the refresh rate of 50 or 60 Hz. For a CRT display, viewers will still experience slight image flickering at the refresh rate of about 60Hz. Nevertheless, while working in 3D page-flipping mode and watched through shutter glasses, the hold-type operation causes a problem.

When the left-image begins to fill out the pixels, the left-eye shutter opens immediately. However, at this moment and the following 1/120 sec., the left-image at the upper part of the screen becomes more and more image area of screen, the right-image at the lower part of the screen becomes less and less image area of screen. That means, the viewer will see both images at the same time through his right eye. This situation will induce strong double image (or crosstalk) and destroy the 3D perception seriously.

1.1 Intelligent and green energy LED backlighting techniques

1. Scanning backlight method. The setup is as following:
   The backlight unit is separated into several regions. Let's take 4 segmented regions as the example as in Fig.1. the pixel response time is less than three fourths of the frame time when the illumination period is one quarter of the frame time. Arranging for the required illumination period to end just before the new image is written into the panel provides the most relaxed requirement for panel response time; i.e., the response time can be longer (Fig. 1). A novel controlled circuit architecture of scanning regions for 120Hz high frequency and high resolution stereoscopic display is shown in Fig.2. Setup all the parameters of scanning backlight method by counting the amount to decide turning time between 4-regions 、 and 2-regions LED backlight type. If counted times equal to 100 then jump to next backlight segmented region.
   For 4-region scanning backlight method, when the panel is filled in segmented regions 1, 2, 3 and 4 by the new image, the backlight lights up are in the corresponding regions 4, 1, 2 and 3. In anticipation of an image for a left eye and right eye is shown in the region 1 of the panel, we turned on region 3 of the backlight unit. Analogize the image shown in region 2 and turned on region 4 of the backlight unit. For 2-region scanning backlight method, when the panel is filled in regions 1 and 2 by the new image, the backlight lights up in the corresponding regions 2 and 1. For avoiding seeing both L-image and R-image at the same time, the backlight regions R1 have to be off until R1 filled up the image. Analogize the backlight regions R2 have to be off until R2 filled up the image.
2. In backlight strobe method as shown in Fig.3, Setup the parameters of backlight strobe method by counting the amount to decide turning on time of full screen (full screen of one frame 1/120sec counted amount equal to 400). Setup the parameters of backlight strobe duty time by counting the amount to decide turning time on full screen backlight regions. If counted times equal to (400 - 400*9/10) then jump to next full screen backlight region. The backlight is turned off when the image data refreshes. The backlight only turns on at the system time, or at most a little bit longer than the system time. But the system time is short compared with the time between two adjacent vertical synchronization signals (less than 10%), the display brightness operated under this method is probably quite small.

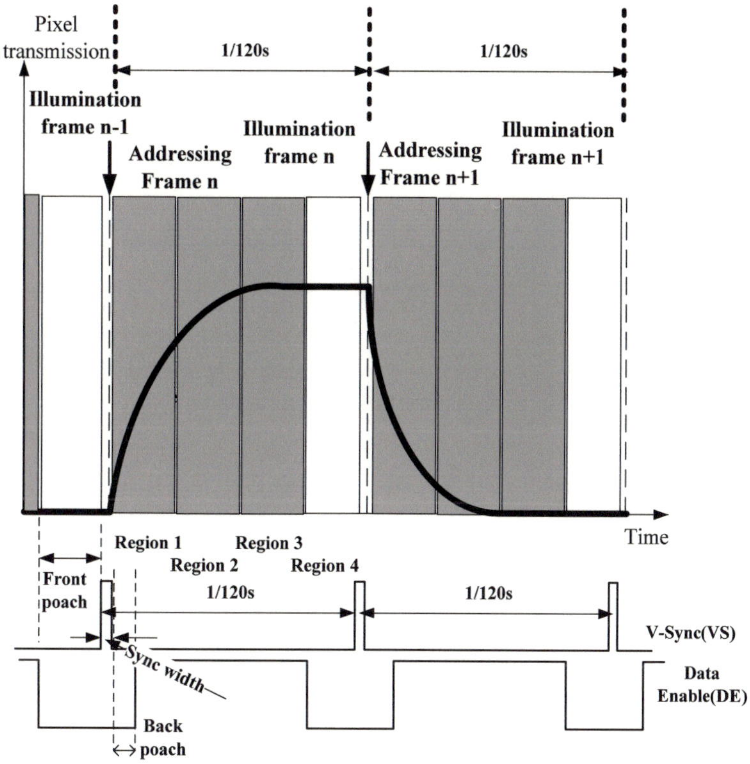

Fig. 1. Schematic diagram of scanning backlight method

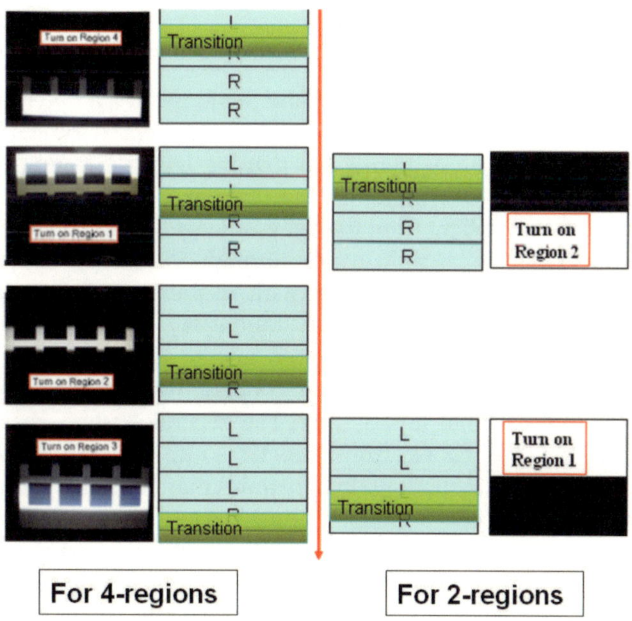

Fig. 2. Synchronization Signal LED backlight architecture

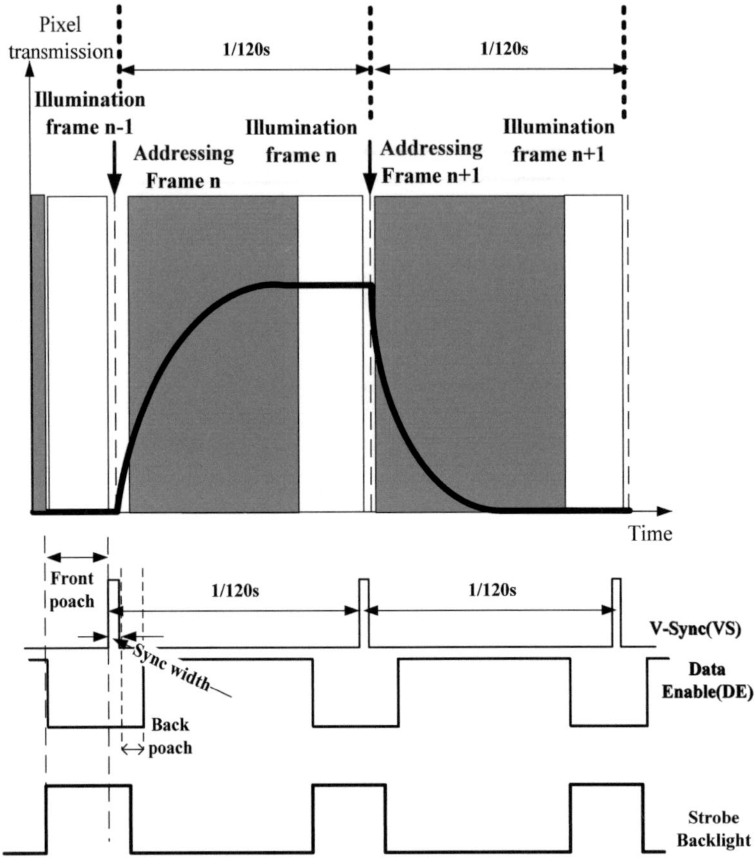

Fig. 3. Time scheme of backlight strobe method

In this research, we have successfully designed and demonstrated a decent performance with 120Hz optimized synchronization signal between LED brightness/darkness flash and adjusted shutter glasses signal. It has been demonstrated that the 120Hz scanning characteristic from upper row to lower row of the horizontally arranged of stereoscopic image. A quadrate image for a left eye is projected by the light from the left eye image file and a circle image for a right eye is projected by the light from the right eye file through a liquid crystal panel.

LED scanning backlight stereoscopic display with shutter glasses is provided to realize stereoscopic image viewing even in a liquid crystal display. In a frame time, some kinds of brightness/darkness characteristic from upper row to lower row of the horizontally arranged rows of LED in the backlight module, cooperating with the scanning of the LCD, to thereby realize an effect similar to scanning. The general strategy that we employ is to integrate all relatively small-signal electronic functions into one ASIC to minimize the total number of the components. This strategy demonstrates that both the cost is lowered and the amount of the printed circuit board area is reduced. Based on this concept, a smart three dimensional multiplexed driver for LED switching chip with more than 640 LEDs are proposed and the circuit architecture is shown in Fig. 4. It is difference from the traditional two dimensional arrays driven by scanning scheme. Three lines are employed to control one LED, including voltage, shift register, and data line. Each LED requires a voltage line for the

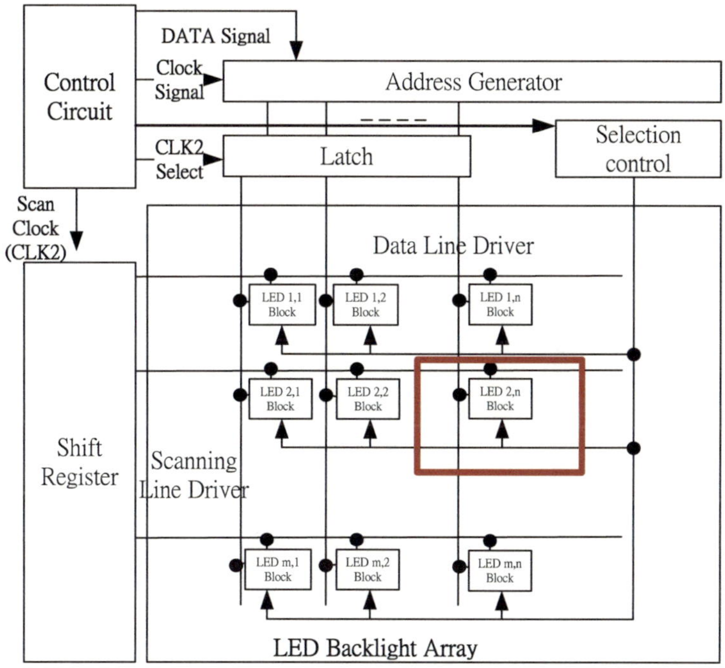

Fig. 4. Block diagram of control algorithm for LED backlight array module

driving current and shares the same ground with the other resistors. The resistors are individually addressable to provide unconstraint signal permutations by a serial data stream fed from the controller. The shift register is employed to shift a token bit from one group to another through AND gates to power the switch of a LED group. The selection of a LED set is thus a combined selection of the shift register for the group and the data for the specific LED. Such an arrangement allows encoding one data line from the controller to provide data to all of the LEDs, permitting high-speed scanning by shortening the LED selection path and low IC fabrication cost from the greater reduction of circuit component numbers.

2. 3D display backlight and application

Throughout the proposed system, lower power consumption are successfully obtained as well as high contrast ratio even with less number of drivers than that of conventional local dimming method. This chapter also contains a new adaptive dimming algorithm and image processing technique for the proposed stereo LCD backlight system.

Recent progress in stereo display research has led to an increasing awareness of market requirements for commercial systems. In particular areas of display cost and software input to the displays are now of great importance to the programme. Possible areas of application include games displays for PC and arcade units; education and edutainment; Internet browsing for remote 3D models; scientific visualisation and medical imaging.

Intelligent and green power LED backlighting techniques of two-dimensional (2D) to three-dimensional (3D) convertible type, shutter glasses type, mult-view time multiplexed naked eye type, and multi-viewer tracking type for stereo liquid crystal displays are shown as follows.

2.1 Two-dimensional (2D) to three-dimensional (3D) convertible display

Convertible two-dimensional-three-dimensional display using an LED array based on modified integral imaging as shown in Fig.5. This type propose a two-dimensional (2D) to three-dimensional (3D) convertible display technique using a light-emitting diode (LED) array based on the principle of modified integral imaging. This system can be electrically converted between 3D and 2D modes by using different combinations of LEDs in the LED array without any mechanical movement. The LED array, which is controlled electrically, is used for backlight, and a lens array is used for making a point light source array with higher density. We explain the principle of operation and present experimental results.

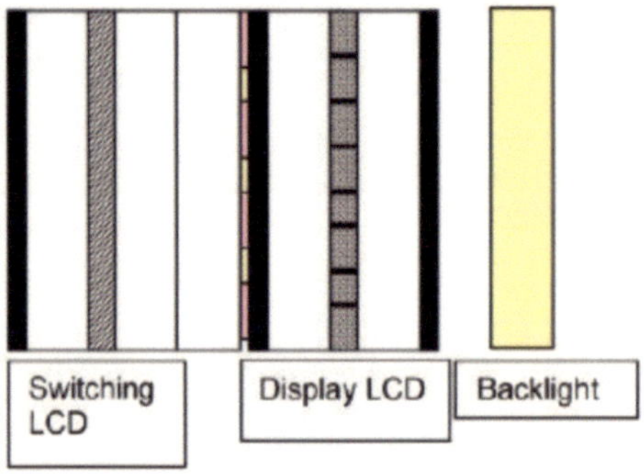

Fig. 5. Display configuration

2.2 Shutter glasses type stereoscopic displays

This type proposes to employ multi-dimensional controller for driving LED backlight scanning in a 120Hz LCD for overcoming the hold-type characteristic of an LCD in time-multiplexed stereoscopic displays. A synchronization signal circuit is developed to connect the time scheme of the vertical synchronization for reducing scanning time. The general strategy is to integrate three dimensional controller and all relatively small-signal electronic functions into one ASIC to minimize the total number of the components. The display panel, LED backlight scanning, and shutter glass signals could be adjusted by vertical synchronization and modulation to obtain stereoscopic images. Each row of LED in a backlight module is controlled by multi-dimensional data registration and synchronization control circuits for LED backlight scanning to flash in bright or dark. LED backlight scanning stereoscopic display incorporated with shutter glasses is provided to realize stereoscopic images even viewed in a liquid crystal display as shown in Fig.6. The eye shutter signal is alternately switched from the left eye to the right eye with 120Hz of LCD Vertical synchronization (V-sync). This kind of low cross-talk shutter glasses stereoscopic display with an intelligent multiplexing control of LED backlight scanning has low cross-talk below 1% through a liquid crystal shutter glasses.

2.3 Multi-view time multiplexed autostereoscopic displays

Three-dimensional displays which create 3D effect without requiring the observer to wear special glasses are called autostereoscopic displays. A number of techniques exist - parallax barriers, spherical and lenticular lenses, the latter being the most common one. Depending on the design parameters, various tradeoffs between screen resolution, number of views and optimal observation distance exist. The most popular ones, so called multiview 3D displays,

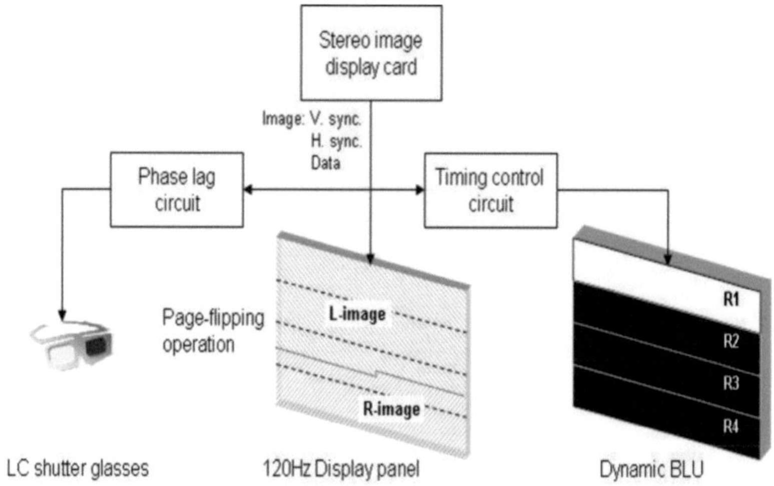

Fig. 6. Schematic diagram of dynamic backlight method

work by simultaneously showing a set of images ("views"), each one seen from a particular viewing angle along the horizontal direction. Such effect is achieved by adding an optical filter, which alters the propagation direction for the information displayed on the screen.

Currently, several 2D/3D switched displays had been proposed such as switched barrier and LC-lens. However, both of the parallax barrier and the cylindrical lens arrays still has the issues of narrow viewing angle and low resolution when displaying the 3D images. Besides, in order to balance the horizontal versus vertical resolution of an autostereoscopic a display, a slanted lens array is used. This causes the high crosstalk of stereo display and the subpixels of a view to appear on nonrectangular grid. This type describes the work aimed at developing optical system; active barrier dynamic backlight slit multi-view full resolution and lower crosstalk 3D panel as shown in Fig.7. The panel of 240Hz displays the corresponding images of the four viewing zones by the same time sequence according to temporal multiplexed mechanism.

All modern multiview displays use TFT screens for image formation. The light generated by the TFT is separated into multiple directions by the means of special layer additionally mounted on the screen surface. Such layer is called "optical layer", "lens plate" and "optical filter". A characteristic of all 3D displays is the tradeoff between pixel resolution (or brightness or temporal frequency) and depth. In a scene viewed in 3D, pixels that in 2D would have contributed to high resolution are used instead to show depth. If the slanted lenticular sheet were placed vertically atop the LCD, then vertical and horizontal resolution would drop by a factor equal to the number of views.

This type addresses the specific technological challenges of autostereoscopic 3D displays and presents a novel optical system that integrates a real-time active barrier dynamic backlight slit system with a naked eyes multi-view stereo display. With 240Hz display and tunable frequency LED backlight slits, only a pair of page-flipped left and right eye images was necessary to produce a multi-view effect. Furthermore, full resolution was maintained for the images of each eye. The loading of the transmission bandwidth was controllable, and the binocular parallax and motion parallax is as good as the usually full resolution multi-view autostereo display.

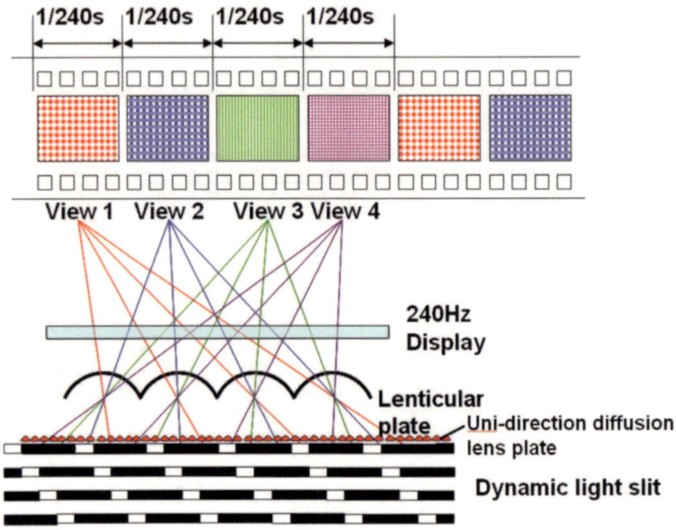

Fig. 7. The structure of the proposed multi-view 3D display

A lenticular-based 3D display directs the light of neighboring sub-pixels into different directions by means of small lenses placed immediately in front of the sub-pixels. In this manner different pictures can be transmitted into different directions. Usually a multitude of directions is chosen, e.g. 4 different views. Two of these views can be seen by the left and right eye respectively, and as such create a stereoscopic (3D) image. Fig.7. shows the structure of the proposed multi-view 3D display. Only one eye individually receives the image at one corresponding viewing zone at its displaying time period, such as 1/240 second. As a result, the 3D image can be created for the viewer by naked eyes. Each view is 60Hz.

2.4 Multi-viewer tracking stereoscopic display

Many people believe that in the future, autostereoscopic 3D displays will become a mainstream display type. Achievement of higher quality 3D images requires both higher panel resolution and more viewing zones. Consequently, the transmission bandwidth of the 3D display systems involves enormous amounts of data transfer. This type integrated a viewer-tracking system and a synchro-signal LED scanning backlight module with an autostereoscopic 3D display to reduce the crosstalk of right/left eye images and data transfer bandwidth, while maintaining 3D image resolution. Light-emitting diodes (LED) are a dot light source of the dynamic backlight module as shown in Fig.8. When modulating the dynamic backlight module to control the display mode of the stereoscopic display, the updating speed of the dynamic light-emitting regions and the updating speed of pixels were synchronal. For each frame period, the viewer can accurately view three-dimensional images, and the three-dimensional images displayed by the stereoscopic display have full resolution. The stereoscopic display tracks the viewer's position or can be watched by multiple viewers. This type demonstrated that the three-dimensional image displayed by the stereoscopic display is of high quality, and analyzed this phenomenon. The multi-viewer tracking stereoscopic display with intelligent multiplexing control of LED backlight scanning had low crosstalk, below 1%, when phase shift was 1/160s.

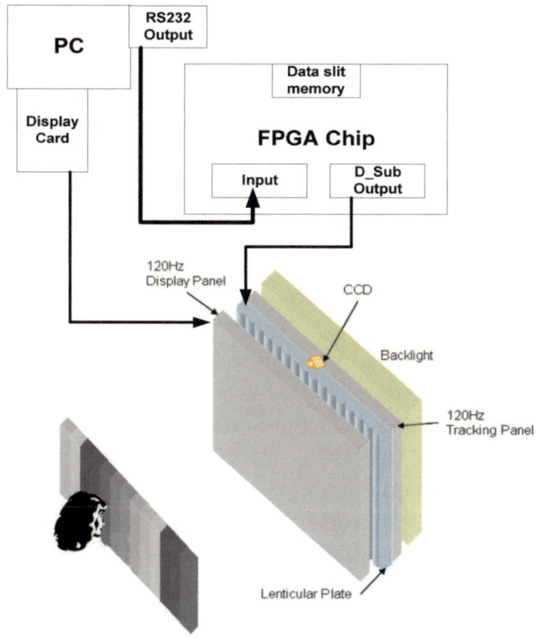

Fig. 8. Viewer-tracking 3D display system

Due to the LCD with physical delay characteristic (low response speed), both images are alternately switched from one to the other by switching the light emitting full panel. Thus, stereoscopic images are shown to a viewer. As a result, all kinds of low cross-talk stereoscopic display with an intelligent multiplexing control LED scanning backlight have low crosstalk below 1% through a liquid crystal display.

Intelligent and green energy LED backlighting techniques of stereo liquid crystal displays have been successfully designed and demonstrated a decent performance of all kinds of stereo types display system with optimized synchronization signal between LED brightness/darkness flash and adjusted driving signal enabled by a 3-Dimensional controlling IC. At high scanning rate from upper row to lower row, the system demonstrated horizontally arranged clear stereoscopic images. This method of backlighting also allows dimming to occur in locally specific areas of darkness on the screen. This can show truer blacks and whites at much higher dynamic contrast ratios, at the cost of less detail in small bright objects on a dark background, such as star fields.

3. Experimental and results

3.1 Two-dimensional (2D) to three-dimensional (3D) convertible display

As LED backlight of flat panel display become large in format, the data and gate lines turn into longer, parasitic capacitance and resistance increase, and the display signal is delayed. Three dimensional architecture of multiplexing data registration integrated circuit method is used that divides the data line into several blocks and provides the advantages of high accuracy, rapid selection, and reasonable switching speed.

The design concept can be easily scaled up for large LED backlight array format TFT-LCD elements system without much change in the terminal numbers thanks to the three dimensional hierarchy of control circuit design, which effectively reduces the terminal numbers into the cubic root of the total control unit numbers and prevent a block defect of the flat panel. The TFT-LCD unit lights, line(s) in the vertical or horizontal axis appear dim, but not completely on or off. These defects are generally the result of a failure in the row (horizontal) or column (vertical) drivers or their connections. The TFT-LCD includes an extension part defect such as an extension piece overlapping with a pixel electrode of boundary pixels at a boundary data line applying a data signal to the boundary pixels.

3.1.1 Experimental results and discussion

A LED backlight of flat panel display with three dimensional architecture of multiplexing data registration integrated circuit having a plurality of scanning electrodes, a plurality of data electrodes extending perpendicularly to the scanning electrodes, and liquid crystal filling a space between the scanning electrodes and data electrodes, pixels being formed at each intersection of the scanning and data electrodes together with the liquid crystal, the display panel being divided into an even row part and a odd row part; a scanning control circuit for scanning the scanning electrodes by sequentially supplying scanning voltages to each scanning electrode and by maintaining the same for a predetermined period, the scanning electrodes located in the even row part of the panel and the scanning electrodes located in the odd row part being scanned separately but simultaneously in the same directions from upper to lower of the panel; an image data control circuit for sequentially supplying image data voltages to the data electrodes in synchronism with scanning of the scanning electrodes, the scanning electrodes are scanned in such a manner that the image data is written on the pixels in a selecting period, the written image data is held on the pixels in a holding period and the image data is eliminated in an eliminating period; In traditional control circuit design for TFT-LCD elements array system, each TFT-LCD element requires one driver switch. As a result, when the TFT-LCD backlight of LEDs' pixels scale up into a large array, the numbers of input/output ports will increase enormously. To handle large array of driving circuits for such large pixels array, 2D circuit architecture was employed for the traditional driving circuit to reduce the IO number from n*n into 2n+1. However, firstly, this reduction still can not meet the requirement for high speed signal scanning with low data accessing points when switch numbers greater than 640×480 pixels. It would be necessary to increase the display frequency to 240 Hz or higher to eliminate flicker. If the display frequency is 240 Hz, a period of time for writing one frame is 4.17 ms. Assuming the number of scanning electrodes is 480, a period of time available for writing one line is only 8.7 microseconds. The number of scanning electrodes has to be larger than 480 to display a high resolution image, making the writing period further shorter. Secondly, no technology is ever completely perfect, of course, and the LCD can still suffer from some defects in the displayed image. In this technology, though, most defects in the basic electronics, such as failure of the backlight or the row or column drivers, result in a completely unusable display, and so when such occur in production they are easily detected and corrected. It is extremely rare for a product to ship with any such problems.

To achieve this, In this study, a three dimensional data registration flat panel display scheme (Fig. 9) to reduce the number of data accessing points as well as scanning lines for large array TFT-LCD element with switch number more than 640×480 is proposed (Fig.10). The total numbers of data accessing points will be $N=3\times\sqrt[3]{Y}+1$, which is 68 for 640×480 switches by the 3D novel design, the scanning time is reduced up to 30% (The scanning speed is also

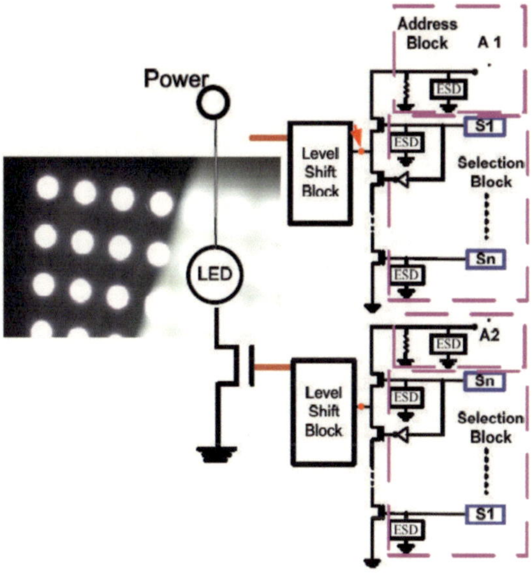

Fig. 9. Photograph of LEDs backlight sets

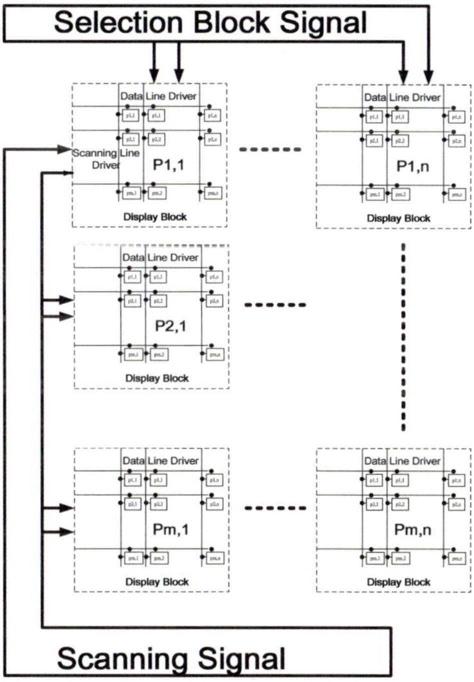

Fig. 10. 3D scanning display block array

increased by 3 times) thanks to the great reduction of lines for 3D scanning, instead of 2D scanning. Fig. 11 is shown pad connections from 1D, 2D, and 3D control circuits. Fig.12. is localized dimming LED backlight.

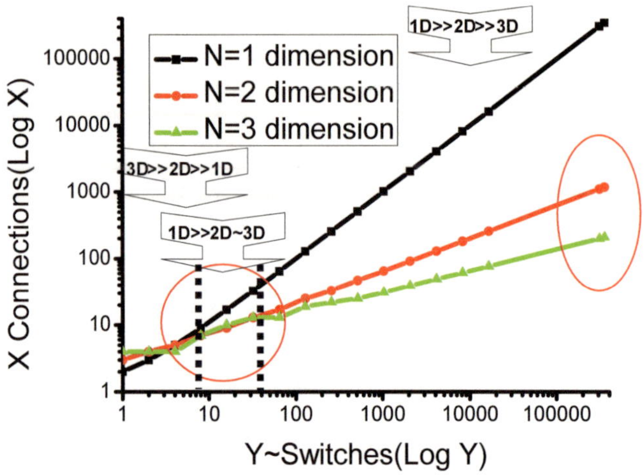

Fig. 11. Pad connections from 1D, 2D, and 3D control circuits

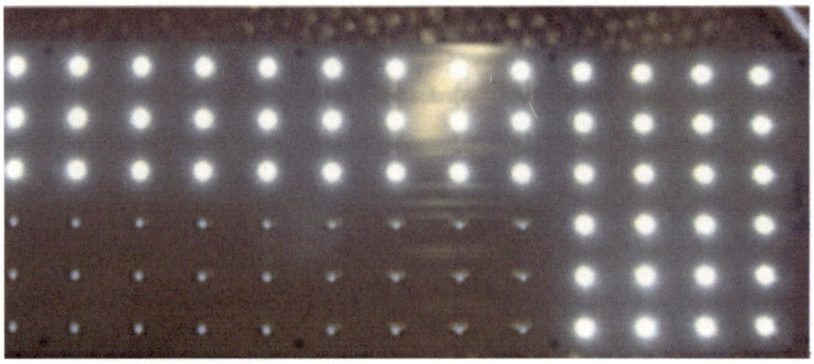

Fig. 12. Localized dimming LED backlight

3.1.2 Localized 2D/3D switchable maked-eye 3D display

The patented 2D/3D control module is made of a low-resolution panel with micro-retarder film, the LCD paneland backlight dynamically controlled form a patented architecture that can switch 2D/3D mode display area, provides three-dimensional image web application. Area by demand, 2D/3D switchable, 2D coexists with 3D on the one screen, 2D area shows small character clearly and 3D area shows multi-view naked-eye stereo image.

For simultaneous display of 2D and 3D information, we have develop an integrates naked-eye 2D/3D display window technology, called integrated 2D/3D windows (i2/3DW), that can display 2D and 3D images with flexibility and best quality on the same screen. As for stereo 3D gaming, there are dialog windows or pop-up windows in the stereo 3D game frames. Without an integral 2D/3D display, 2D texts in the 3D mode windows appear in broken and blurred characters. This situation is very much annoying especially for small fonts. But, with i2/3DW's localized 2D/3D switchable display technology, 3D gaming becomes true joy because 2D texts will be as clear as they are on a 2D screen while the 3D game scenes would still be the same fascinating as on a 3D device. Fig. 13 shows the construction of an i2/3DW display made according to technology from ITRI. It comprises three "primary "component layers." The first at left is a conventional liquid crystal display panel (LCD panel). Table 1 shows the specifications of the image liquid crystal panel.

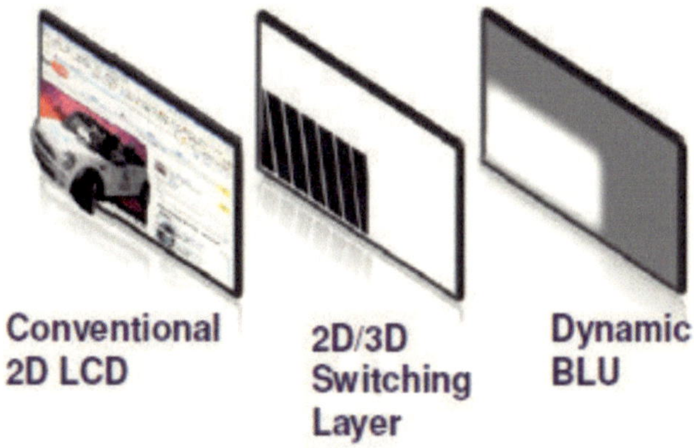

Fig. 13. The structure of 2D/3D switching design for integrated 2D/3D display

Property	Specification
Resolution	1680 x 1050, SXGA+
Panel Diagonal	22 inch diagonal
Speed	60Hz
contrast	800 : 1

Table 1. Image Liquid Crystal panel specification

The third at right is a dynamic back-light unit (DBLU), one that is similar, for example, to an LED matrix-based BLU now seen in many laptop PC display panel but with brightness of each (or at least groups) of the LED elements in the matrix separately controllable. There is a second component layer, the 2D/3D switching component, inserted between the first and third that is responsible for the automatic switching of individual pixels in the first between its 2D and 3D display mode. This technology uses a microretarder-based switching device to partially switch various parts of the display screen between 2D mode and 3D mode. Structurally, a microretarder plate has an interleaved pattern of half-wavelength-retardation and zero-retardation stripes. Working together with the microretarder is a liquid crystal (LC) switching panel inserted between the microretarder and the polarized backlight of the device. Each "pixel" of the LC switching panel functions as the switching unit cells between 2D and 3D mode regions. As a general rule of thumb, the smaller these "pixels" are, the smaller the 2D/3D switching cells can be. Though, for any practical application, the number of units required for the 2D/3D switching LC panel lands in where barely enough but with the lowest cost case, for example, 16 by 10.

3.2 Shutter glasses type stereoscopic displays

In such a shutter-glasses type stereoscopic display, the display signal (including vertical synchronization, horizontal synchronization and data) is sent from the display card to the LCD panel. The switching of the shutter glasses is driven by the vertical synchronization signal from the display card. Due to the possible phase lag between the shutter glasses and

the LCD panel, a phase lag circuit is set between these two devices. The scanning or the strobe of the dynamic backlight is driven by the same vertical synchronization signal.
Two experimental setups are used to implement the two dynamic backlight methods.

1. In scanning backlight method, a novel controlled circuit architecture of scanning regions for 120Hz high frequency. Setup all the parameters of scanning backlight method by counting the amount to decide turning time between 4 、and 2 LED backlight regions. If counted times equal to 100 then jump to next backlight region. For 4-region scanning backlight method, when the panel is filled in regions 1, 2, 3 and 4 by the new image, the backlight lights up in the corresponding regions 3, 4, 1 and 2. In anticipation of an image for a left eye and right eye is shown in the region 1 of the panel, we turned on region 3 of the backlight unit. Analogize the image shown in region 2 and turned on region 4 of the backlight unit. For 2-region scanning backlight method, when the panel is filled in regions 1, and 2 by the new image, the backlight lights up in the corresponding regions 2, and 1. For avoiding seeing both L-image and R-image at the same time, the backlight regions R1 have to be off until R1 filled up the image. Analogize the backlight regions R2 have to be off until R2 filled up the image.
2. In backlight strobe method, Setup the parameters of backlight strobe method by counting the amount to decide turning on time of full screen (full screen of one frame 1/120sec counted amount equal to 400). Setup the parameters of backlight strobe duty time by counting the amount to decide turning time on full screen backlight regions. If counted times equal to (400 - 400*9/10) then jump to next full screen backlight region. The backlight is turned off when the image data refreshes. The backlight only turns on at the system time, or at most a little bit longer than the system time. But the system time is short compared with the time between two adjacent vertical synchronization signals (less than 10%), the display brightness operated under this method is probably quite small.

In order to tell which method is better for a shutter-glasses stereoscopic display, three experiments are done. They are 4-region scanning backlight, 2-region scanning backlight and backlight strobe methods. One of the most important properties of the 3D display, the crosstalk, is used in these experiments to tell which method is better. The test pattern is in a video stream form, which is like the left diagram.
The crosstalk is given by equations (1) and (2):

$$C_L = \frac{BW - BB}{WB - BB} \tag{1}$$

and

$$C_R = \frac{WB - BB}{BW - BB} \tag{2}$$

Where

WB represents a video stream with all-white as left-eye images, all-black as right-eye images),
BW represents a video stream with all-black as left-eye images, all-white as right-eye images),
BB represents a video stream with all-black for both left and right eyes.
C_L and C_R represent the crosstalk experienced by the left eye and right eye

The CS-100 Spot Chroma Meter is used in this research to measure all the luminance values. The images are displayed in page-flipping mode using the resolution and color-depth set in Stereo/Page-flip Setup as shown in Fig.14 and Fig.15.

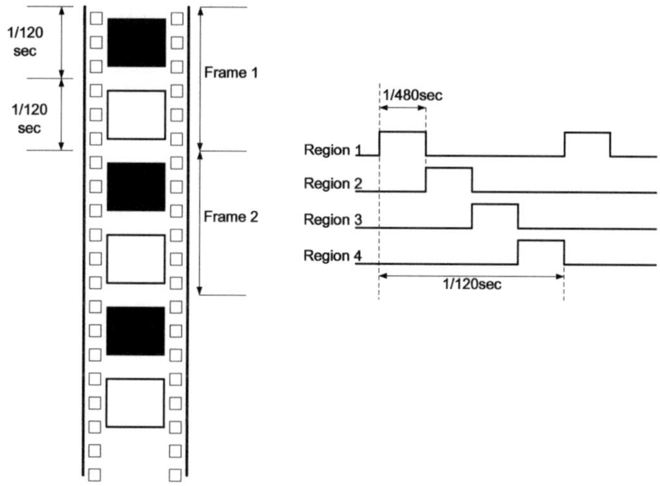

Fig. 14. LED scanning backlight duty cycle

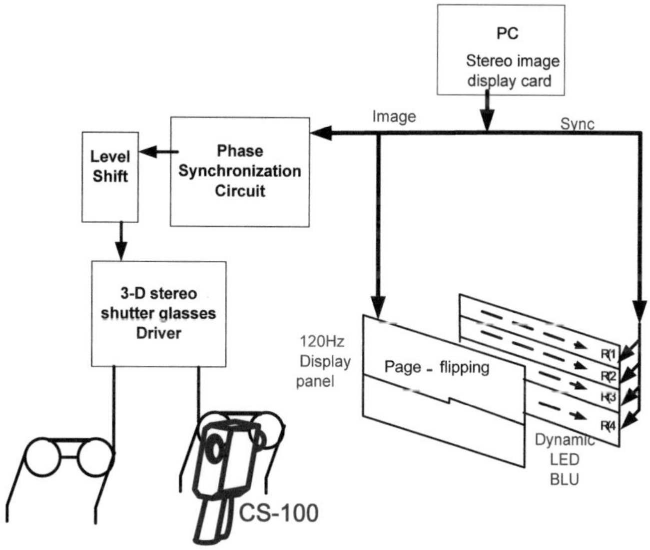

Fig. 15. The CS-100 measurement stereoscopic display system

The scanning backlight method turns on several (e.g., 2 or 4) horizontal regions of the backlight in turn, corresponding with the fill-out of the LCD panel. The backlight strobe method is to synchronously apply a control signal to the whole backlight to provide flashing effect rather than scanning. Both methods can control the brightness of the backlight module by adjusting the duty cycle of the control signal (Fig.14). Table 2 shows the specifications of the image liquid crystal panel.

Property	Specification
Resolution	1680 x 1050
Panel Diagonal	22 inch diagonal
Speed	120Hz
contrast	1,000:1 (20,000:1 'MEGA' Dynamic Contrast)
Response time (G2G)	5ms (2D), 3ms (3D)

Table 2. Image Liquid Crystal panel specification

3.2.1 Results and discussion

There is a phase difference between the vertical synchronization signal and the shutter glasses switching time. A phase lag circuit is applied to correct the difference. The waveform shown in Fig. 16 is an adjusted result.

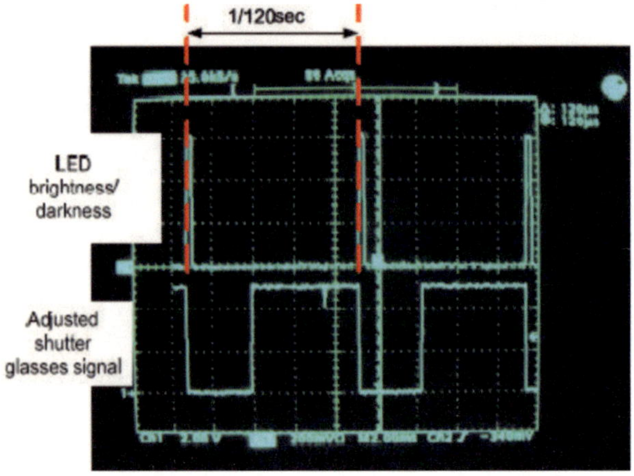

Fig. 16. Optimized synchronization signal

The synchronized shutter glasses can be used to separate the left-eye image and the right-eye image. However, even the block and transparent function of the shutter glasses is perfect, if the synchronization is not exact, or the response time of the liquid crystal is not fast enough to operate with 100 to 120 Hz, the viewer will still experience serious crosstalk. Therefore, reduction of the crosstalk is very important while making a stereoscopic display. The observation of cross-talk reduction effect special pattern is clearly shown in Fig.17. The display content is a video stream with small squares as the left-eye image and small circles for the right eye.

The luminance of the 3D LCD is measured and recorded as WB, BW and BB charts. The measurement distance is 1 meter and the recording unit is "nit". The crosstalk is calculated by the equations 1 and 2. As a result, the luminance of a 4R scanning backlight display for one viewing zone is 28.825 nits, that of a 2R scanning backlight display for one viewing zone is 57.25 nits, and that of backlight strobe display is 28.45 nits.

The crosstalk under different scanning conditions is calculated from data, including the cases of 4-Region and 2-Region scanning backlight method, and backlight strobe method (1 Region). The crosstalk of 4R and 2R scanning backlight displays are 3.2% and 5.08% respectively, and that of backlight strobe display is 1.68%.

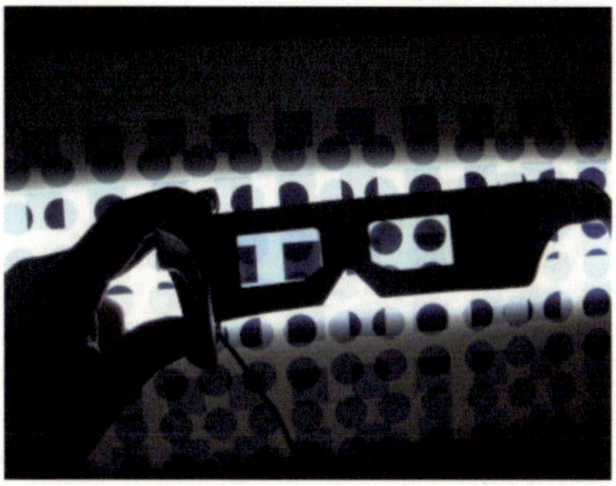

Fig. 17. Special pattern with eye shutter.

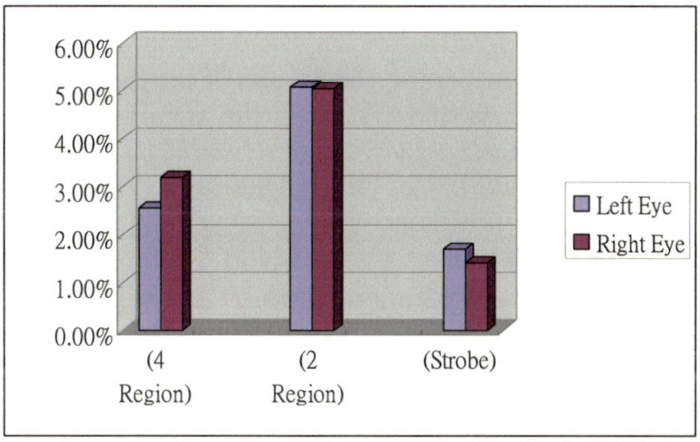

Fig. 18. Results of crosstalk calculation of different dynamic backlight methods

According to the results of and 18, the brightness of the 4R scanning backlight and the widened backlight strobe is about half of the 2R scanning backlight. But the crosstalk performance of the widened backlight strobe crosstalk is the best of three methods. Only about 1.68%. Although the 4R scanning backlight display crosstalk is higher than widened backlight strobe, it still performs better than the 2R scanning backlight one.

3.3 Multi-view time multiplexed autostereoscopic displays

Multi-view displays use TFT screens for image formation . The light generated by the TFT is split into multiple directions by the means of special optical layer (called also lens plate or optical filter) mounted in front of the TFT. The intensity of the light rays passing through the filter changes as a function of the angle, as if the light is directionally projected. There are two important points to note when considering multi-view screens: (1)They are not directly compatible with standard stereo-scopic footage or software. This is because, rather than display two distinct views to the viewer (as with most other stereoscopic displays), they provide up to 5, allowing the viewer to walk around the screen whilst maintaining the 3D effect. (2)Due to the display showing 4 or 5 views simultaneously, each view contains only 1/4 or 1/5 of the standard resolution of the display panel. An optimal lens, designed to handle a single situation, would be shaped to contain only as many distinct views as local participants; maximizing each views resolution would require a lens width sufficient to cover the same number of subpixels as views.

Our current prototype system uses a lenticular lens multiview technique with a time-multiplex autostereoscopic display based on active directional backlight (active dynamic backlight). Throughout the proposed system, lower power consumption are successfully obtained as well as high contrast ratio even with less number of drivers than that of conventional local dimming method. This architecture also contains a new adaptive dimming algorithm and image processing technique for the proposed stereo LCD backlight system. Recent progress in stereo display research has led to an increasing awareness of market requirements for commercial systems. In particular areas of display cost and software input to the displays are now of great importance to the program. Possible areas of application include games displays for PC and arcade units; education and edutainment; Internet browsing for remote 3D models; scientific visualisation and medical imaging. Intelligent and green power LED backlighting techniques of two-dimensional (2D) to three-dimensional (3D) convertible type, shutter glasses type, multi-view time multiplexed naked eye type, and multi-viewer tracking type for stereo liquid crystal displays are applied for 3D display system.

According to the time sequence for turning the groups of the light source, multiple viewing zones at multiple directions are created. To meet the requirements of different one-eye images, we propose that the real-time active barrier dynamic backlight slit system on stereo-display. To confirm our design workable, we did the optical simulation using Advanced Systems Analysis Program (ASAP) software. The detector is set at the convergent point, and the intensity profiles of the four views are shown in Fig. 19. The intensity profiles are evolving every 1/240 sec, and the separation of peaks is about 60 mm, quite close to the design value, 65 mm. The small inaccuracy resulted from the absorption of black matrix, and it makes the dead zone explicit. Setting the pixel size larger, or the black matrix smaller would improve the results. On the other hand, the center-viewing group has the least crosstalk, and side lobe groups have larger crosstalk, especially when viewing groups departs from center very much. The increase of crosstalk arises from the abbreviation when light is not close to the optical axis. Thus, the profiles of the four views confirm our design workable. Table 3 shows the specifications of the image liquid crystal panel.

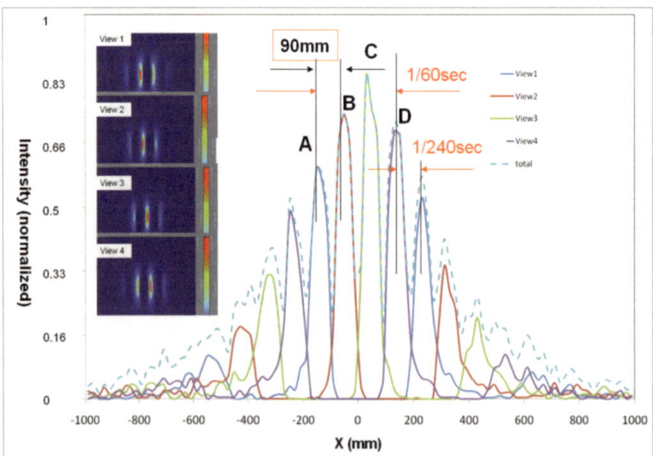

Fig. 19. The optical simulation using ASAP software

Property	Specification
Resolution	1024 x 768
Panel Diagonal	6.0 inch diagonal
Speed	240Hz
contrast	1000 : 1

Table 3. Image Liquid Crystal panel specification

3.3.1 Experimental results

The photographs of displayed images used the luminance meter (Konica Minolta CS-200) as shown in Fig. 20. The distance from the optical sensor to the center of LED backlight panel is in 60 cm, which is the normal range of distance for watching a 3D computer monitor. Measurements of angle are done from observation point -50 degree to observation point +50 degree; view 2 is the central view. The illuminance meter has an analog output to the oscilloscope and the illuminance signal can be recorded and processed by a computer.

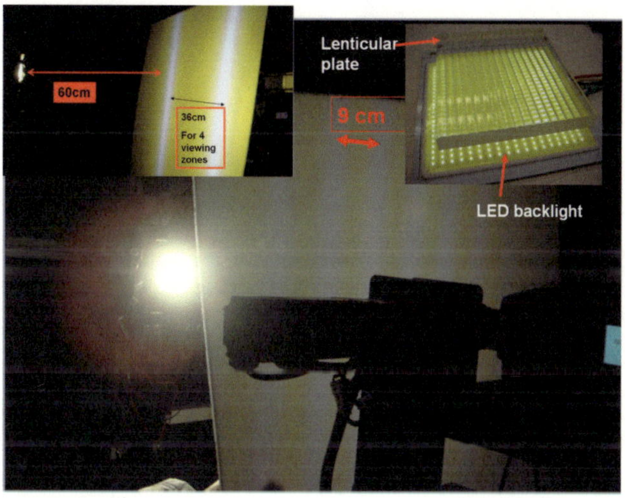

Fig. 20. Four views backlight

In this research, we observed backlight light stain structure for 3D image display based on lenticular lens array. In Fig.21, the photo is illustrated for four viewing zones 1-4 located at the viewing location. Each viewing zone uses 1/240 second to display one image. The light source at specific location is grouped corresponding to each lenticular lens of the lenticular lens array. For the four viewing zones, each lenticular lens has four groups 1-4 of light sources corresponding to four viewing zones 1-4. The four groups of light are sequentially turned on for 1/240 second. The group 1 of light source is turned on, and then the group 2 of light source is turned on next for 1/240 second. Likewise, the groups 3 and 4 of light source are sequentially turned on for 1/240 second. Generally, the multiple viewing zones equally shares 1/60 second for one image frame. The viewable zone area from first viewing zone to be contiguous to second viewing zone is 90 mm, and for 4 viewing zone of viewable area is 360 mm(The separation of viewing zones is about 90 mm, and overall width of viewing group is 360 mm) as shown in Fig. 20.

Yellow stripe pattern is created by phosphor of yellow color. White LEDs are blue LED chips covered with a phosphor that absorbs some of the blue light and fluoresces with a broad spectral output ranging from mid-green to mid-red. So, the backlight modular was taken on yellow stripe.
The configuration of uni-direction diffusion lens plate is shown in Fig. 21(b). The panel of 240Hz displays the corresponding images of the four viewing zones by the same time sequence according to temporal multiplexed mechanism. The uni-direction diffusion lens plate can condense the light individually belonging to each the lenticular lens at transverse direction. The lenticular lenses of the lens array receive the light and deflect the light into each viewing zone in a time sequence, respectively.

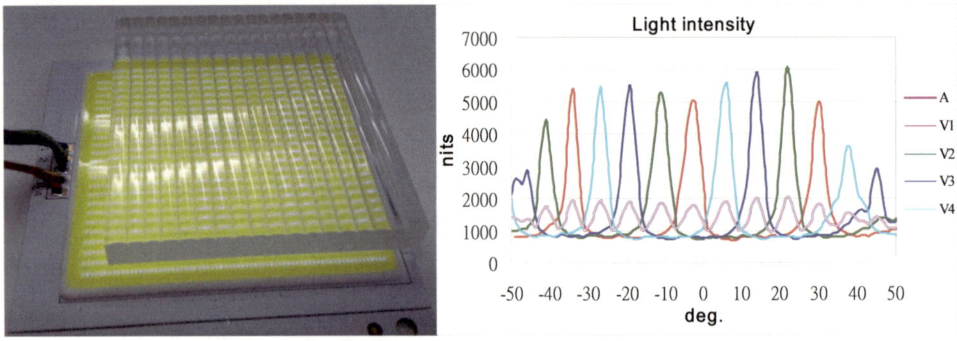

(a) Lenticular/LED

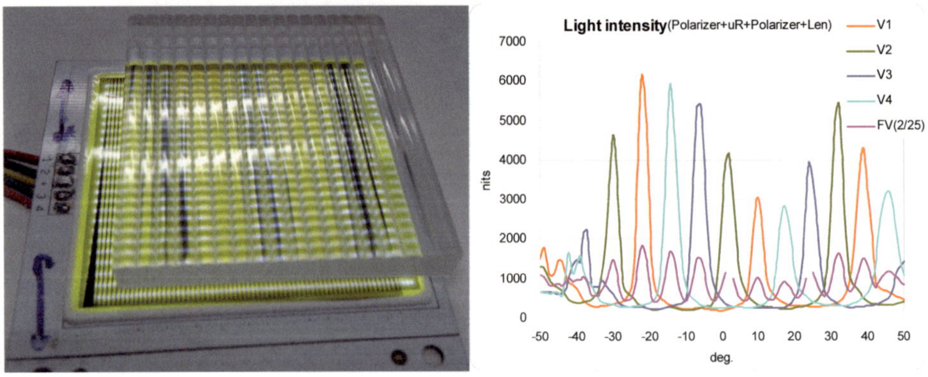

(b) Lenticular/Optical film/LED

Fig. 21. The crosstalk under different observation scanning angles

According to the time sequence for turning the groups of the light source, multiple viewing zones at multiple directions are created. To meet the requirements of different one-eye images, we propose that the real-time active barrier dynamic backlight slit system on stereo-display. The center viewing group has the least crosstalk, and side lobe groups have larger crosstalk, especially when viewing groups departs from center very much. The crosstalk under different observation scanning angles is showed from data in Fig. 21, including the cases of 4-views field scanning. The crosstalk of view 1 is about 5% respectively, the results are better than slanted lenticuler lens type.

3.4 Multi-viewer tracking stereoscopic display

This study integrated an autostereoscopic display with a viewer-tracking system. Fig. 22. illustrates the basic structure of the display and table 4 shows the specifications of the image liquid crystal panel. In the proposed structure, a retarder inserted between the image panels rotated the light beam at 90°; simultaneously, a lenticular plate adjusted the light direction to show the light slit from the tracking display. Retarder film is a clear birefringent material that alters the phase of a polarized beam of light. A quarter wave plate can convert linearly polarized light (oriented at 45° from the direction of the fast/slow axis) into circularly polarized light. Conversely, the wave plate can convert a circularly polarized beam into linearly polarized light.

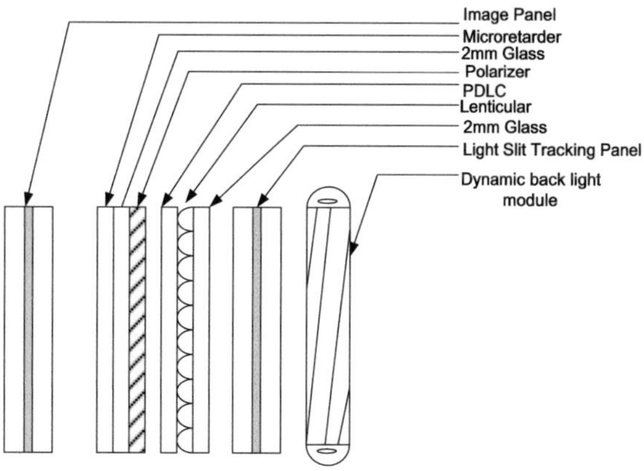

Fig. 22. The structure of the proposed viewer-tracking display panel

Property	Specification
Resolution	1920 x 1080
Panel Diagonal	23.6 inch diagonal
Speed	120Hz
contrast	1000 : 1
Response time (G2G)	2ms (3D)

Table 4. Image Liquid Crystal panel specification

In this study, when the polarization direction of the incident light formed an included 45° angle with the optical axis of the retarder, the polarization of the light passing through the $\lambda/2$ retardation regions rotated by 90° and became orthogonal to the polarization of the light passing though the 0° retardation regions. The molding method fabricated the lenticular plate with polymeric film as the substrate material. One of the light slit pattern pairs adjusted the direction of light from the tracking panel to the viewer's eyes through the lenticular plate.

In this display, the PDLC panel played an important role in the function of the 2D/3D switch. When the PDLC panel was turned to clear state, the microretarder interacted with the polarizers to form a parallax barrier pattern as shown in Fig.23, making the display autostereoscopic. In a case where the PDLC panel is in a diffusive state, the light passing

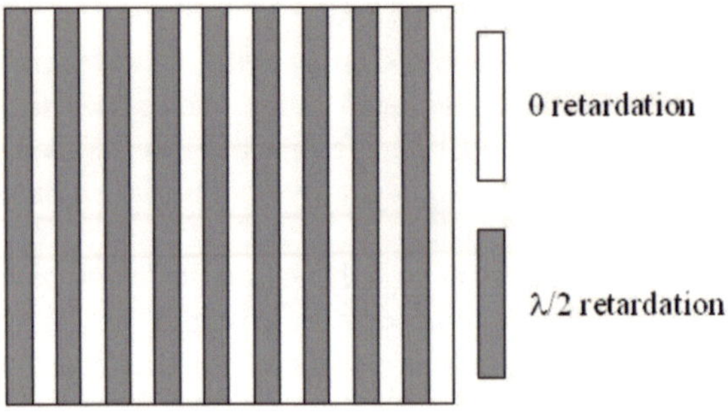

Fig. 23. The pattern of a microretarder

through the PDLC destroys the polarization. The microretarder then loses its function as a parallax barrier, and the display becomes a general 2D display.

This study developed autostereoscopic display apparatus and a display method. The autostereoscopic display apparatus included a display panel, a backlight module, a tracking slit panel and an optical lens array. In a frame time, the display panel and tracking panel share the same synchronization signal for the display panel. The tracking panel controls the light of the backlight module. The tracking panel features tracking slit patterns and switches the slit patterns according to the synchronization signal. Until all screen data is updated, the backlight module is inactive during the frame time. A light provided by the part of the backlight regions passes through the tracking slit set, optical lens array, and the display panel in such a way that each eye separately perceives images. As shown in Fig. 24, when the viewer moves to the left, the tracking slit set changes its pattern to display the correct image.

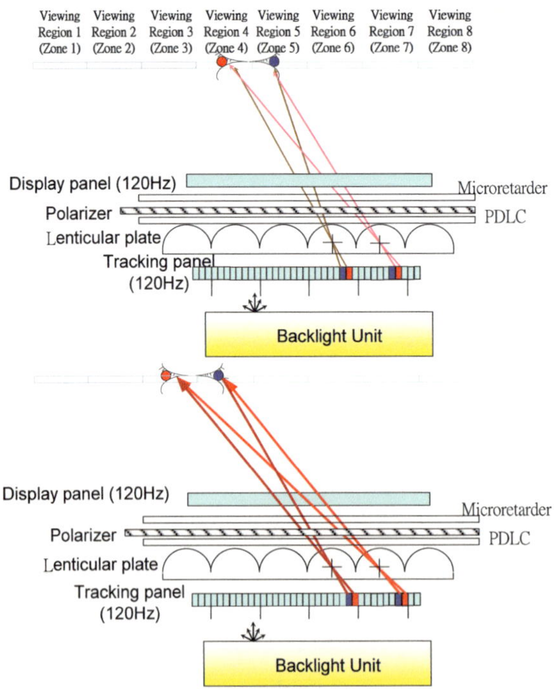

Fig. 24. Relations between viewer and tracking panel

The autostereoscopic display integrated a webcam as the real-time detection device for tracking of the viewer's head/eye positions, so that the display showed left and right eye images correctly. The computer vision-based tracking method detects viewer's eyes over a specific range and under conditions of low and fluctuating illumination. By capturing the image of the viewer in front of the display, the viewer's position is calculated and the related position data is transferred to the field programmable gate array (FPGA) controller through RS232. When the viewer recognizes that he/she is standing at the borders of the viewing zones, analyzing the captured viewer images determines the border positions of the viewing zones. The resulting eye reference pattern allows the tracker to locate the viewer's eyes in live video images. If an observer moves away from his original position, the tracking slit will vary its pattern according to the viewer's new position. The viewer still perceives two eye images separately before exceeding the webcam detection range. Fig. 25 shows the viewer-tracking system.

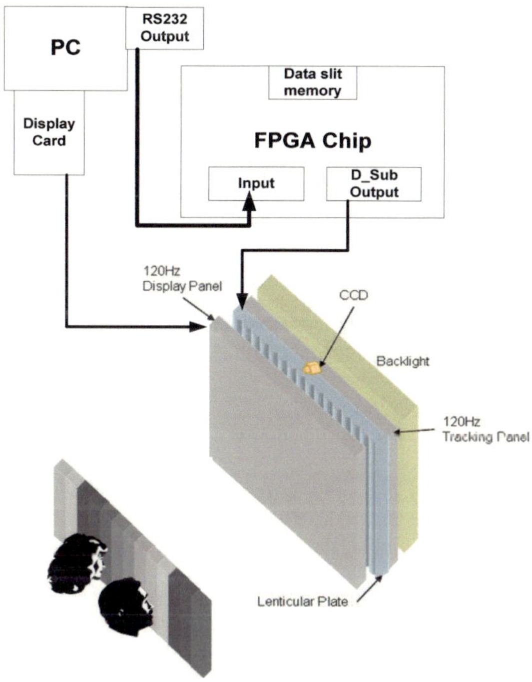

Fig. 25. Viewer-tracking 3D display system

This research addresses the specific technological challenges of autostereoscopic 3D displays and presents a novel system that integrates a real-time viewer-tracking system with an autostereoscopic display. Our successfully designed prototype utilized a FPGA system to synchronize between a display panel and tracking slit panel. With 120Hz display and tracking panels, only a pair of page-flipped left and right eye images was necessary to produce a multi-view effect. Furthermore, full resolution was maintained for the images of each eye. The loading of the transmission bandwidth was controllable, and the binocular parallax and motion parallax is as good as the usually lower resolution multi-view autostereo display.

(B) LED Backlight architecture

Many types of LED backlights are applied to 2D or 3D displays. To date, research on 3D display systems has generally focused on providing uniform, collimated illumination of the LCD, rather than addressing low crosstalk issues. This study investigated the method of using an autostereoscopic multi-viewer tracking 3D display with a synchro-signal LED scanning backlight module to reduce the crosstalk of right eye and left eye images, enhancing data transfer bandwidth while maintaining image resolution. Fig. 26A is a schematic view of a stereoscopic display. Fig. 26B is a block diagram illustrating the stereoscopic display; the stereoscopic display can track the viewer's position and be watched by multiple viewers.

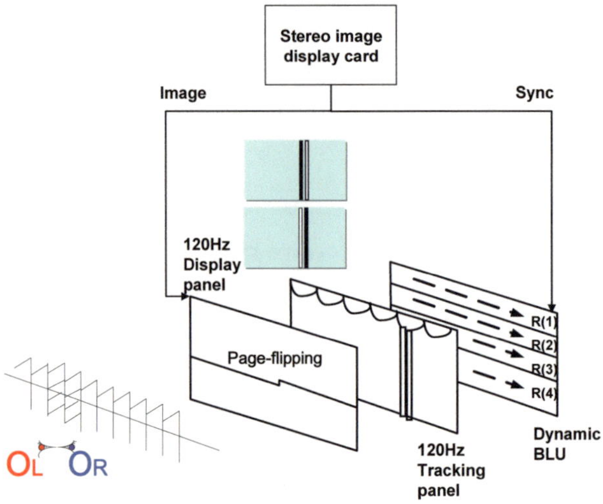

Fig. 26A. The schematic view illustrating a stereoscopic display.

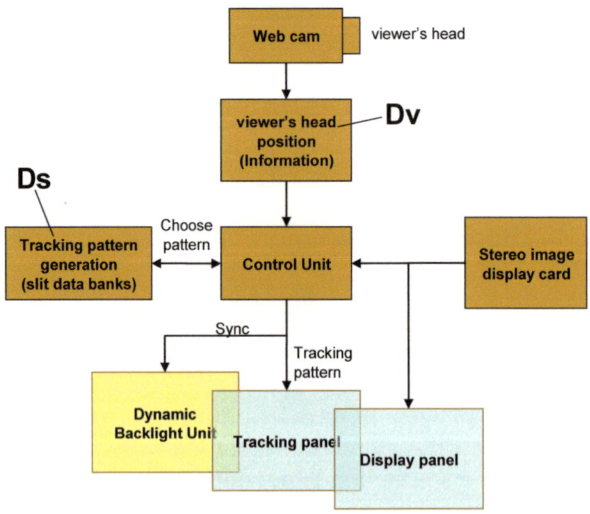

Fig. 26B. The block diagram illustrating the stereoscopic display

The backlight module of the stereoscopic display is a dynamic backlight module featuring many light-emitting regions R(1)~R(4). Fig. 26A excludes the control unit and optical lens array. In the stereoscopic display, the graphic card outputs and transmits the vertical synchro-signal to the control unit. After receiving the synchro-signal, the control unit outputs the synchro-signal to control (turn on or off) the light-emitting regions R(1)~R(4).

To meet the requirements of different one-eye images, we propose that the dynamic LED backlight tracking panel has have many backlight slit sets. According to the position information of viewer O and the vertical synchro-signal, one of the slit sets of the tracking panels is selected and turned-on. Each slit set includes either left or right eye slits. Light emitted from the dynamic backlight module passes through the either left or right eye slit and the display panel, and projects onto one eye of viewer O. Similarly, light emitted from the dynamic backlight module passes through the either left or right eye slit and the display panel, and projects onto the other eye of viewer O. In this way, the pair images are projected to the two eyes of viewer O, who can see accurate three-dimensional images. For example, light emitted from the dynamic backlight module passes through the left eye slit of the slit set and the display panel, and projects onto the left eye O_L of viewer O. Similarly, light emitted from the dynamic backlight module passes through the right eye slit of the slit set and the display panel, and projects onto the right eye O_R of viewer O. The one-eye slits are stripe-shaped and the lengths of the one-eye slits are approximately equal to the longitudinal length of the display panel.

When the display panel displays an image based on the vertical synchro-signal, the slit set of the tracking panel is enabled. Meanwhile, pixels in the updated region of the display panel display a left-eye image, but pixels in the non-updated region of the display panel still display the previous right-eye image. Light passing through the slit set of the tracking panel and the non-updated region of the display panel can be projected onto left eye O_L of viewer O (i.e. a crosstalk phenomenon) if no alternative methodology is applied. This research proposes using a dynamic backlight module to suppress the crosstalk. The light-emitting regions R1~R4 of the dynamic backlight module are separately controlled according to the vertical synchro-signal.

During a frame period, the light-emitting regions R(1) and R(2) corresponding to the updated region are turned on and the light-emitting regions R(3) and (4) corresponding to the non-updated region are turned off. In this way, only the light-emitting regions R(1) and R(2) provide light, so that no light passes through the slit set of the tracking panel and the non-updated region of the display panel. This reduces the crosstalk phenomenon of the stereoscopic display system.

As shown in Fig. 26A and Fig. 26B, the display method of the stereoscopic display comprises the following steps:

First, slit data banks (Ds) corresponding to the many viewing angles of the stereoscopic display apparatus is established. Next, the control unit receives information (Dv) on the position of the viewer. The control unit compares the position information and the slit data banks stored in advance. Meanwhile, the control unit outputs the vertical synchro-signal from the graphic card to control the output mode of the dynamic backlight module and operation mode of the tracking panel. The display panel is driven to display images (i.e. image updating) according to the vertical synchro-signal output from the graphics card. Many of the light-emitting regions (R(1)~R(4)) of the dynamic backlight module are stripe-shaped and the light-emitting regions R(1)~R(4) extend across the slits of the tracking panel. The extending direction of the light-emitting regions R(1)~R(4) is perpendicular to the extending direction of the slits of the tracking panel. Many of the light-emitting regions (R(1)~R(4)) of the dynamic backlight module are array in an arrayed manner.

3.4.1 Crosstalk analysis

To avoid ghost images, the backlight modular provides backlight control signals which are dependent on the position of an associated part of the panel. The system is provided for controlling synchronization timing between backlighting and pixel refresh, in dependence of a location of a section within the display panel. The backlight unit is separated into several regions. Let's take 4 regions as the example, the pixel response time is less than three fourths of the frame time when the illumination period is one quarter of the frame time. Optical sensor and CS-100 Spot Chroma Meter of luminance crosstalk measurement of the 4-regions 、 2-regions scanning and strobe backlight method without lenticular. Frame sequential(page flip, temporal multiplexed) process, the process is referred to as alternate frame sequencing.

Crosstalk is a critical factor determining the image quality of stereoscopic displays. Also known as ghosting or leakage, high levels of crosstalk can make stereoscopic images hard to fuse and lack fidelity. Crosstalk is measured by displaying full-black and full-white in light-emitting regions R(1)~R(4) of the display system without lenticular and using an optical sensor to measure the amount of leakage between channels.

For example, the optical sensor is placed at the left eye position (either behind the left eye of 3D glasses, or in the left eye viewing zone for an autostereoscopic display) and measurements are taken for the four cross-combinations of full-white and full-black in the left and right eye-channels. An additional reading is also taken with the display in the off state. These readings can then be used in the crosstalk equations described above. This metric can be called black-and-white crosstalk and this metric is often used because maximum crosstalk occurs when the pixels in one eye-channel are full-black and the same pixels in the opposite eye-channel are full-white. According to the results, the brightness of the 4R scanning backlight and the widened backlight strobe is about half of the 2R scanning backlight. But the crosstalk performance of the widened backlight strobe crosstalk is the best of three methods. Only about 1.68% left. Although the 4R scanning backlight display crosstalk is higher than widened backlight strobe, it still performs better than the 2R scanning backlight one.

In the study, the CS-100 Spot Chroma Meter was used to measure the brightness of the backlights, which can be controlled using the duty cycle of backlight signal, as shown in Fig. 27. Moreover, photodiode s3072 was used to measure the optic characteristics of the display device.

To view the correct image from the tracking display, the synchronization relationship between image display and backlight requires calibration. Fig. 28 shows that the V-sync signal exceeds the backlight signal in 1/160s. If the V-sync signal triggers the backlight signal directly, the observer sees three white regions and one black region (not fully white or fully black). As the human eyes determine light source, a vertical signal must trigger the first region backlight (the dotted line of Fig. 28). Fig. 29 is the crosstalk of right eye and left eye under three different brightness conditions. The phase of the V-sync signal exceeds the phase of the backlight signal in 1/480s. It is too dark if the duty cycle is lower than 50%, so the three chosen duty cycles all exceeded 50%. According to Fig.29, the differences in brightness do not significantly affect the crosstalk. The performances of both eyes were approximately in agreement. The experiment selected maximum brightness.

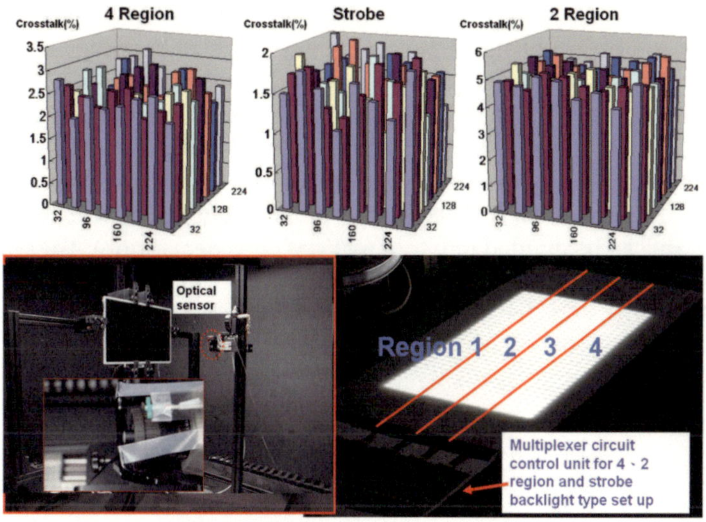

Fig. 27. Optical sensor and CS-100 Spot Chroma Meter of luminance crosstalk measurement of the 4-regions , 2-regions scanning and strobe backlight method without lenticular

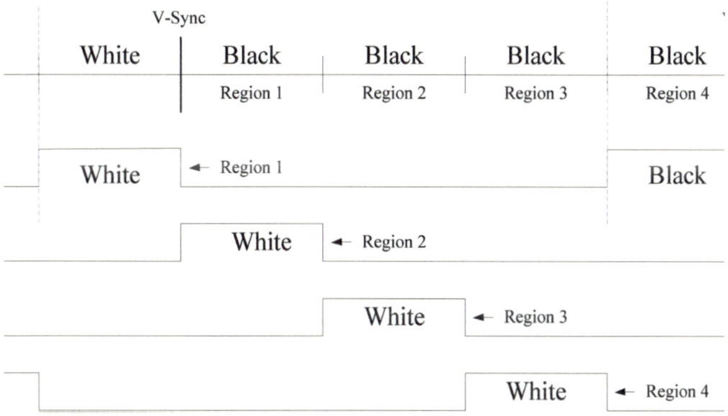

Fig. 28. The synchronization relationship between image display and backlight (from top to bottom)

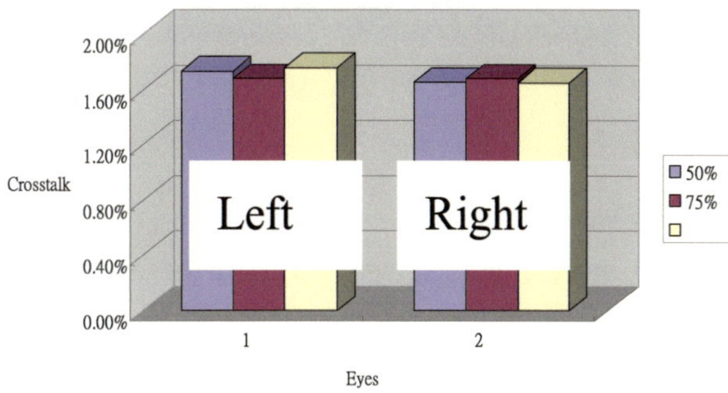

Fig. 29. The crosstalk under different brightness conditions

Fig. 30 shows the crosstalk of the right eye and left eye with different phase shifts between the V-sync signal and backlight signal, where the duty cycle of backlight signal is 100%. The lowest crosstalk only occurs when phase shifts are 1/160s, not both 1/480s and 1/160s. Here, light leaking to other regions and the response time of the liquid crystal affect the crosstalk (Fig.31). Fig. 31 is the response reaction of the liquid crystal from full black to full white. The horizontal axis is time (5 ms per grid) and the vertical axis is voltage (20 mV per grid). One display frame is 1/120 second, approximately equal to 8 milliseconds. The response waveform can be divided into four sections (2ms per section). The waveform of the liquid crystal still rises (section II) when the phase of V-sync signal exceeds the backlight signal in 1/480s (≒2ms); here, the phase does not reach a bright state. But, the phase shifts of 1/240s (≒4ms) and 1/160s (≒6ms), located at region III and region IV, respectively, gradually near the bright state; this explains the difference in crosstalk.

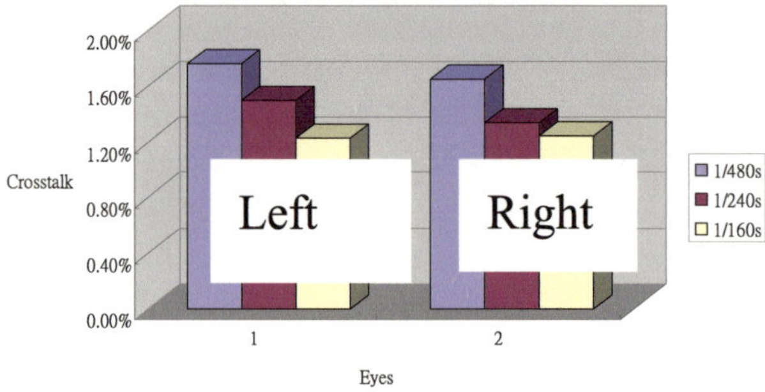

Fig. 30. The crosstalk under different phase shift conditions

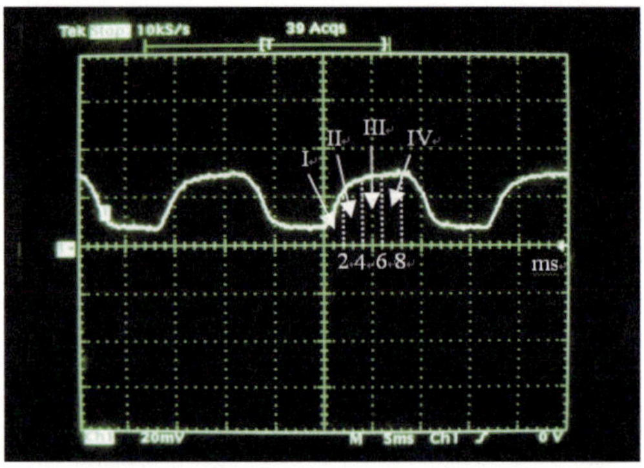

Fig. 31. The response time of liquid crystal from dark state to bright state

3.4.2 Measured results

(A) The optical properties measurement

Detailed and quantitative measurements were used in the autostereoscopic display. To measure the borders and the performance of the viewing zones, a luminance meter, Minolta CS-200, was located at the designed viewing distance (630mm from the display), which reduced the display panel's optical interference during measurement. Only the backlight module, including a backlight, tracking panel, and lenticular plate, was used in optical luminance measurement experiments. In this backlight structure, no additional brightness was lost in the optical path.

Fig. 32 shows the results of the luminance intensity experiment. The entire measuring process was completed in a darkroom, which provided measurements of fairly high quality. In the luminance intensity experiment, only 40 degrees on both sides of the center of the backlight module was measured; the intensity value was captured every 0.5 degree. When measuring, only one viewing zone of the tracking panel was switched on. The luminance meter was used to scan the viewing zones horizontally. The maximum peak luminance value of the viewing zone was approximately 514 cd/m^2 when CS-200 detected the luminance intensity value near the center of the backlight module. The minimum peak luminance intensity value of the backlight was approximately 364 cd/m^2 at the edge of the viewing group. The luminance intensity range of the viewing group in front of the backlight module ranged between 264 and 514 cd/m^2. The intersection point between two adjacent luminance intensity curves may determine the borders of the viewing zones.

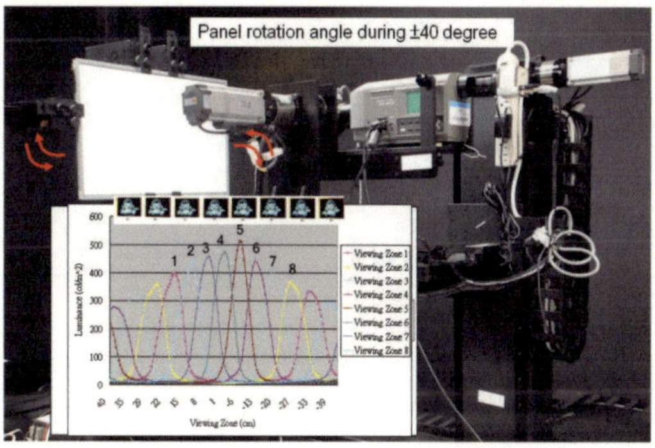

Fig. 32. Luminance intensity distribution of lenticular-type BLU

After luminance intensity measurement of viewing zones 1 to 8 was completed, data could be combined to yield a luminance intensity distribution figure to verify the optical design parameters. The peak luminance intensity value of viewing zone 4 is in front of the center of the backlight module. The tracking panel was slightly misaligned with the lenticular plate. Due to the light intensity distribution of the tracking panel and the entrance angle difference of the light path between the backlight and the lenticular lens, a stronger luminance intensity curve was measured in the central part of the viewing group. The viewing group of the 3D display system was approximately 53 cm wide at a viewing distance of 63 cm, indicating that each viewing zone is 6.625 cm wide on average.

In building a viewer-tracking-based autostereoscopic display, the viewer's position and border positions of the viewing zones are the key parameters. To accurately define the viewer's position, the black and white pictures for both eyes were displayed as calibration images. The positions of the borders in the viewing zones could be determined by analyzing the images of the viewer captured while the viewer reported to be at the border positions.

(B) Motion parallax function result

For a 3D display to simulate the natural vision of human beings, both binocular parallax and motion parallax are required. For a multi-view autostereoscopic display system, viewers can see stereoscopic images with binocular parallax and motion parallax within a group of viewing zones. However, a high number of viewing zones is necessary to achieve smooth motion parallax for a sufficiently large view. This normally causes significant reduction of image resolution. While maintaining good image resolution, we implemented smooth motion parallax by adopting viewer tracking function and real-time image updating in a two-view autostereoscopic display system.

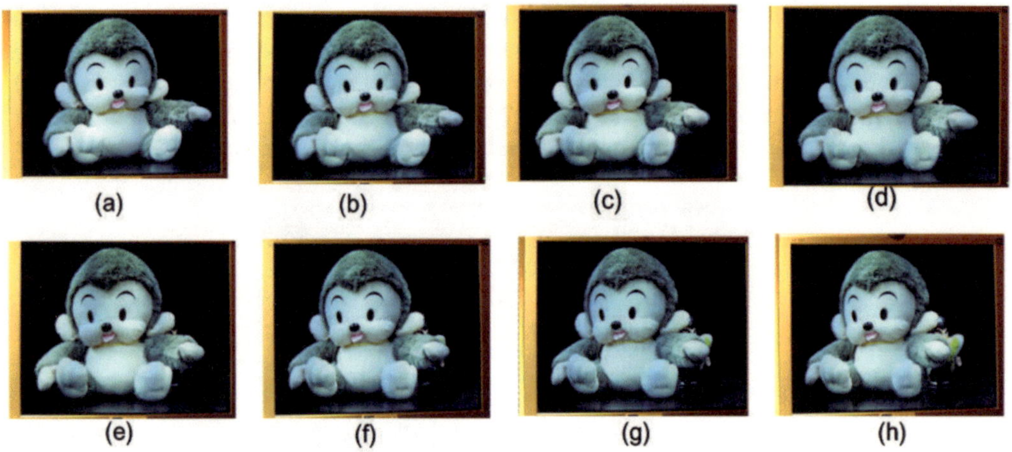

(a) 35° image content, (b) 40° image content, (c) 45° image content, (d) 50° image content,
(e) 55° image content, (f) 60° image content, (g) 65° image content, (h) 70° image content.

Fig. 33. The viewer-tracking-based 2D/3D switchable autostereoscopic display

This study used an ad-boost algorithm capable of evaluating important features to quickly track viewers. If the viewer's eyes were detected in specific viewing zones in front of the display, the viewer's position would determine the images of the corresponding viewing angles shown. When the viewer's eyes move inside the same viewing zones, full stop needs moving, the images for the new viewing angles are fed into the same viewing zones. The viewer experiences the motion parallax because he/she sees different images from different angle.

When the viewer's eyes continue to move and finally cross the border of the viewing zones, the images of the new angle are reversed left-and-right and presented in real-time on the display.

In this study, tracking stability was good with viewing angles ranging from -15 degrees and 15 degrees. The refresh rate achieved 30 frames per second when the resolution of the capturing image was set to 160×120. To match the resolution of the display, the resolution of image content for each eye was 840×1050. The image content was rendered from the 3D model built from 3D Max or directly captured using cameras. For the webcam coordinates, the accuracy of one pixel was about 6.25 mm according to the viewing angle of the webcam indicated and the designed viewing distance in the autostereoscopic display. Therefore, the rendering or capturing angle was set to 0.5 degree according to the accuracy of one pixel. As shown in Fig. 33, the 2D/3D switchable auto- stereoscopic display correspondingly provides about 160 pairs of stereo images to the viewer moving in the viewing angle of the system. The viewer is consistently able to experience the reality of motion parallax.

4. Conclusion

The design of the three dimensional hierarchy with control circuit for large LED backlight array, which effectively reduces the terminal numbers into the cubic root of the total control unit numbers and prevent a block defect of the flat panel. The display panel is divided into many scanning block parts, each part is separately and simultaneously scanned in the same directions to write images on the pixels on the respective scanning electrodes. These defects are generally the result of a failure in the row (horizontal) or column (vertical) drivers or their connections. We have reached the advantages of high accuracy, rapid selection, and reasonable switching speed flat panel.

Several shutter-glasses type stereoscopic displays have been measured to analysis difference of their 3D performance. The less the backlight regions are, the brighter the display with scanning backlight method is. Therefore, a 2R scanning backlight is brighter than a 4R one. Nevertheless, due to better separation of a 4R scanning backlight, the crosstalk of it is less than that of a 2R scanning backlight. However, from the other aspect, the uniformity of a scanning backlight method is usually not as good as than backlight strobe method. For a higher luminance and lower crosstalk, it is suggested to combine the 4R scanning backlight method and backlight strobe method. In this way, a 120Hz LCD can be made a very good performance stereoscopic display with shutter glasses.

In full resolution multi-view autostereoscopic display research, we have successfully designed and fabricated the optical system, high density active barrier dynamic LED backlight, the slit pitch is 700um, and the LED chip size is 10×23mil for full resolution multi-view autostereoscopic display. From the measurement results, the dynamic LED backlight optical system can yield ideal parabolic curvature and the crosstalk is lower than 5%. Besides, the lenticular lenses of the lens array optical system was successfully received the light and deflected the light into each viewing zone in a time sequence, which could be one of the candidates for future full resolution time-multiplexed 3D applications.

A viewer-tracking-based auto- stereoscopic display of a synchro-signal LED scanning backlight system that can correspondingly send different pairs of stereo images based on the viewer's position. Additionally, an 8-view autostereoscopic display was implemented with full resolution in the display panel, achieving high 3D image quality in the preliminary configuration. Further modifications, e.g. higher precision for viewer-tracking positioning and design of more viewing zones in the display system, may improve system performance.

5. References

P.de Greef and H. Groot Hulze, "Adaptive Dimming and Boosting Backlight for LCD-TV System,"SID Symposium Digest Tech Paper 38, 1332-1335(2007).

E.H.A. Langendijk, R. Muijs, and W. van Beek."Quantifying Contrast Improvements and Power Savings in Displays with a 2D-Dimming Backlight",IDW(2007).

P. J. Bos, K. Roehler/Beran "The pi-cell: a fast liquid-crystal optical-switching device" Mol. Cryst. Liq. Cryst., 113, 329-339,(1984).

P. Bos "Stereoscopic imaging system with passive viewing apparatus", US Patent 4,719,507,(1988).

L.Lipton, J.Halnon, J.Wuopio, B. Dorworth "Eliminating pi-cells artifacts", Sterescopic Displays and Virtual Reality Systems VII, Vol. 3295, 264-270, (2000).

C. vanBerkel and J.A. Clarke "Characterisation and Optimisation of 3D-LCD Module Design" Proc SPIE Vol. 3012, (1997).

S. Pastoor and M. Wopking 3-D displays: a review of current technologies. Displays 17, (1997).

L. Lipton Synthagram: autostereoscopic display technology. Proceedings of the SPIE, Vol. 4660, (2002).

S.S. Kim, K.H. Cha, J.H. Sung," 3-D Display" SID'02 Digest, pp.1422,(2002).

Hamagishi, M. Sakata, A. Yamashita, K. Mashitani, E. Nakayama, S. Kishimoto, and K. Kanatani, "New stereoscopic LC displays without special glasses," Asia Display '95, pp. 791–794 (1995).

Roese, "Stereoscopic electro-optic shutter CRT displays—a basic approach," in Processing and Display of Three-Dimensional Data II, J. J. Pearson, ed., Proc. SPIE 507, 102–107 (1984).

Bohm, J. Frank, and L. Joulain, "An LCD helmet mounted display with stereoscopic view for helicopter flight guidance," in Helmet- and Head-Mounted Displays V, R. Lewandowski, L. Haworth, and H. Girolamo, Eds., Proc. SPIE 4021, 66–78 (2000).

Jian-Chiun Liou and Fan-Gang Tseng, "120Hz Display with Intelligent LED Backlight Enabled by Multi-Dimensional Controlling IC , Displays, Vol. 30, No. 3, pp. 147-153, 2009.

P. Surman, K. Hopf, I. Sexton, W.K. Lee, R. Bates, "Solving the 3D problem - The history and development of viable domestic 3-dimensional video displays", In (Haldun M. Ozaktas, Levent Onural, Eds.), Three-Dimensional Television: Capture, Transmission, and Display (ch. 13), Springer Verlag, 2007.

C. van Berkel, D. Parker and A. Franklin, "Multiview 3D LCD," in Proc. SPIE Vol. 3012, pp. 32-39, 1996.

A. Schmidt and A. Grasnick, "Multi-viewpoint autostereoscopic displays from 4D-vision", in Proc. SPIE Photonics West 2002: Electronic Imaging, vol. 4660, pp. 212-221, 2002.

C. van Berkel, "Image preparation for 3D-LCD," in Proceedings of SPIE, 1999, vol. 3639, pp.84-91,(1999).

J. Konrad and P. Angiel, "Subsampling models and anti-alias filters for 3-D automultiscopic displays", IEEE Trans. Image Processing, vol.15, no.1, pp. 128-140, Jan. 2006.

Xiaofang Li, Qiong-hua Wang, Yuhong Tao, Dahai Li, and Aihong Wang,"Cross-talk reduction by correcting the subpixel position in a multiview autostereoscopic three-dimensional display based on a lenticular sheet", Chinese Optics Letters, Vol. 9, Issue 2, pp. 021001- (2011).

Jian-Chiun Liou, Kuen Lee, Chun-Jung Chen, "Low Cross-Talk Multi-Viewer Tracking 3-D Display of Synchro-Signal LED Scanning Backlight System", IEEE/OSA Journal of Display Technology, VOL. 7, NO. 8, AUGUST 2011,pp.411-419 (2011).

J.-Y. Son, V. V. Saveljev, Y.-J. Choi, J.-E. Bahn, and H.-H. Choi, "Parameters for designing autostereoscopic imaging systems based on lenticular, parallax barrier and IP plates," Opt. Eng., vol. 42, no. 11, pp.3326–3333 (2003).

M.Sakata, G.Hamagashi, A.Yamashita, K.Mashitani, E.Nakayama "3D displays without special glasses by image-splitter method", pp.48-53, Proc. 3D Image Conference 1995, Kogakuin University, 6/7 July (1995).

E. Kurutepe, M. R. Civanlar, and A. M. Tekalp, "Interactive transport of multi-view videos for 3DTV applications,"Journal of Zhejiang University SCIENCE A: Proc. Packet Video Workshop 2006, vol. 7, no. 5, pp. 830–836(2006).

Y. Huang and I. Essa. "Tracking multiple objects through occlusions." In IEEE Conf. on Computer Vision and Pattern Recognition, pages II: 1051-1058 (2005).

F. Jurie and M. Dhome. "Real time tracking of 3D objects with occultations." In Int'l Conf. on Image Processing, pages I: 413-416 (2001).

Thomas B. Moeslund, Adrian Hilton, and Volker Kruger. "A survey of advances in vision-based human motion capture and analysis." Computer Vision and Image Understanding, 104(2):90-126 (2006).

J. Pan and B. Hu. "Robust occlusion handling in object tracking." In IEEE Workshop on Object Tracking and Classification Beyond the Visible Spectrum, pages 1-8 (2007).

Rajwinder Singh Brar, Phil Surman, Ian Sexton, Richard Bates, Wing Kai Lee, Klaus Hopf, Frank Neumann, Sally E. Day, and Eero Willman," Laser-Based Head-Tracked 3D Display Research", Journal of Display Technology, Volume: PP , Issue: 99 , Page(s): 1 - 14(2010).

Jian-Chiun Liou and Fo-Hau Chen, "Design and fabrication of optical system for time-multiplex autostereoscopic display", Optics Express, Vol. 19, Issue 12, pp. 11007-11017 (2011).

Won-Sik Oh; Daeyoun Cho; Kyu-Min Cho; Gun-Woo Moon; Byungchoon Yang; Taeseok Jang;"A Novel Two-Dimensional Adaptive Dimming Technique of X-Y Channel Drivers for LED Backlight System in LCD TVs" ,Journal of Display Technology, Volume: 5 , Issue: 1 , Page(s): 20 - 26(2009).

U. Vogel, L. Kroker, K. Seidl, J. Knobbe, C. Grillberger, J. Amelung, and M. Scholles, "OLED backlight for autostereoscopic displays," SPIE, vol. 7237, pp. 72370U-1–9, (2009).

N. Raman and G. J. Hekstra, "Content Based Contrast Enhancement for Liquid Crystal Displays with Backlight Modulation," IEEE Trans. Consumer Electron., vol. 51, no. 1, pp. 18-21, Feb. (2005).

M. Doshi and R. Zane, "Digital architecture for driving large LED arrays with dynamic bus voltage regulation and phase shifted PWM," in Proc. IEEE Appl. Power Electron. Conf. (APEC), pp. 287-293 (2007).

S.-Y. Yseng et al., "LED Backlight Power System with Auto-tuning Regulation Voltage for LCD Panel," in Proc. IEEE Appl. Power Electron. Conf.(APEC), pp. 551-557 (2008).

Y. Hu and M. M. Javanovic, "LED Driver With Self-Adaptive Drive Voltage,"IEEE Transactions on Power Electronics, Vol. 23, No 6, pp. 3116-3125,Nov.(2008).

J. Son and B. Javidi, "Three-Dimensional Imaging Methods Based on Multiview Images," J. Display Technol. Vol.:1 Issue:1, pp.125-140 (2005).

8

Gas Safety for TFT-LCD Manufacturing

Eugene Y. Ngai[1] and Jenq-Renn Chen[2]
[1]*Chemically Speaking LLC, New Jersey*
[2]*Department of Safety, Health and Environmental Engineering, National Kaohsiung First University of Science & Technology, Kaohsiung*
[1]*USA*
[2]*Taiwan*

1. Introduction

The fabrication of TFT-LCD panel includes the growing and etching of a-silicon and silicon nitride films. These fabrication processes utilize significant amounts of silane, phosphine, ammonia, chlorine, boron trichloride, nitrogen trifluoride, fluorine, and hydrogen which are highly flammable, reactive, corrosive, and/or toxic. The amount used and scale of the supply system for these gases are far larger than for other tech industries such as the semiconductor and photovoltaic industries. Accidental leaks and fires of these gases are the major safety concern in the TFT-LCD fabs.

This paper first reviews the hazardous properties of the gases used in the TFT-LCD manufacturing processes. The best practices for handling these hazardous gases are then described. Finally, the past incidents and emergency response actions are also reviewed.

2. Hazardous properties of specialty gases

The TFT-LCD manufacturing processes utilize significant amount of gases for thin-film deposition and etching. These include silane, phosphine, ammonia, and hydrogen for polycrystalline silicon and silicon nitride thin film deposition in a plasma-enhanced chemical vapor deposition (PECVD) reaction chamber. In addition, nitrogen trifluoride or fluorine is used in the PECVD chamber cleaning. Chlorine and sulfur hexafluoride are used in the dry etching of thin film. The hazards of these gases can be classified as pyrophoric, flammable and oxidizing gases. There are other bulk gases such as nitrogen, argon and helium which are used for inerting or purging and will not be discussed here owing to their low risks.

2.1 Silane

Silane or silicon tetrahydride (SiH_4) is the most common silicon source used in TFT-LCD manufacturing. It is a highly flammable gas and has a very wide flammable range, from 1.37% to 96%. Silane is colorless and odorless although it has been reported to have a prudent odor and a reported time weighted threshold limit value (TLV-TWA) of 5 ppm by ACGIH. However, this data is based on analogy of another hydride gas Germane rather than the actual toxicological data. In fact, silane has no odor and its toxicity is low with a high median lethal concentration (LC_{50}) of 9,600 ppm for rat at 4 hours exposure. The prudent odor comes from the impurity of trichlorsilane during early silane production.

Silane is used as a compressed gas. However, silane has a critical temperature of -3.4°C. It is possible that liquefaction may occur in cold storage or during expansion cooling from compressed sources. A silane supply system must be designed properly to prevent unexpected pressure surge from liquefaction and vaporization.

Silane is also a pyrophoric gas that normally ignites upon contact with air. Silane has a reported autoignition temperature of -50~-100°C. Autoignition of silane in air has also been reported down to -162°C, depending on the oxygen concentration of the mixture (Baratov et al., 1969). However, the autoignition temperature only denotes the temperature above which a given fuel-air mixture will autoignite in a closed vessel. It however does not refer to details of the autoignition process and how autoignition is affected by flow. The major potential hazard of silane is however not in its pyrophoricity but rather in its unpredictable ignition behavior in accidental releases. A silane release from a pressure source has been known not to lead to spontaneously ignition (Koda, 1992) and frequently delayed ignition occurs when the release is shut-off resulting in a "pop". In a semi-confined space with gas accumulation, the pop can lead to a gas explosion with significant damage (Ngai et al., 2007).

The mechanism of delayed ignition has been studied by Tsai et al. (2011) which demonstrated that silane release without prompt ignition was most likely caused by quench of the reactive kernel from the flow strain or scalar dissipation accompanied by the large velocity and concentration difference between the release gas and the ambient air. With diminishing release velocity, the flow strain reduces along with quench of the reactive kernel and ignition then occurs at a critical exit velocity. The ignition at reducing velocity may ignite the released gas and create a significant combustion or explosion. The critical exit velocity of indefinitely delayed ignition lies between 0.3 m/s to 4.3 m/s for vent diameter of 2.03 ~4.32 mm as shown in Figure 1. These velocities are at least two orders of magnitude smaller than the velocity from a pressurized source which in most cases reaches sonic velocity. Furthermore, Figure 1 also highlights that the critical velocity is also decreases with decreasing vent size.

There are factors other than vent size affecting the critical exit velocity. Among them, the temperature, moisture and silane combustion powder are the most noticeable factors. At higher temperature, the reactive kernel is stronger and thus the critical exit velocity is larger (Liang et al., 2010). Removing the moisture in the air also results in significantly higher critical exit velocity indicating an inhibitory role of moisture on silane ignition (Liang et al., 2010). The inhibitory action of moisture on the silane autoigniton in air is also in consistent with other fuel such as hydrogen and methane. Silane combustion produces white to brown powders as shown in Figure 2. These powders are also known to promote ignition but its exact role remains to be studied.

The presence of a critical exit velocity for prompt ignition of silane release has important implications in the safety of silane operations. First of all, almost all current silane uses in the semiconductor, TFT-LCD, and photovoltaic industries are supplied in the form of pressurized cylinders with pressure up 12.5 MPa. The high pressure silane is then regulated to 0.8 MPa for the supply tube into the cleanroom and then further regulated down to 0.3~0.45 MPa before feeding into the process chamber. Thus, almost all potential leak points in silane utilization have pressure and possible leak velocity higher than the reported critical velocity for prompt ignition. In addition, the potential leak points are the cylinder valve, tubing, and tube connections (Huang and Ngai, 2006). The potential release size will be much smaller than 2.03 mm, except for a catastrophic full bore tube rupture. Thus, delayed ignition should be considered as the usual case rather than a rare case in most silane operations.

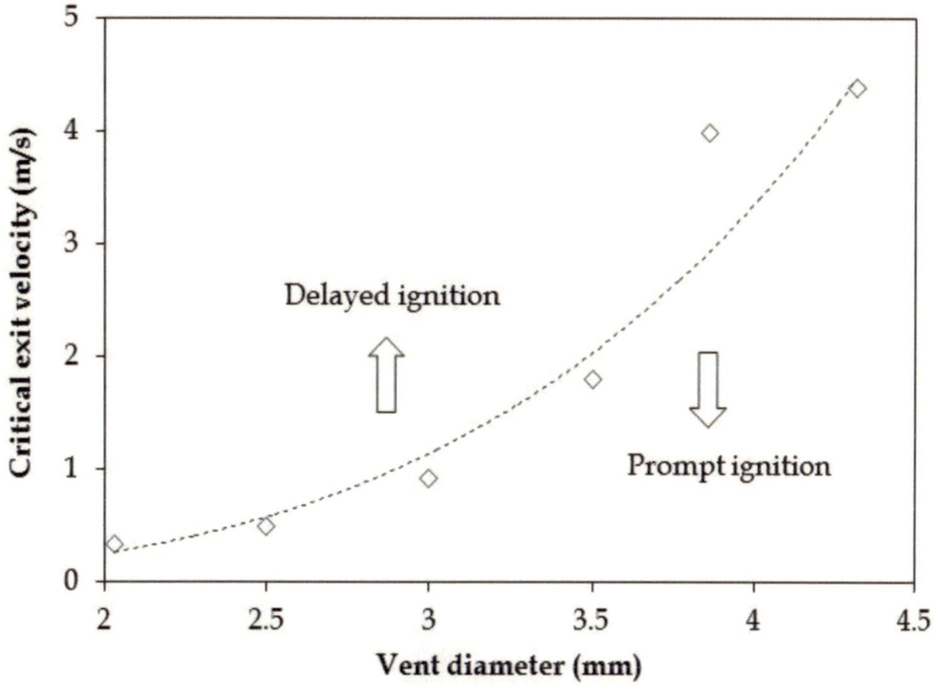

Fig. 1. Critical exit velocity for delayed/prompt ignition as a function of vent diameter.

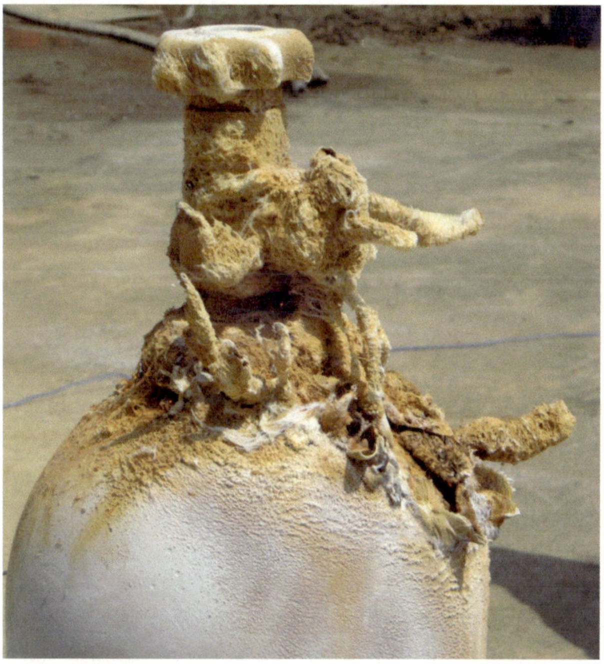

Fig. 2. Typical silane combustion powder.

2.2 Phosphine

Phosphine (PH_3) is a highly toxic and pyrophoric liquefied compressed gas. Phosphine is used primarily as a phosphorus dopant (N-type dopant) in the TFT LCD industry. It is typically supplied as mixture with hydrogen.

Pure phosphine is a colorless, pyrophoric liquefied gas. Although pure samples of phosphine are odorless, the gas often has an odor of garlic or decaying fish due to the presence of substituted phosphine and diphosphine (P_2H_4). It is a highly toxic gas with a PEL of 300 ppb, an LC_{50} of 20 ppm for rat at 1 hour exposure.

Phosphine has a low auto-ignition temperature <0°C so it may ignite spontaneously on contact with air. Autoignition of phosphine released into air is also subjected to effects by temperature, moisture, and flow strain just like silane. For example, autoignition of phosphine is reported to occur only in dry air but not in moist air (Brittion, 1990). However, there is still a lack of detailed studies on the exact conditions for these effects. It also has an extremely wide flammable range of 1.6% - 95% in air. It burns in air with an orange flame forming white phosphorous pentoxide as a byproduct as shown in Figure 3(a), which is a severe respiratory irritant. Under certain conditions, phosphine will form red phosphorus as a byproduct in a fire as shown in Figure 3(b).

(a) (b)

Fig. 3. Typical phosphine combustion flame and powder. (a) Orange flame and white phosphorous pentoxide powder (b) Red phosphorus powder.

Acute exposures to phosphine cause respiratory tract irritation that attacks primarily the cardiovascular and respiratory systems causing peripheral vascular collapse, cardiac arrest and failure, and pulmonary edema. Inhalation is the primary exposure route, and there is no known dermal exposure route. Symptoms may include restlessness, irritability, drowsiness, tremors, vertigo, double vision (diplopia), ataxia, cough, dyspnea, retrosternal discomfort, abdominal pain, and vomiting. There is no antidote for phosphine exposures; treatment is supportive. Guidance on proper medical treatment is available from the US Health

Department Agency for Toxic Substances and Disease Registry (ATSDR) sheet on phosphine.

Transportation regulations prohibit the supply of pure phosphine in cylinders larger than 50 liters since it is highly toxic. As a dopant, phosphine is supplied to the process as a low concentration mixture in order to precisely control the diffusion rate for the TFT-LCD processes. As a mixture, the gas is no longer classified as highly toxic, therefore it is supplied in bulk high cylinders of 450 liters or larger.

Some gas suppliers are also supplying gas mixing systems which will dynamically blend the phosphine from a pure source cylinder with hydrogen. A single cylinder will supply 200 cylinders of 0.5% concentration.

2.3 Hydrogen

Hydrogen is a colorless, odorless, tasteless, highly flammable gas. It is also the lightest weight gas. It is much less dense than air and can disperse rapidly or accumulate in the upper sections of enclosed spaces

Hydrogen has a wide flammability range, 4% to 75% in air, and the small amount of energy required for ignition necessitate special handling to prevent the inadvertent mixing of hydrogen with air. Sources of ignition such as sparks from electrical equipment, static electricity, open flames, or extremely hot objects should be eliminated. Hydrogen and air mixtures within the flammable range can explode and may burn with an almost invisible flame. Although hydrogen is not pyrophoric like silane or phosphine, hydrogen release from a tube or a crack under high pressure may lead to spontaneous ignition. This is attributed to the low ignition energy, high blow-off velocity and the adiabatic compression from the shock wave of the release (Dryer et al., 2007)

Gaseous hydrogen may be supplied in cylinders or in tubes that are designed and manufactured according to applicable codes and specifications for the pressures and temperatures involved. The pressure rating and internal volume of a container determines the quantity of hydrogen it can hold. Cylinders may be used individually or can be manifolded together to allow for a larger gas storage volume. Tubes are mounted on truck-trailer chassis or in ISO frames for transportation and are referred to as tube trailers or tube modules, respectively.

2.4 Ammonia

Ammonia (NH_3) is a toxic and corrosive liquefied compressed gas. It is used to grow a silicon nitride layer in the TFT-LCD industry. Pure ammonia is a white liquefied gas. It is a strongly irritating gas. It is a toxic gas with a PEL of 50 ppm and an LC_{50} of 7,338 ppm. Ammonia is lighter than air (specific gravity: 0.59) and is highly soluble in water. Ammonia is thermally stable.

Ammonia has a very narrow flammable range of 16% to 25% and minimum ignition energy (MIE) 40,000 times (680 mJ) higher than that of hydrogen (0.017 mj). The common flammable gases such as propane have a MIE of 0.25 mJ which is still many times less than amonnia. The high lower flammable limit (LFL) and lighter than air vapor density makes it unlikely for an ammonia release to reach concentrations high enough to burn in open areas. It is thus classified as non-flammable for transportation purposes. The national building and fire regulations however classify it as flammable in storage and use.

As a flammable gas, ammonia is reactive with oxidizer gases such as nitrous oxide or oxygen. Since ammonia has an LC_{50} higher than 5,000 ppm (threshold for toxic gas), the United States classifies it as a non-toxic non-flammable gas. All other countries classify ammonia as a toxic gas. Ammonia is a severe irritant with a sharp pungent odor that will have an immediate effect on moist tissues, eyes, upper respiratory system and skin. Guidance on proper medical treatment is available from the US Health Department Agency for Toxic Substances and Disease Registry (ATSDR) sheet on Ammonia.
Ammonia is very corrosive to low alloy carbon steel, causing stress corrosion. Ammonia is also corrosive to viton, zinc and copper. Bulk ammonia is supplied in low or high pressure bulk containers (250 liter - 950 liter). Some facilities use 40,000 lb ISO Modules that are sited like the nitrogen trifluoride or silane modules for continuous supply. It is vaporized for use on demand.

2.5 Nitrogen trifluoride

Nitrogen trifluoride (NF_3) is a colorless, odorless, nonflammable, oxidizing compressed gas. Nitrogen trifluoride is the most common reactor cleaning gases used in the TFT-LCD industry because of its advantages, such as high etching rates, high selectivity, carbon-free etching, and minimal residual contamination. It is first decomposed in a remote plasma chamber into a reactive fluorine before fed to the reaction chamber where it is used to react solid deposits on the reactor walls to form gaseous byproducts such as silicon tetrafluoride.
Nitrogen trifluoride is not toxic, it has an LC_{50} of 6,700 ppm (1 hr. rat). It is not hazardous by skin contact and is a relatively minor irritant to the eyes and mucous membranes. Exposure of rats to a 100 ppm concentration for 7 hrs per day, 5 days per week over a 19 week period caused minor pathological changes to the liver and kidney. No other effects were noted. The ACGIH established the current TLV-TWA of 10 ppm based on 1/10 of this value. Some countries classify nitrogen trifluoride as a toxic gas because it has a TLV-TWA <200 ppm.
Under ambient temperatures and pressures, nitrogen trifluoride is inert. However it can become an extremely reactive under certain conditions. It is a strong oxidizer that is violently explosive at elevated pressures with flammable gases such as hydrogen or silane. Purification of nitrogen trifluoride using a solid absorbent media such as molecular sieve is to be avoided. Fires and explosions have occurred.
Nitrogen trifluoride is reported to be a significant greenhouse warming gas with a Greenhouse Warming Potential (GWP) >10,000 as compared to carbon dioxide. Properly design remote plasma cleaning system will react 100% of the nitrogen trifluoride, so it has a GWP comparable to fluorine under these conditions.

2.6 Chlorine

Chlorine (Cl_2) is a toxic and corrosive liquefied compressed gas. It is also an oxidizing gas. Chlorine is used primarily as an etching gas in the TFT-LCD industry. Pure chlorine is a yellow-green liquefied gas. It is a strongly irritating and highly toxic gas with a PEL of 0.5 ppm and an LC_{50} of 293 ppm. Chlorine is much heavier than air (specific gravity: 2.49) and is slightly soluble in water. Chlorine is thermally stable.
Chlorine is a severe irritant with a pungent odor that will have an immediate effect on moist tissues, eyes, upper respiratory system and skin. Chronic exposure will corrode teeth, in

some cases it will cause flu-like symptoms. Guidance on proper medical treatment is available from the US Health Department Agency for Toxic Substances and Disease Registry (ATSDR) sheet on chlorine.
Chlorine is an oxidizer gas that is equivalent to a 60% concentration of pure oxygen or almost 3 times that of air. As an oxidizer chlorine is reactive with flammable gases such as silane or hydrogen. In some cases it can be explosive, even in concentrations as low as 3% when mixed with these gases. Pure chlorine has also autoignited flammable materials.
Similar to ammonia, bulk chlorine is supplied in low or high pressure bulk containers (250 liter - 950 liter). It is vaporized for use.

2.7 Fluorine

Fluorine is a highly toxic gas with extreme oxidation and corrosion potentials. It has a sharp, pungent odor that can be detected by most people at very low levels (0.1 ppm). It has a TLV-TWA of 1 ppm and an LC_{50} of 185 ppm (1 hr. rat). Chronic fluorine exposures at low levels may cause fluorosis or abnormal calcium accumulation in bone structure. Fluorine is highly reactive with moisture, forming hydrofluoric acid which is extremely corrosive to human tissue. Fluorine will have an immediate effect on the eyes and respiratory system.
Medical treatment of exposure is specific and specialized. Releases into air can form varying concentrations of hydrofluoric acid, depending on conditions. When the aqueous hydrofluoric acid concentration is below 50%, it has delayed (up to six hours) symptoms of exposure. In a fire, fluorine will react to form other toxic fluoride compounds.
Fluorine is the most reactive of all elements and the most powerfully know oxidizing agent. Fluorine is able to react with almost all elements and compounds depending on pressure and temperature, with the exception of lighter noble gases (e.g. helium, argon), inorganic fluorides of the highest valency, and perfluorinated organic compounds. It reacts with many organic substances even at room temperatures and often accompanied by combustion and possible explosion (European Industrial Gases Association, 2011). Therefore special steps, such as special cleaning and passivation procedures, must be taken to protect systems.
Fluorine reacts vigorously at ambient temperatures with most metals. The reaction intensity depends upon the surface area of the metal. Powdered metals or fine wires may react violently. Most inorganic materials react with fluorine; water forms hydrofluoric acid, and salts convert to fluorides. Contact with organic materials generally results in ignition or violent explosion.
Due to its extreme reactivity, national transportation regulations typically limit the total amount of fluorine in a 50 liter cylinder to 3 kg or to a pressure of 30 bar. Pure fluorine cylinders can be fairly reactive at 30 bar. Some users use a mixture which is less reactive. Testing by Air Products and Chemical Inc. (2004) has shown that fluorine is 2.5 times more reactive than oxygen, when it is mixed in a 20% concentration with nitrogen it will not ignite a ¼" carbon steel rod at 13,790 kPa.
Testing by Ngai (2005) has shown that pure fluorine even at a pressure as low as 1 atm (101 kPa) can cause 316 stainless steel tubing to burn. Pure fluorine did not react immediately with the raw chicken. To better simulate human skin, some human hair was placed on the chicken skin. This immediately ignited and started the skin on fire as shown in Figure 4. The reaction was so hot that it ignited the stainless steel tubing, which sprayed the chicken with molten stainless steel. This continued to react until the fluorine flow was stopped (Ngai, 2005).

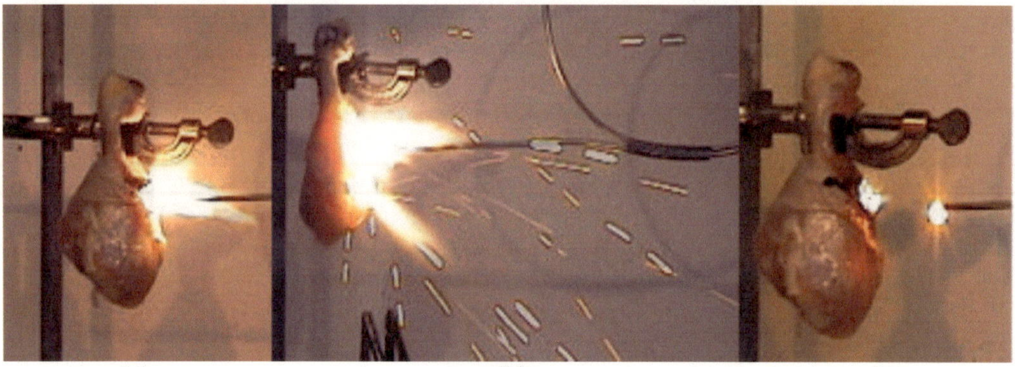

(a) (b) (c)

Fig. 4. Reaction of fluorine and chicken skin. (a) The hair on the skin reacts with the fluorine and ignites the chicken skin. (b) Skin burns hot enough to ignite the stainless steel tubing, which is sprayed at the chicken. (c) The tubing continues to react until the flow is stopped.

Despite its toxicity and reactivity, fluorine is being increasingly used as a reactor cleaning gas owing to less energy requirement, lower Greenhouse Warming Potential, and faster clean time (Stockman, 2009). To supply pure fluorine, small generators are being installed and managed by gas suppliers. These limit the amount of fluorine stored. A typical fluorine generator is an electrolytic cell which dialysis hydrogen fluoride to form fluorine and hydrogen (Stockman, 2009).

3. Best practice for handling

3.1 Silane

The first step in safe handling of silane is to comply with relevant regulatory & industrial standards such as Compressed Gas Association (CGA), Semiconductor Equipment and Materials International (SEMI), National Fire Protection Association (NFPA), Factory Mutual (FM), European Industrial Gas Association (EIGA), Asian Industrial Gas Association (AIGA), Transportable Pressure Equipment Directive (TPED), Transportation of dangerous goods (ADR), SEVESO, ATEX, etc. The installation and operation practices are however not necessarily detailed in the above codes or standards. Rufin (2011) has presented a comprehensive list of best practices for silane supply system. The key features are summarized below.

From the perspectives of preventing silane explosion, the most effective measure is preventing the accumulation of unignited silane through the use of forced ventilation or open air storage. It is also important to detect an unignited release by installing a gas sensor. Flame (UV/IR) detector is also needed for detecting leaks with autoignition. Sprinklers are needed for cooling the cylinders and system. Operators must be also be protected by barriers and flame resistant protection suit.

Other minor but equally important issues include:

- Welding for all indoor distribution systems
- Joule-Thomson compensation for high flow >50 L/min
- Automation and remote control
- Prevention of "domino effect" by steel plate or firewall between sources

- Normally closed pneumatic shut off valves
- Dedicated purge sources with automated purge cycles
- High pressure test and He outboard leak test
- Emergency remote shut down

For high flow rate greater than 50 L/min, large pressure drop will produce liquid silane mixture causing failure and leaks mostly in regulators. Thus, use of 2 regulators in series and heaters to compensate Joule-Thomson expansion cooling and prevent condensation.

3.2 Fluorine

As fluorine reacts with almost everything, corrosion of materials in contact with fluorine is dependent upon the fluoride formed with the material. If the fluoride is volatile or in the form of loose flakes, new material will constantly be exposed to the fluorine and will be continuously eroded until it burns through the material. For this reason material selection is extremely important. Materials that form a good fluoride coating will be protected from further attack.

Systems for fluorine service must not only be fabricated of the proper material, the design of the system must consider the following factors:

- Adiabatic compression - Adiabatic compression and heating is caused by the rapid pressurization of a system where the gas absorbs the energy and the gas temperature rises. This heating occurs at the point of compression or the point where the flow of gas is stopped, such as at a valve or regulator seat. Depending on the material in use where the hot gas impinges, the heat may be sufficient to ignite the material.
- High flow – High flow velocity increases the potential for a particle to impinge onto a surface with sufficient force to ignite it. A common area where this occurs is at a direction of flow change such as an elbow.
- Reactive materials - If the valve or regulator has a seat made of a nonmetallic material, it may be more prone to ignition because nonmetals typically have lower ignition temperatures than metals.
- Drying – removal of moisture from the system surface is essential
- Cleaning - to remove contaminants such as oil and particles from the system. These can be ignited easily generating sufficient heat to propagate a reaction to the system.
- Passivation – Fluorine passivation of metals is a pretreatment of the components in order to stabilize the wetted surfaces of the system by formation of a self-limiting protective fluoride film on the surface of the metal. The typical film thickness range from thickness of 5 Å (0.0005 microns) to 50 Å (0.005 microns) depending on the metal (Air Force Rocket Propulsion Laboratory, 1967).
- Friction - from a component that is malfunctioning or operating poorly can generate heat. Friction between two materials generates fine particles, which may ignite from the heat generated.

3.3 Other specialty gases

Silane and fluorine represent the two extremes, the flammable and the oxidizer, of the specialty gases used in TFT-LCD manufacturing. All other gases bear similar but less extensive practices and protection measures as these two gases. For example, gas detectors are needed for all gases except for nitrogen trifluoride. Flame (UV/IR) detectors and sprinklers are needed for all flammable gases. High pressure test and He outboard leak test with emergency remote shut down are also typical practices for all gases.

4. Incidents and emergency response

4.1 Silane

There are numerous silane leaks in the forty years of silane use (Huang and Ngai, 2006) mostly resulted from poor procedures, practices, and equipment. The reactions from the leaks were reported to be a fire, minor "pops" to explosions. The "pops" can occur when a small amount of Silane trapped behind the valve cap or pigtail is released rapidly without prompt ignition. In a few cases they have been severe enough to cause eardrum rupture if the amount of accumulated silane is significant. Thus, the past silane leaks do support the fact that a leak with indefinitely delayed ignition is commonly observed in silane operations.

Although the release with indefinitely delayed ignition were mostly small and had finite impact, the unignited release may turn into large release with significant impact. The following major incidents were known where silane was released without immediate ignition and exploded after a delay:

- 1976, Germany, 1 fatality
- 1989, Japan, 1 fatality & 1 injury
- 1990, Japan, 1 fatality & 3 injuries
- 1992, US, 1 injury
- 1996, Japan, 1 fatality
- 2003, US, no injuries
- 2005, Taiwan, 1 fatality
- 2007, India, 1 fatality

Among these incidents, the incident in Taiwan in 2005 has been described in detail by Chen et al. (2006), Chang et al. (2007) and Peng et al. (2008). The cause of leak is a failed silane cylinder valve with loosened valve stem retainer. Silane leak from such a loosened retainer has been shown to have delayed ignition by Ngai et al. (2007). The operator in this case did not notice the unignited silane release from the loosened retainer and continued to turn the valve handwheel until eventually the retainer, valve stem, and valve diaphragm were all detached. The rapid silane release with ignition at end of release resulted in a fatal vapor explosion.

Emergency response to a silane incident is the most challenging one among all gas incidents (Ngai, 2011). The greatest danger is a leak and no fire. As ignition is inevitable at the end of leak, a pop or explosion should be expected. Use surveillance camera to make initial assessment if possible. Remotely review the gas detector response over time on a graph to insure that it is not sensor drift or electronic spike. Evacuation should be done first if there is a gas detector alarm but without visual confirmation of flame. The responders should wear full personal protection equipment, including fire proof suit, glove, hardhat, ear plug etc. and approach the leak with great care. The leak could ignite as the pressure decays in the source cylinder or an air flow disturbance is created. Assess the leak rate with a portable gas detector through the gas cabinet duct. For a small leak, it might be safe to remotely shutoff the supply but expect ignition and a pop at shutoff. Slow shutoff of the valve will minimize the amount of silane when it ignites. A leak with a gas detector reading below the maximum scale, normally about 50 ppm, may be considered as a small leak. For a larger leak, activate sprinkler or apply water spray from a distance to the leak point to disperse the gas cloud and reduce the extent of potential explosion. Always prepare for possible explosion.

If the leak is burning, cool the adjacent area with a water spray. Do not spray water directly onto the fire as this could put it out and create an explosive environment. The heat could melt the fusible metal in the pressure relief device (PRD). The PRD will not activate unless the internal pressure reaches 27.6 MPa (4000 psig). Water spray from the deluge system onto the tubes will prevent this from ever reaching this temperature. A leak on fire can plug from the solids formed. Solids can also trap silane which can still be reactive. Even after a fire and the cylinder has vented to zero gauge pressure, there is still one atmosphere of gas in the cylinder. Emergency response teams typically will dilute the remaining silane by pressurizing the leaking cylinder with nitrogen to 0.79 MPa (100 psig) and venting 3 times. This will dilute the silane concentration to 2000 ppm. At this concentration and pressure, the cylinder can be shipped back as a non-hazardous package.

4.2 Hydrogen

The authors are aware of six hydrogen fires with one of the fires resulting in rupture of the hydrogen tube. The fire resulted from the detachment of an aging copper tube to the cylinder pack in a hydrogen filling station. The hydrogen ignited during tube detachment from a filling manifold and resulted in jet fire which impinged on a nearby tube trailer. The tube trailer was parked without protection of sprinkler system and two tubes were ruptured in less than 5 minutes of flame impingement. The most important emergency response after the tube explosion was continuous cooling and venting of hydrogen in the remaining tubes.

4.3 Nitrogen trifluoride

There have been reported incidents of nitrogen trifluoride with flammable oils used in vacuum pumps. Systems for nitrogen trifluoride must be designed similar to the requirements for fluorine in 3.2. Adiabatic compression and heating can initiate the decomposition reaction of nitrogen trifluoride into reactive fluorine. Incidents have occurred with improperly designed or cleaned systems during routine activities such as pressurizing the system. Internationally the maximum fill limits for nitrogen trifluoride cylinders is 0.5 kg/liter. At 21°C the cylinder pressure will be 10.78 MPa. Quickly opening the cylinder valve into a system with a small dead volume will cause adiabatic compression and heating to 307°C. In the presence of a reactive nonmetal component or a contaminant, this can initiate the decomposition reaction. Given sufficient energy and pressure, the reaction can cause the metal system to also become involved. Figure 5 shows a burnt valve in a nitrogen trifluoride line caused by adiabatic compression.

4.4 Chlorine

Chlorine is very corrosive to carbon steel and stainless steel. With the moisture in the air it forms hypochlorous (HClO) and hydrochloric (HCl) acids which will corrode the steel as shown in Figure 6. A corroded connection generally leads to leak. A good design suply system should pick up the leak by detector and shutdown the system. If the system is not

Fig. 5. A burnt valve in a nitrogen trifluoride line caused by adiabatic compression.

Fig. 6. Green hypochlorous and hydrochloric acids from chlorine corrosion.

designed properly, the chlorine may continuous to leak and affect the cleanroom. The author is aware of a chlorine leak from a tube connection in an IC fab that was not detected until one hour later. The chlorine cloud speaded and significantly affected the cleanroom and production tools.

4.5 Other specialty gases

The authors are not aware of any phosphine, ammonia, or fluorine fatalities or fires related to semiconductor or TFT-LCD uses.

5. Conclusions

The TFT-LCD manufacturing processes utilize significant amount of gases ranging from highly flammable to strong oxidizer gases. Special cares must be taken in handling these specialty gases to prevent fire, explosion and toxic release.

This paper has reviewed the hazardous properties and potential incidents of these specialty gases. The best practices for handling these gases are then described. The emergency response actions are also reviewed. It is sincerely hope that the current summary may bring an integrated view on the complicated and delicate specialty gas supply system, and preventing future incidents from occurring.

6. References

Air Force Rocket Propulsion Laboratory (1967). Halogen Passivation Procedural Guide, Technical Report AFRPL-TR-67-309, December 1967.

Air Products and Chemical Inc. Safetygram #25: 20% Fluorine in Nitrogen, June 2004.

Baratov, A. N.; Vogman, L. P. & Petrova, L. D. (1969). Explosivity of monosilane-air mixtures. *Combustion, Explosion and Shock Waves*, Vol. 5, No. 4, (September 1969), pp. 592–594, ISSN 0010-5082.

Britton, L. G. (1990). Combustion hazards of silane and its chlorides, *Plant/Operations Progress*, Vol. 9, No. 1, (January 1990), pp. 16-38, ISSN 0278- 4513.

Chang, Y. Y., Peng, D. J., Wu, H. C., Tsaur, C. C., Shen, C. C., Tsai, H. Y., & Chen, J. R. (2007). Revisiting of a silane explosion in a photovoltaic fabrication plant, *Process Safety Progress*, Vol. 26, No. 2, (June 2007), pp. 155-157, ISSN 1066-8527.

Chen, J. R. , Tsai, H. Y., Chen, S. K., Pan, H. R., Hu, S. C., Shen, C. C., Kuan, C. M., Lee, Y. C., & Wu, C. C. (2006). Analysis of a silane explosion in a photovoltaic fabrication plant, *Process Safety Progress*, Vol. 25, No. 3, (September 2006), pp. 237-244, ISSN 1066-8527.

Dryer, F. L., Chaos, M., Zhao, Z., Stein, J. N., Alpert, J. Y. & Homer, C. J. (2007). Spontaneous ignition of pressurized release of hydrogen and natural gas into air. *Combustion Science and Technology*, Vol. 179, No. 4, (April 2007), pp. 663–94, ISSN 0010-2202.

European Industrial Gases Association (2011). Code of Practice: Compressed Fluorine and Mixtures with Inert Gases, European Industrial Gas Association, Globally Harmonized Document, IGC 140/11, 2011

Huang, K. P. P. & Ngai, E. Y. (2006). Silane Safety and Silane Incidents, *Silane Safety and ER Seminar*, Kaohsiung, Taiwan, May 17-19.

Koda, S. (1992). Kinetic aspects of oxidation and combustion of silane and related compounds, *Progress in Energy Combustion Science*, Vol. 18, No. 6, (November 1992), pp. 513-528, ISSN 0360-1285.

Liang, R. L., Wu, S. Y., Tseng, N. J., Ku, C. W., Tsai, H. Y. & Chen, J. R. (2010). Effects of Temperature and Moisture on the Ignition Behavior of Silane Release into Air, *8th International Symposium on Hazards, Prevention, and Mitigation of Industrial Explosions*, Yokohama, Japan, September 6-10 , 2010.

Ngai, E. Y. (2005). Release Testing of Reactive Fluoride Reactor Cleaning Gases, *Semicon China*, Shanghai, China, March 14 & 15, 2005.

Ngai, E. Y. (2011). Silane Emergency Response, *Silane Safety Seminar - Safe Handling and Use of Silane Gas*, Taichung City, Taiwan, January 20, 2011.

Ngai, E. Y., Huang, K. P. P., Chen, J. R., Shen, C. C., Tsai, H. Y., Chen, S. K., Hu, S. C., Yeh, P. H., Liu, C. D., Chang, Y. Y., Peng, D. J., & Wu, H. C. (2007). Field tests of release, ignition and explosion from silane cylinder valves, *Process Safety Progress*, Vol. 26, No. 4, (December 2007), pp. 265-282, ISSN 1066-8527.

Rufin, D. (2011). Distribution Systems Best Practices, *Silane Safety Seminar - Safe Handling and Use of Silane Gas*, Taichung City, Taiwan, January 20, 2011.

Stockman, P. (2009). Going Green with On-Site Generated Fluorine: Sustainable Cleaning Agent for PECVD Processes, *EHS Workshop for Photovoltaic Manufacturers, Intersolar 2009* , Munich, Germany, May 28 2009.

Tsai, H. Y., Wang, S. W., Wu, S. Y., Chen, J. R., Ngai, E. Y. & Huang, K. P. P. (2010). Experimental Studies on the Ignition Behavior of Pure Silane Released into Air, *Journal of Loss Prevention in the Process Industries*, Vol. 23, No. 1 (January 2010), pp. 170-177, ISSN 0950-4230.

Portable LCD Image Quality: Effects of Surround Luminance

Youn Jin Kim
Samsung Electronics Company, Ltd.
Korea

1. Introduction

Liquid crystal display (LCD) has become a major technology in a variety of display application markets from small sized portable displays to large sized televisions. Portable LCD devices such as smart phones and mobile phones are used in a diverse range of viewing conditions. We usually experience images on a mobile phone with a huge loss in contrast under bright outdoor viewing conditions; thus, viewing condition parameters such as surround effects, correlated colour temperature and ambient lighting have become of significant importance. (Katoh et al., 1998; Moroney et al., 2002) Recently, auxiliary attributes determining the mobile imaging were examined and the surround luminance and ambient illumination effects were considered as the first major factor. (Li et al., 2008) Surround and ambient lighting effects on colour appearance modelling have been extensively studied to understand the nature of colour perception under various ambient illumination levels (Liu & Fairchild, 2004, 2007; Choi et al., 2007; Park et al., 2007); thus, this study intends to figure out characteristics of the human visual system (HVS) in *spatial frequency domain* by means of analysing the contrast discrimination ability of HVS. In consequence, we propose an image quality evaluation method and a robust image enhancement filter based on the measured contrast sensitivity data of human observers under various surround luminance levels.

The former is to quantify the observed trend between surround luminance and contrast sensitivity and to propose an image quality evaluation method that is adaptive to both surround luminance and spatial frequency of a given stimulus. The non-linear behaviour of the HVS was taken into account by using contrast sensitivity function (CSF). This model can be defined as the square root integration of multiplication between display modulation transfer function (MTF) and CSF. It is assumed that image quality can be determined by considering the MTF of an imaging system and the CSF of human observers. The CSF term in the original SQRI model (Barten, 1990) is replaced by the surround adaptive CSF quantified in this study and it is divided by the Fourier transform of a given stimulus.

The latter is a robust image enhancement filter which compensates for the effects of surround luminance on our contrast perceiving mechanism. Precisely, the surround luminance adaptive CSF is used as a guide for determination of the adaptive enhancement gain in the proposed algorithm.

2. Measuring and modelling of the surround adaptive CSF

This study examined the effects of surround luminance on shape of spatial luminance CSF and reduction in brightness of uniform neutral patches shown on a computer controlled display screen is also assessed to explain the change of CSF shape. Consequently, a large amount of reduction in contrast sensitivity at middle spatial frequencies can be observed; however, the reduction is relatively small for low spatial frequencies. In general, effect of surround luminance on the CSF appears the same to that of mean luminance. Reduced CSF responses result in less power of the filtered image; therefore, the stimulus should appear dimmer with a higher surround luminance.

2.1 Backgrounds

The CSF represents the amount of minimum contrast at each spatial frequency that is necessary for a visual system to distinguish a sinusoidal grating or Gabor patterns over a range of spatial frequencies from a uniform field. Physiologically, both parvocellular (P) and magnocellular (M) cells have receptive fields organised into two concentric antagonistic regions: a centre (on- or off-) and a surrounding region of opposite sense. This arrangement is common in vertebrates. The receptive fields of small bistratified cells appear to lack clear centre-surround organization. (Dacey & Lee, 1994) The distributions of sensitivity within centre and surround mechanisms are usually represented by Gaussian profiles of a ganglion cell's receptive field. The spatial properties of the visual neurons are commonly inferred from a neuron's spatial modulation transfer function (van Nes & Bouman, 1967) or contrast sensitivity function (Enroth-Cugell & Robson, 1966) measured with grating patterns whose luminance is modulated sinusoidally. In practice, monochromatic patterns in which luminance varies sinusoidally in space are used. CSFs typically plot the reciprocal of the minimum contrast that is also referred to as threshold and provide a measure of the spatial properties of contrast-detecting elements in the visual system. (Campbell and Green, 1965) It is believed that CSF is in fact the envelope of the sensitivity functions for collections of neural channels that subserve the detection and discrimination of spatial patterns. (Braddick et al., 1978; Graham, 1980)

The first measurement of luminance CSF for the human visual system (HVS) was reported by Schade (Schade, 1956) in 1956 and the luminance CSF has been extensively studied over a variety of research fields - such as optics, physiology, psychology, vision and colour science - and the same basic trends were observed. Luminance CSF exhibits a peak in contrast sensitivity at moderate spatial frequencies (~ 5.0 cycles per degree; cpd) (Campbell & Green, 1965) and falls off at both lower and higher frequencies; thus, generally shows band-pass characteristics. The fall-off in contrast sensitivity at higher spatial frequency can be explained by spatial limitations in the retinal mosaic of cone receptors. The reduction in contrast sensitivity at lower spatial frequencies requires further neural explanations. (Westland et al., 2006) Centre-surround receptive fields are one possible reason for this low-frequency fall-off. (Wandell, 1995)

CIE technical committee (TC) 1-60 (Martinez-Uriegas, 2006) has recently collected luminance CSF measurement data from various literatures. (Campbell & Robson, 1968; Watson, 2000; Martinez-Uriegas et al., 1995; Barten, 1999) Those data were measured using different experimental contexts; for instance, Campbell and Robson used Garbor patches and the others used sinusoidal gratings. The all data were normalised to unity at the maximum contrast sensitivity of each data set for a cross-comparison on a single plot. Consequently, they corresponded to one another and their trends are remarkably similar; therefore, they could be accurately fit by a single CSF model (Barten, 1999) in spite of the significant difference in conditions, methods and stimulus parameters.

The CSF model used was originally proposed by Barten as a function of spatial frequency and dependent on a field size (or viewing angle in degree) and mean luminance of the sinusoidal grating stimulus. As the mean luminance of the sinusoidal grating stimulus is decreased, the following variations occur (See Fig. 1). The contrast sensitivity at each spatial frequency decreases, and the maximum resolvable spatial frequency decreases. In addition, the shape of luminance CSF changes; the peaks in the functions shift toward lower spatial frequencies, broaden, and eventually disappear. (Rohaly & Buchsbaum, 1989; Patel, 1966; de Valois et al., 1974)

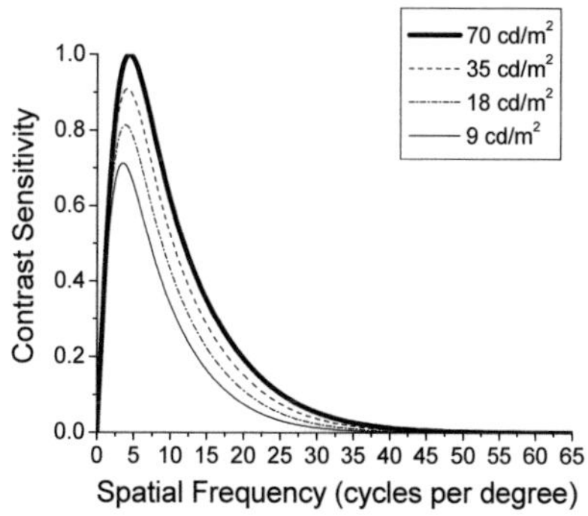

Fig. 1. Predicted CSF by Barten's model with various mean luminance levels for a field size of 5 degrees. As the mean luminance of the sinusoidal grating stimulus is decreased, contrast sensitivity at each spatial frequency decreases, and the maximum resolvable spatial frequency decreases as well. The peaks in the functions shift toward lower spatial frequencies and broaden.

The wealth of data in the literature also reports a variety of changes in CSF shape with senescence, (Owsley et al., 1983; Tulunay-Keesey et al., 1988; Higgins et al., 1988; Rohaly & Owsley, 1993; Pardhan, 2004) eccentricity (Rovamo et al., 1978; Koenderink et al., 1979; Wright & Johnston, 1983; Johnston, 1987; Snodderly et al., 1992) and degree of adaptation to noise (Farchild & Johnson, 2007) in a given stimulus. Briefly, luminance CSFs for older subjects exhibit losses in contrast sensitivity at the higher frequencies, although much of the loss is attributed to optical factors. (Owsley et al., 1983; Burton et al., 1993) Sensitivity to the local contrast at the peripheral region can be measured by instructing the observer to fixate on a marker whilst the actual object is placed at some distance from the marker. The distance is usually expressed in an angular measure called 'eccentricity' and the contrast sensitivity is measured as a function of eccentricity. With increasing eccentricity, capillary coverage increases up to 40%. (Snodderly et al., 1992) Fairchild and Johnson (Farchild & Johnson, 2007) found the fact that the adapted luminance CSF relates to the reciprocal of the adapting stimulus' spatial frequency. However, surround effects on the luminance CSF in spatial frequency domain appears to be less well investigated so far. Cox et al. (Cox et al., 1999) measured the effect of surround luminance on CSF and visual acuity using computer-generated sinusoidal gratings under a surround levels up to 90 cd/m^2 for the purpose of ophthalmic practice in 1999. In consequence, reduced contrast sensitivity was measured under the highest surround luminance (90 cd/m^2) and the optimal surround level was found to be at 10 ~ 30% of mean luminance of a target stimulus. Precisely, contrast sensitivity increases when luminance of the surround increases from 0 to 10 ~ 20% of that of stimulus; however once the surround luminance exceeds the optimal level contrast sensitivity suddenly falls off.

Recently, portable display devices such as mobile phones and portable media players are viewed in a diverse range of surround luminance levels and we usually experience images on a mobile phone display with a huge loss in contrast under bright outdoor viewing conditions. Ambient illumination and surround have been thought of as the first major factor among the mobile environmental considerations (Li et al., 2008); therefore, it is worthy to measure the changes in luminance CSF shape under highly bright surrounds as a simulation of outdoor sunlight. In two psychophysical experiments we examined luminance CSFs under different surround luminance levels are estimated and change in brightness of uniform neutral patches shown on a computer controlled display screen is observed. In specific, Experiment 1 is conducted to measure the compound results of contrast threshold perception and physical contrast of a display resulted from the increase of ambient illumination. The former could be attributed to simultaneous lightness contrast (Palmer, 1993) between stimuli on a display and surround luminance so may cause change in CSF. The latter is usually decreased by the surface light reflections off the front of the monitor screen referred to as viewing flare. In addition, a more psychophysics for variation in brightness is carried out to support and justify the surround effects in Experiment 2.

In this study, MTF of the display used is computed for each surround condition and divides the results from Experiment 1 in order to deduct the display's resolution term as well as effects of viewing flare. Because resolution of the display device used may limit the detectable contrast sensitivity of a human observer, the display factor should be discounted. In an equation form, let $F(u, v)$ represent MTF of a display which comes from the Fourier transformed line spread function (LSF). If the image from the display is filtered by CSF denoted by $H(u, v)$, the Fourier transform of the output $\psi(u, v)$ can be given by (Barten, 1990, 1999)

$$\psi(u,v) = H(u,v)F(u,v) \quad (1)$$

where u and v are spatial frequency variables.

Therefore, CSF $H(u, v)$ can be estimated by deducting MTF $F(u, v)$ in linear system (See eq. (2)). Viewing flare is an additional luminance across the whole tonal levels from black to white and increases the zero frequency response only. More detailed discussions are followed in Results section.

$$H(u,v) = \psi(u,v)/F(u,v) \quad (2)$$

2.2 Methods

2.2.1 Apparatus

A 22.2-inc. Eizo ColorEdge221 liquid crystal display (LCD) was used to present experimental stimuli such as sinusoidal gratings and uniform neutral patches. Spatial resolution of the LCD is 1920 × 1200 pixels and the bit depth was 8 bits per channel. The maximum luminance producible is approximately 140 cd/m^2 in a dark room and the black level elevates up to 1 cd/m^2 due to the inherent leakage light problem of typical LCDs. The display was illuminated by using an EVL lighting colourchanger 250 light source in a diagonal direction. The ambient illuminance levels could be adjusted by changing the distance between the display and light source. Two particular illuminance levels, i.e. 7000 and 32000 lx, were achieved when the distance settings from the display are respectively 270 and 135 cm. The white coloured wall located behind the display was used as surround. In our previous works, (Kim et al., 2007, 2008) illuminance of few real outdoor viewing situations was measured. The lower level (7000 lx) is for simulating 'overcast' and the higher one (32000 lx) for 'bright' outdoor sunlight conditions. Note that the light source illuminates not only the surround region but also the display screen. The physical contrast loss, which can be caused by the light reflection from the screen (See Table 1), is deducted by using MTF of the display for each viewing condition. More details about this viewing flare compensation will be discussed later in Results section.

	Dark	Overcast	Bright
L_{max} (cd/m^2)	140	147	154
L_{min} (cd/m^2)	1	8	15
Viewing Flare (cd/m^2)	0	7	14
Michelson Contrast (Mc)	0.986	0.897	0.828
Relative Mc to Dark	1	0.910	0.840
Surround luminance (cd/m^2)	0	1500	7000

Table 1. Breakdown of each viewing condition

Table 1 provides measured maximum and minimum luminance levels of the display for each viewing condition along with the viewing flare, absolute Michelson contrast (Mc), relative Mc to dark and surround luminance. Viewing flare can be estimated by the additional luminance increase due to the ambient illumination. As surround is changed from dark to overcast to bright, mean of evenly sampled 8 luminance values across the surround wall behind the display increases from 0 to 1500 to 7000 cd/m^2. The amount of viewing flare also increases, so Michelson contrast levels (See eq. (3)) are respectively decreased to 0.897 and 0.822 for overcast and bright as given in Table 1.

$$\textit{Michelson Contrast} = (L_{max} - L_{min}) / (L_{max} + L_{min}) \quad (3)$$

where L is luminance and maxima and minima are taken over the vertical position of the sinusoidal grating stimulus pattern.

The temporal stability of the light source was measured every 20 seconds continuously for 30 minutes from the cold start and the results are depicted in Fig. 2. Crosses are measured data points for overcast and open circles are for bright. The illuminance level became stable after approximately 3 minutes for both cases. The stabilised illuminance values for the two lighting conditions were fluctuating around 6000 ~ 8000 lx for overcast and 31000 ~ 37000 lx for bright and their mean could be found near 7000 and 32000 lx.

2.2.2 Experiment 1: Compound results of contrast threshold perception and physical contrast variation

Experiment 1 is conducted to measure compound results of contrast threshold perception and physical contrast reduction caused by increase of ambient illumination. The former is

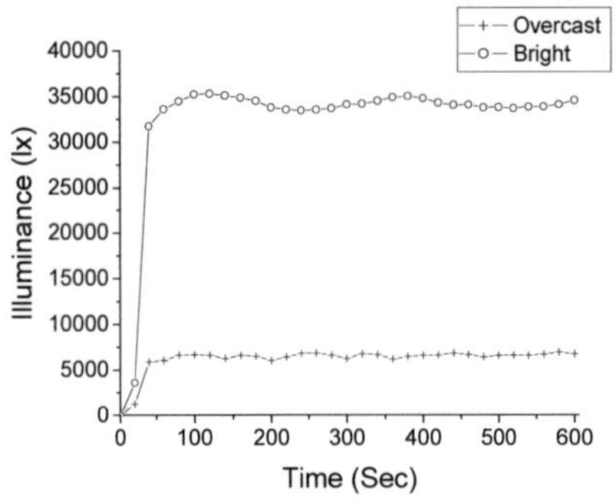

Fig. 2. Temporal illuminance measurement of the simulated outdoor sunlight using EVL lighting colourchanger 250. The temporal stability of the light source was measured every 20 seconds continuously for 30 minutes from the cold start. Crosses are measured data points for overcast and open circles are for bright. The illuminance level became stable after approximately 3 minutes for both cases.

affected by the level of surround luminance and the latter relates to the amount of viewing flare that was provided in Table 1. A sinusoidal grating pattern, of which contrast modulation gradually varies, is displayed on the display. Along the vertical axis of the screen, contrast becomes the highest in the bottom and lowest in the top of the pattern as can be seen in Fig. 3. This sinusoidal grating pattern (**Q**) was produced by means of the product of a non-linear gradient function along the vertical axis (**M**) and a one-dimensional sinusoidal function of spatial frequency across the horizontal axis (**F**). Practically, those functions can be discretely sampled and expressed by

$$\mathbf{Q} = \mathbf{M}\mathbf{F}^{T} \tag{4}$$

where $\mathbf{F}^T$ denotes transpose of **F**.

The compound effects of contrast threshold perception and physical contrast were measured at 11 spatial frequencies: 1, 2, 3, 4, 5, 6, 7, 13, 23, 32 and 65 cpd. The first 7 spatial frequencies (1 to 7 cpd) are sampled at the low spatial frequency area with steps of 1 cpd in order to accurately measure the peak sensitivity and the sharp fall-off of CSF. Two middle spatial frequencies, 13 and 23 cpd, where the gradual fall-off after the peak can be observed, are also selected. The highest spatial frequency sampled in this study is 65 cpd for predicting the maximum resolvable frequency.

Fig. 3. Example of sinusoidal grating stimulus. Along the vertical axis of the screen, contrast becomes the highest in the bottom and lowest in the top of the pattern.

In total, 6 observers (4 females and 2 males) participated in Experiment 1 and their ages ranged from 26 to 38. They were required to identify vertical positions of the sinusoidal pattern, by double-clicking a wireless mouse, when the contrast becomes just indistinguishable. This experimental technique emulates a method suggested by (Kitaguchi & MacDonald, 2006). We implemented a software using Microsoft foundation class in Visual C++ 6.0 to display sinusoidal patterns, to read the coordinates of double-clicked vertical position by the observer and to calculate the contrast. Technically, contrast can be defined as Michelson contrast (See eq. (3)) and it is usually converted into sensitivity unit that is the reciprocal of contrast threshold as given in eq. (5).

$$\text{Sensitivity} = 1 \,/\, \text{Threshold} \tag{5}$$

Each sinusoidal pattern is displayed on the LCD monitor in a random order. Under the dark surround condition, the procedure was repeated for 5 times and the results were averaged to obtain contrast threshold values. The same procedure was also applied for the other viewing conditions: overcast and bright. The sequence of these psychophysical sessions for the viewing conditions was also randomly decided for each observer. In order to assure maximum observer adaptation to the viewing condition including the LCD monitor white point and ambient illumination level, observers were given 30 seconds adaptation period (Fairchild & Reniff, 1995) prior to each session. Precisely, observers were instructed to stare at a full white patch displayed on the LCD monitor screen under a certain ambient illumination. The distance between an observer and LCD was set to be 3 m in order to minimize the quantization error of the 8-bit display used. The total number of psychophysical assessments collected for data analysis was 990 (11 stimuli × 5 repeats × 6 observers × 3 viewing conditions).

2.2.3 Experiment 2: Magnitude estimation of brightness

Experiment 2 aims to measure the change in brightness (Blakeslee et al., 2008) of a series of neutral colours shown on an LCD under varied ambient illumination levels and to find out whether the brightness change can affect the contrast threshold perception of human observers. The brightness / lightness distinction may not always be clear to subjects. (Arend & Spehar, 1993a, 1993b; Rudd & Popa, 2007) Lightness means perceived reflectance as a surface property, while brightness is even more ambiguously defined as the perceived luminance of a light source or subjective correlate of luminance. (Rudd & Popa, 2007) We decided to adopt brightness because the all test stimuli used are shown on a monitor rather than reflective colours.

Nine neutral patches were uniformly sampled across a 8-bit RGB scale from 0 to 255 with steps of 51 and each of the neutral colours was displayed at a time on the whole LCD screen. Five observers (2 females and 3 males) participated in this experiment in total and their ages ranged from 26 to 38. The apparent brightness of a full white patch displayed on the LCD monitor screen of which the RGB values are (255, 255, 255) was assigned as an arbitrary brightness magnitude value of 100. Prior to the brightness estimations, observers were required to memorize the white patch on the monitor in a dark room and judge a brightness ratio of each of the rest of test neutral colours at a time not only under dark but also under the other two ambient illumination conditions: overcast and bright. Observers were given the following written-instruction. *"Please estimate the level of perceived brightness according to the reference patch of which its perceived brightness is assigned as 100."* Each observer repeated all judgments five times in a random order and its mean opinion score (MOS) (ITU-R, 2002) was collected for data analysis.

The sequence of the experiment for those ambient illumination conditions was also randomly decided for each observer and a mid-gray of which RGB value is (128, 128, 128) was shown to the observer as a transient patch whilst changing stimulus. The transient patch is usually displayed to prevent from any illusions while the scene is changed. In the field of image quality, this illusion artefact is referred to as image sticking. (Lee et al., 2009) In order to assure maximum observer adaptation to the viewing condition including the LCD monitor white point and ambient illumination level, observers were given the 30 seconds adaptation period (Fairchild & Reniff, 1995) prior to each session. They were allowed to look back into the reference white patch under dark viewing condition but re-adaptations were performed when viewing condition is altered. The total number of psychophysical assessments used for data analysis was 675 (9 stimuli × 5 repeats × 5 observers × 3 viewing conditions).

2.2.4 Experiment 3: Statistical analysis for observer variation

Variation between observers was evaluated in terms of three test methods: ITU-R BT 500-10, a modified version of coefficient of variance (CV) (Luo et al., 1991) and the Pearson correlation. First, ITU-R BT 500-10 method (ITU-R, 2002) rejects observations which are statistically incoherent with the other observers and show unusual peakedness of the probability distribution of a real-valued random variable. It should be ascertained whether the distribution of an observer's data is normal, using the Kurtosis test. Second, CV is often used as a measure of the 'observer accuracy' which represents the mean discrepancy of a set of psychophysical data obtained from a panel of observers from their mean value. This term has been widely used in colour appearance and difference studies (Luo et al., 2001, 2006) and usage of it was also verified in image quality studies. (Kim et al., 2008, 2010a, 2010b) The original CV is a normalised measure of dispersion for a repeated measurement but was applied to measure the degree to which a set of data points varies in this study. The CV is normally displayed as percentage and, for a perfect agreement between them, equals to 0. Third, Pearson correlation reflects the degree of linearity in the relationship between a pair of variables (e.g. x and y). It is defined to be the sum of the products of the standard score of the two variables divided by the degree of freedom. When the variables are perfectly linearly related, their Pearson correlation is +1.

2.3 Results

2.3.1 Observer variation

Performance of the observers participated in Experiments 1 and 2 was evaluated using the three statistical test methods previously introduced and so obtained results are summarised in Table 2. Basically, the all observations can be accepted by ITU-R BT 500-10 method and CV values ranged from 19 to 25 in Experiment 1 which can be within the acceptable level for observer accuracy. (Kim et al., 2008, 2010a, 2010b; Luo et al., 2006) Even lower CV values were measured in Experiment 2 (13 ~ 18) because of the simplicity of magnitude estimation technique. Pearson correlations for the all assessments are larger than 0.98 meaning the strong linearity between the mean and each observation. Especially for Experiment 2, since brightness estimates are known for their subject variability, the individual data are also illustrated along with their mean for each viewing condition in later section.

Experiment	Method	Dark	Overcast	Bright
1	ITU-R BT 500-10	All passed	All passed	All passed
	CV	20	19	25
	r	0.983	0.996	0.994
2	ITU-R BT 500-10	All passed	All passed	All passed
	CV	13	18	17
	r	0.993	0.994	0.991

Table 2. Observer variation test results

2.3.2 Compound results of contrast threshold perception and physical contrast

In Experiment 1, the compound results of contrast threshold perception and physical contrast loss were achieved. They resulted from the increase of ambient illumination level causing both increase of surround luminance and viewing flare. The measured data were converted into the sensitivity unit using eq. (5), which is denoted as *psi* (ψ), in eqs. (1) through (2). Figure 4 depicts those ψ data for the three viewing conditions. Every data point was normalised to the unity at the maximum value obtained in dark (288) and adjacent data are linearly connected. Consequently, as the viewing condition changes from dark to overcast to bright, the data moved toward zero in general. The shape of the all three plots appears typical band-pass and the spatial frequency where the maximum contrast sensitivity occurred was moved toward a lower frequency, i.e. from 5 to 4 cpd. The compound effects of surround luminance and viewing flare on the contrast threshold perception and physical contrast loss seem to be similar to that of mean luminance as previously reported by the wealth of data in the literature as discussed in Introduction section. Error bars represent standard errors that can be defined as standard deviation divided by square root of number of observations.

2.3.3 Deriving the display MTF

It is often assumed that the point spread function (PSF) of a majority of commercial LCD monitors is a rectangle function, *rect*(*x*), (Barten, 1991; Sun & Fairchild, 2004) because the shape of a single pixel in LCDs is rectangular as illustrated in Fig. 5 (a). The rectangle function can be defined as

$$rect(x) = 1 \qquad \text{if} \qquad |x| \leq n/2$$

otherwise

$$rect(x) = 0 \tag{6}$$

where n is the size of a pixel of an LCD in visual angle. (Sun & Fairchild, 2004)
Magnitude of the Fourier transform of the rectangle function can be expressed as shown in eq. (7).

$$\begin{aligned} MTF(u) &= \left|\Im\left[rect(x)\right]\right| \\ &= \left|\frac{\sin \pi n u}{\pi u}\right| \end{aligned} \tag{7}$$

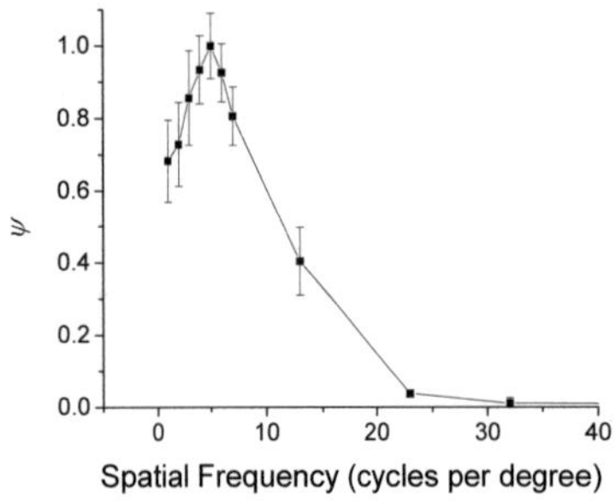

(a) Dark

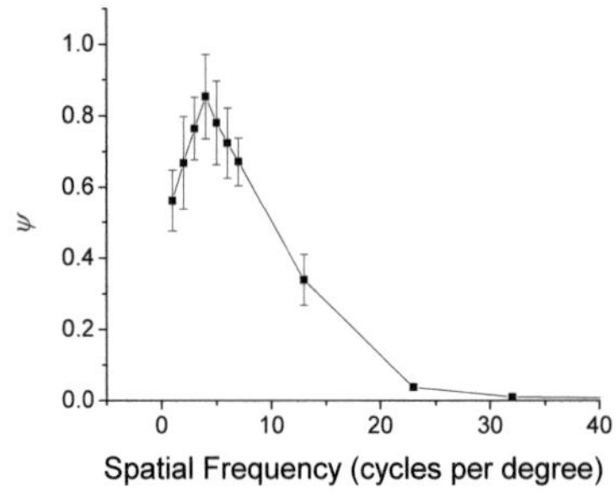

(b) Overcast

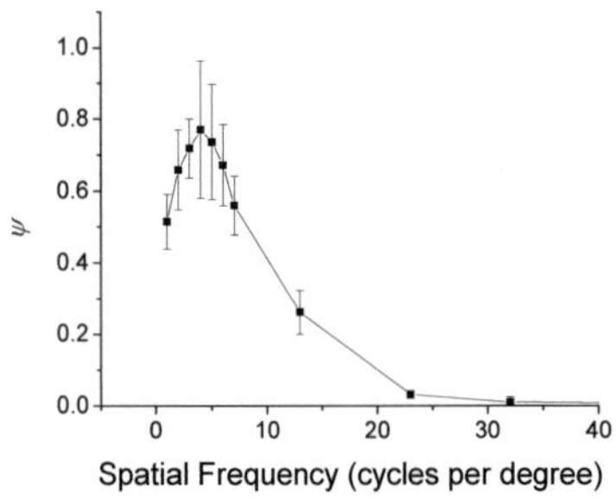

(c) Bright

Fig. 4. Measured data from Experiment 1 under 3 different ambient illumination conditions with linear interpolation for (a) dark (b) overcast and (c) bright. As the viewing condition changes from dark to overcast to bright, the data moved toward zero in general. The shape of the plots appears typical band-pass and the spatial frequency where the maximum contrast sensitivity occurred was moved toward a lower frequency, i.e. from 5 to 4 cpd.

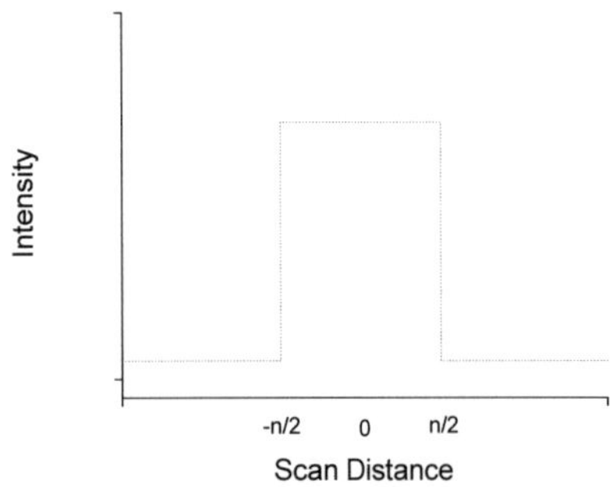

(a)

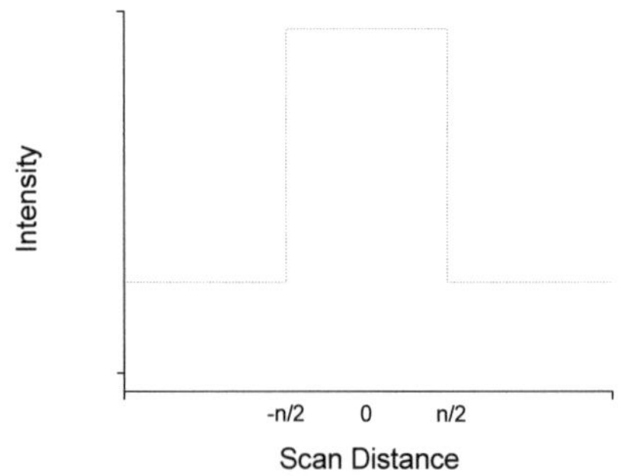

(b)

Fig. 5. (a) The original PSF and (b) viewing flare added PSF. Viewing flare is an additional luminance across the whole tonal levels from black to white and increases the zero frequency response only.

Then eq. (7) is divided by n, because $MTF(u)$ should equal to 1, so sinc function can be used as the MTF of LCDs.

$$MTF(u) = \left|\frac{\sin \pi n u}{\pi n u}\right| = \left|\operatorname{sinc}(\pi n u)\right| \quad (8)$$

where $\Im[\cdot]$ denotes the Fourier transform of the argument.

Viewing flare can be defined as the additional luminance due to surface reflections off the front of a display caused by ambient illumination. It boosts the PSF by a constant offset level as illustrated in Fig. 5 (b); thus, the zero frequency response (or *dc* component) is increased only and other frequency responses remain the same if the signal is transformed into Fourier domain. When the MTF is normalised at the maximum, $MTF(0) = 1$ and $MTF(u>0)$ is multiplied by a weighting factor a for $u > 0$ as shown in eq. (9).

$$MTF_i(u) = \alpha MTF_0(u) = \alpha\left|\operatorname{sinc}(\pi n u)\right| \quad (9)$$

where i represents the amount of viewing flare. For instance of this, MTF_0 shows the MTF for dark viewing condition so MTF_i is the MTF for a viewing condition where the amount of viewing flare is i cd/m². The weighting factor a refers to the ratio of zero frequency response between $MTF_0(u)$ and $MTF_i(u)$ as given in Eq. (10). Practically, mean value of the PSF can be simply used instead of calculating zero frequency response of the MTF in Fourier domain therefore a values should be identical to the relative Michelson contrast to the dark viewing condition as can be expected (See Table 1).

$$\alpha = \frac{MTF_0(0)}{MTF_i(0)} = \frac{(L_{Max,0} + L_{Min,0})/2}{(L_{Max,i} + L_{Min,i})/2} \quad (10)$$

The estimated MTF of the LCD monitor used in this study is presented in Fig. 6 (See the solid line). Single-pixel size of the LCD is set to be 0.00474° in visual angle unit. The estimated MTFs for the higher illumination levels are shown in Fig. 6 as well represented by dashed and dotted lines.

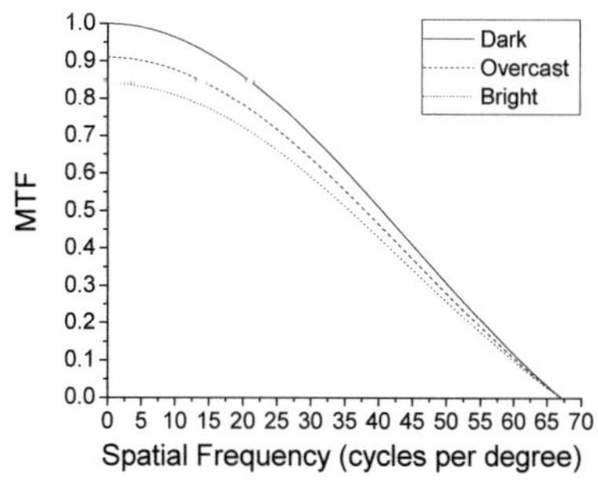

Fig. 6. MTF of the LCD used in this study and the approximated MTFs under two different levels of viewing flare. Single-pixel size of the LCD is set to be 0.00474° in visual angle unit. The compensation factors (a) for viewing flare for the three viewing conditions are listed in Table 3.

	Dark	Overcast	Bright
φ	1	0.534	0.191

Table 3. The surround luminance effect function (φ)

2.3.4 Estimating CSF by compensating for MTF

As given in eqs. (1) through 2 in Introduction section, CSFs for the three viewing conditions can be estimated by dividing ψ measured in Experiment 1 by the corresponding MTFs as illustrated in Fig. 7. Data points for dark are linearly interpolated and represented by solid lines and dashed lines for overcast and dotted lines for bright. As can be seen, they show band-pass characteristics and the peak contrast sensitivity for dark is observed at 5 cpd but it moves to 4 cpd for overcast and bright. The peak-shift appears more obvious compared to Fig. 4. However, it is not quite easy to yield significance of the shift on the sampling frequency of 1 cpd. A large amount of reduction in contrast sensitivity at middle frequency area ($4 < u < 13$) can be observed; however, little reduction in contrast sensitivity is found for lower frequencies ($u < 4$). Because the MTF converges to zero at near the maximum spatial frequency we sampled (68 cpd) so contrast sensitivity at 65 cpd is not investigated in the current section due to the limited display resolution.

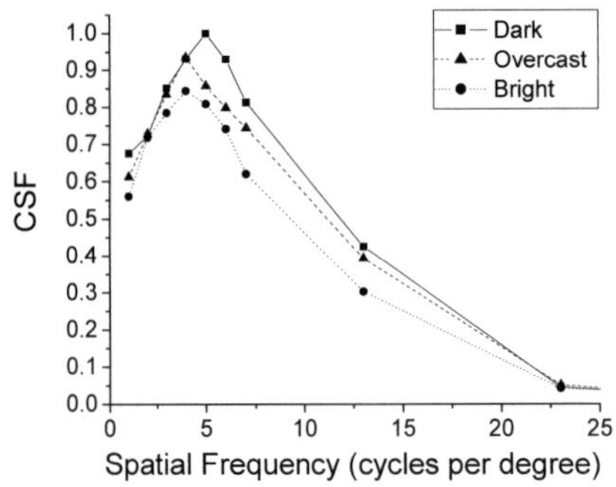

Fig. 7. Estimated CSF data points under 3 different surround luminance levels with linear interpolation. The all three plots show band-pass characteristics and the peak spatial frequency for dark is 5 cpd but moves to 4 cpd for overcast and bright. A large amount of reduction in contrast sensitivity at middle frequency area ($4 < u < 13$) can be observed; however, little reduction in contrast sensitivity is found for lower frequencies ($u < 4$).

Figure 8 illustrates the ratio of the area covered by the three linearly interpolated plots previously shown in Fig. 7. The area of a function or a filter correlates to the power of a filtered image. Area of each plot is normalised at the magnitude of the area for dark viewing condition. As can be seen, about 7 and 15 % of the loss in power was occurred under overcast and bright, respectively due to the increase of surround luminance. The amount of power loss caused by the reduction in contrast sensitivity can be analogous to that of Michelson contrast reduction. As given in Table 1, Michelson contrast decrease reaches up to approximately 10 and 18 % respectively for overcast and bright. It yields to the fact that the amount of physical contrast reduction is larger than that of power loss in CSF. In order

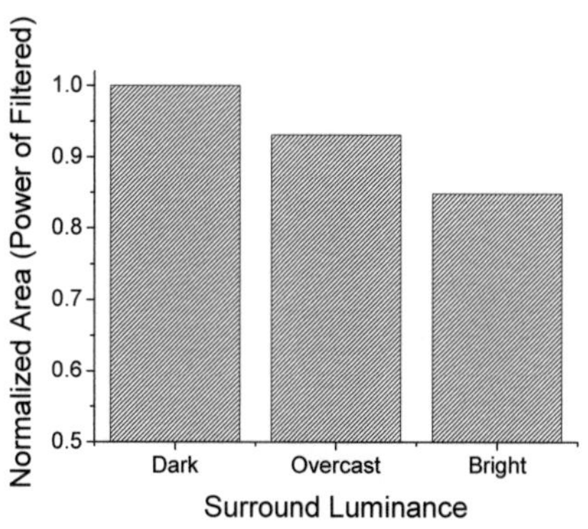

Fig. 8. Ratio of area of *psi* functions given in Figs. 4 (a) through (c). The area of a function or a filter correlates to the power of a filtered image. As can be seen, about 15 and 23% of the loss in power was occurred under overcast and bright, respectively due to the increase of ambient illumination.

to statistically verify the surround luminance and spatial frequency effects on the shape in CSF, two-way analysis of variance (ANOVA) was performed with surround luminance and spatial frequency as independent variables and contrast sensitivity as the dependent variable. Significant effects could be found for both surround luminance and spatial frequency. Their P values were less than 0.0001. A value of $P < 0.05$ was considered to be statistically significant in this study.

Generally, effect of surround luminance on the luminance CSF appears the same to that of mean luminance as previously discussed in Fig. 1. Because CSF response correlates to the filtered light in the ocular media, smaller CSF responses across the spatial frequency domain result in less power of the filtered image; thus, less amount of light can be perceived by the visual system. Therefore, the stimulus should appear darker under a higher surround luminance which can be verified through another set of experiments. The subsequent section discusses the results from Experiment 2.

2.3.5 Change in brightness caused by surround luminance

The mean perceived brightness magnitudes of the nine neutral colours for the 5 observers are drawn in Fig. 9. The abscissa shows measured luminance of the neutral patches shown on an LCD. The ordinate represents their corresponding perceived brightness magnitudes. The filled circles indicate dark, empty circles for overcast and crosses for bright. Data points are linearly interpolated. As can be seen, the all data points for overcast and bright are underneath data points for dark which means that their perceived brightness is decreased in general, as the ambient illumination and surround luminance increase in spite of the additional luminance increase by viewing flare. Similar results of brightness reduction between the surround and focal area can also be found in other works. (Wallach, 1948; Heinemann, 1955) Since brightness estimates are known for their subject variability, the individual data are also illustrated along with their mean for each viewing condition in Fig. 10. Filled circles show mean of the 5 observers and error bars show 95% confidential interval. As the all observations were accepted by the three observer variability tests in Table 2, the all brightness estimates follow the same trends. No particular outliers can be observed.

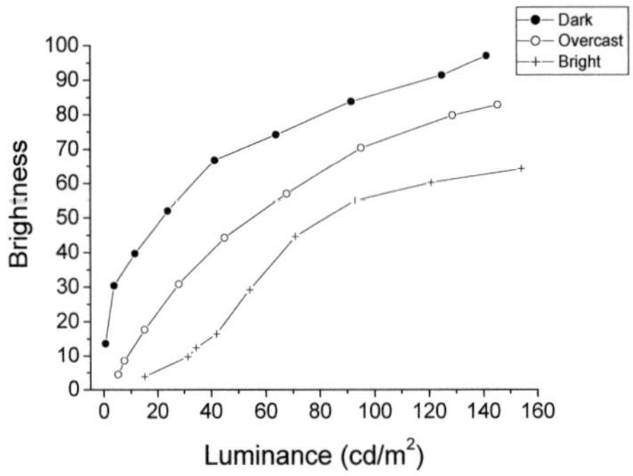

Fig. 9. Luminance vs. brightness under varied ambient illumination levels. The all data points for overcast and bright are underneath data points for dark which means that their perceived brightness is decreased in general, as the ambient illumination and surround luminance increase in spite of the additional luminance increase by viewing flare.

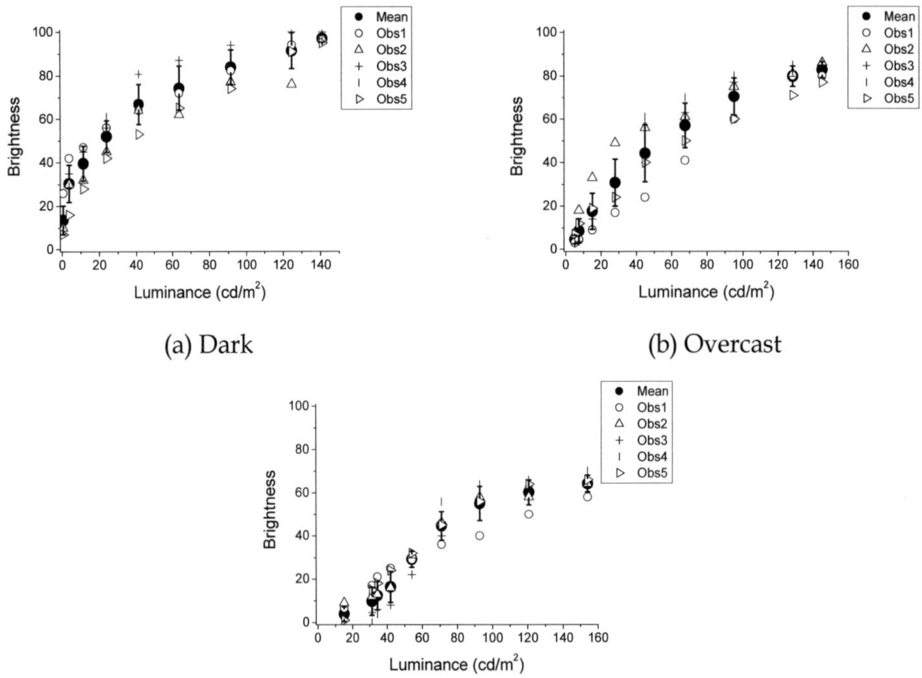

(a) Dark

(b) Overcast

(c) Bright

Fig. 10. Individual brightness estimates for (a) dark (b) overcast and (c) bright. Brightness estimates are known for their subject variability but the all brightness estimates follow the same trends. No particular outliers can be observed. Error bars show standard error.

The precise relation between perceived brightness and stimulus luminance has been extensively studied using reflective colour samples. Traditionally, there are two most frequently cited explanations. (Jameson & Hurvich, 1961) One of them is called *law of retinal stimulus*. It is intuitively expected that, if the amount of light falling on a given stimulus is increased, the intensity of the retinal light image could be increased and the HVS could perceive its increased brightness. All of the stimuli should appear lighter with the aid of increased luminance from ambient illumination. The other most frequently cited explanation for the relation between perceived brightness and stimulus luminance is *law of brightness constancy* (Wallach, 1948; Woodworth & Schlosberg, 1954; Jameson & Hurvich, 1959, 1961). This phenomenon is based on neural processing after light rays pass through ocular media in the HVS. There are some examples that apparent brightness of visually perceived objects is relatively constant in real world: white snow always appears bright but black coal looks very dark regardless a range of illuminance. Specifically, although the coal in the high illumination may actually reflect more intensity of light to the eye than does the snow at the low illumination. According to this theory, the relative brightness between with and without ambient illumination should be constant. However, our experimental results showed reduction in perceived brightness under ambient illumination and neither of the two traditional phenomena could predict this situation. One of the possible reasons for this is that the lighter surround makes the focal area appears darker and this phenomenon is referred to as simultaneous lightness contrast. (Palmer, 1999) The neural contrast mechanism that makes the low-luminance areas appear darker in bright environments more than compensates for the reduced physical contrast caused by intraocular scatter. (Stiehl et al., 1983; Wetheimer & Liang, 1995)

2.4 Summary

This section examined the variation in shape of spatial luminance CSF under different surround luminance levels and reduction in brightness of uniform neutral patches shown on a computer controlled display screen is also assessed to explain change of CSF shape. In specific, Experiment 1 was conducted to measure the compound results of contrast threshold perception and physical contrast decrease of a display resulted from the increase of ambient illumination. The former is found to be attributed by simultaneous lightness contrast (Palmer, 1999) between stimuli on a display and surround luminance so yields to cause the change in CSF shape. The latter is usually decreased by the surface light reflections off the front of the monitor screen referred to as viewing flare. Through a set of brightness magnitude estimations in Experiment 2 the surround luminance effects on the CSF and brightness reduction assumption could be justified. The viewing flare and display terms were successfully deducted by using MTF. Consequently, a large amount of reduction in contrast sensitivity at middle frequency area ($4 < u < 13$) can be observed; however, little reduction in contrast sensitivity is found for lower frequencies ($u < 4$). They show band-pass characteristics and the spatial frequency where the maximum contrast sensitivity occurs moves from 5 to 4 cpd when surround luminance increases from dark to overcast to bright. However, it is not quite easy to yield significance of the shift on the sampling frequency of 1 cpd. Generally, effect of surround luminance on the luminance CSF appears the same to that of mean luminance. Because CSF response can correlate to the filtered light in the ocular media, smaller CSF responses across the spatial frequency domain result in less power of the filtered image; thus, less amount of light can be perceived by the visual system. Therefore, the stimulus should appear dimmer under a higher surround luminance. The power loss in CSF reaches up to 7 and 15 % respectively for overcast and bright. Analogously, the Michelson contrast decrease was 10 and 18 % for overcast and bright. It yields to the fact that the amount of physical contrast reduction

is larger than that of power loss in CSF. The statistical significance of the surround luminance and spatial frequency effects on the shape in CSF, two-way ANOVA was performed and significant effects could be found for both parameters.
The results, which can be obtained from Experiments 1 and 2, are applicable to various purposes. Since CSFs have been widely used for evaluating image quality by predicting the perceptible differences between a pair of images (Barten, 1990; Daly, 1993; Zhang & Wandell, 1996; Wang & Bovik, 1996) surround luminance effects on CSF can be very useful for this application. Furthermore, the results can also be applied to simulate the appearance of a scene (Peli, 1996, 2001) and evaluate the visual performance of the eye. (Yoon & Williams, 2002)

3. Evaluating image quality

This section intends to quantify the effects of the surround luminance and noise of a given stimulus on the shape of spatial luminance CSF and to propose an adaptive image quality evaluation method. The proposed method extends a model called square-root integral (SQRI). The non-linear behaviour of the human visual system was taken into account by using CSF. This model can be defined as the square root integration of multiplication between display modulation transfer function and CSF. The CSF term in the original SQRI was replaced by the surround adaptive CSF quantified in this study and it is divided by the Fourier transform of a given stimulus for compensating for the noise adaptation.

3.1 Backgrounds

3.1.1 Adaptation to spatial frequency of the stimulus

On spatial frequency adaptation, (Fairchild & Johnson, 2007) proposed adjusting two-dimensional CSF based on the degree of a given image's blurness. (Goldstein, 2007) demonstrates spatial frequency adaptation effect as shown in Fig. 11. The left pair consists of patterns having different spatial frequency. Spatial frequency of the upper pattern shows lower than that of the lower pattern. However, the other pair on the right-handed side has two patterns showing the identical spatial frequency. After staring at the bar on the left pair of patterns for a while, the other pair on the right handed side appear to shift in spatial frequency in directions opposite the adapting stimuli (the left pair).
More precisely, a half of the foveal area of the viewer is adapted to the lower frequency of the upper pattern, while the other half of the foveal area is adapted to the higher frequency of the lower pattern. After adapting to the spatial frequency of those stimuli, although the two identical patterns were assessed, the upper right and lower right patterns should appear to show higher and lower spatial frequencies, respectively. Consequently, the adapted contrast sensitivity of the HVS can be related to the reciprocal of the adapting stimulus' spatial frequency as given by (Fairchild & Johnson, 2007)

$$CSF_a(u) = \frac{CSF(u)}{img(u)+1} \tag{11}$$

where $img(u)$ is Fourier transform of a given image.

3.1.2 Square-root integral

The SQRI method (Barten, 2006) can be defined as the square root integration of multiplication between display MTF, i.e., $MTF(u)$ and CSF, the reciprocal of contrast threshold function $M_t(u)$ as

$$SQRI = \frac{1}{\ln 2}\int_0^{u_{\max}} \sqrt{\frac{MTF(u)}{M_t(u)}}\frac{du}{u} \tag{12}$$

where u_{max} is the maximum spatial frequency to be displayed.

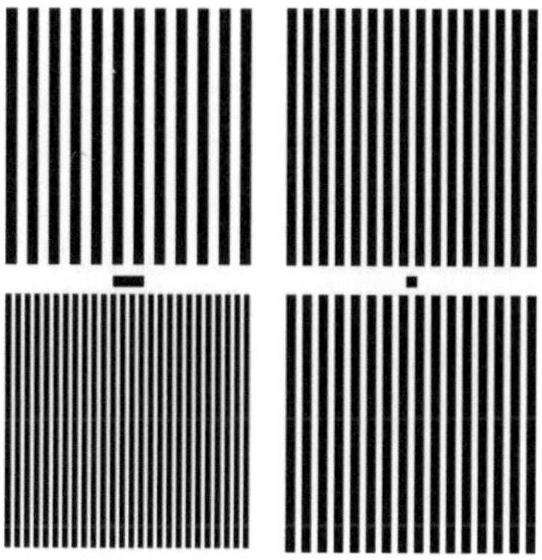

Fig. 11. Demonstration of spatial frequency adaptation

3.2 Modelling the effects of surround luminance

The surround luminance effects on CSF are quantified in this section. In order to compensate for the effects, a weighting function φ was multiplied to the adapting luminance that is denoted as L in (Barten, 1990). Precisely, as previously mentioned in Background section, brightness of a stimulus can be affected by surround luminance increase so a function φ should be multiplied to L. For each surround, the following optimisation process was carried out.

Step 1. A CSF curve is predicted using Barten's model under a given surround condition. The adapting luminance can be obtained by measuring the mean luminance between black and white patches of the display.

Step 2. The predicted CSF curve is adjusted by changing the value of φ so that its maximum contrast sensitivity value can match that of the measured CSF data in (Kim & Kim, 2010) under the given surround condition. Note: in case the surround is dark, φ should equal to one.

Consequently, the maximum contrast sensitivity value of the adjusted CSF curve for overcast could match that of the measured CSF data points when φ equals to 0.534. In the case of bright, φ is found to be 0.339. Table 3 lists the obtained optimum φ values for the three surrounds along with their measured surround luminance levels. The relation between φ against the corresponding surround luminance (L_S) can be modelled by an exponential decay fit as given in eq. (13) and also illustrated in Fig. 12. Its exponential decaying shape appears similar to that of the image colour-quality model (Kim et al., 2007) that predicts the overall colour-quality of an image under various outdoor surround conditions. In addition, the change in "clearness," which is one of the psychophysical image quality attributes, caused by the illumination increase could also be modelled by an exponential decay function as well. (Kim et al., 2008)

$$\varphi = 0.17 + 0.83e^{-10^{-4}\cdot L_S/0.18} \tag{13}$$

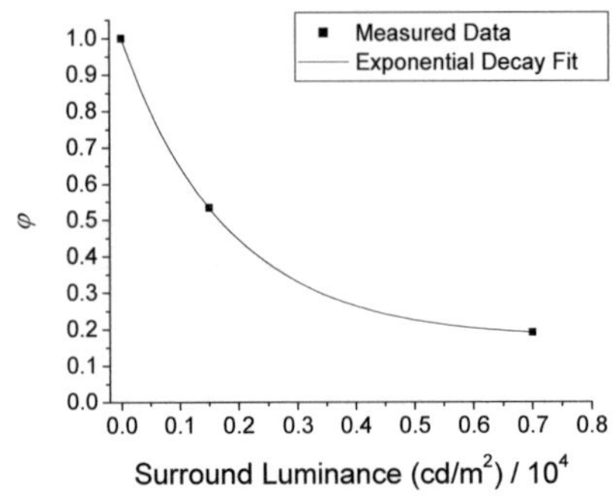

Fig. 12. Relation between the surround luminance factor (φ) and the normalised surround luminance ($L_S / 10^4$)

3.3 Proposed method: Adaptive SQRI

The proposed method - adaptive SQRI ($SQRI_a$) - can be expressed as eq. (14). The $M_t(u)$ in the original SQRI (see eq. (12)) is replaced by $M_{ta}(u)$ which represents the inverse of the adaptive CSF denoted as $CSF_a(u)$.

$$SQRI_a = \frac{1}{\ln 2}\int_0^{u_{\max}} \sqrt{\frac{MTF(u)}{M_{ta}(u)}}\frac{du}{u} \tag{14}$$

where u denotes the spatial frequency and $1/M_{ta}(u)$ is

$$\frac{1}{M_{ta}(u)} = CSF_a(u) = \frac{au\exp(-bu)\sqrt{(1+c\exp(bu))}}{(k \times img(u)+1)}$$

The numerator of CSF_a shows the surround luminance adaptive CSF; a, b, and c are

$$a = \frac{540\left(1+0.7/\varphi L\right)^{-0.2}}{1+\dfrac{12}{w\left(2+u/3\right)^2}}$$

$$b = 0.3\left(1+100/\varphi L\right)^{0.15}$$

$$c = 0.06$$

where the adapting luminance L is the mean luminance between white and black on the display under a given surround luminance and φ is a weighting function for the surround luminance effect as previously given in eq. (13).

As (Fairchild & Johnson, 2007) found the reciprocal relation between the adapted contrast sensitivity of the HVS and the adapting stimulus' spatial frequency, as shown in eq. (11), CSF_a is divided by Fourier transform of the given image. The denominator of the CSF_a shows amplitude of the Fourier transformed image information, $img(u)$. A constant k is multiplied to the magnitude of img(u) for normalisation as

$$k = 10^4 \times \frac{1}{\max(|img(u)|)} \tag{15}$$

Since the denominator of $SQRI_a$ is Fourier transform of a given image, the model prediction can be proportional to the inverse of the image's spatial frequency. In order to attenuate any unwanted spatial frequency dependency of the image, the model prediction should be normalised by that of a certain degraded image expressed as

$$nSQRI_a = \frac{SQRI_a(Original)}{SQRI_a(Degraded)} \tag{16}$$

where $nSQRI_a$ denotes a normalised $SQRI_a$ prediction and $SQRI_a$ (*Original*) and $SQRI_a$ (*Degraded*) respectively represent $SQRI_a$ predictions for a given original image and its degraded version.

The degraded image can be defined as an image of which its pixel resolution is manipulated to a considerably lower level, i.e., 80 pixels per inc. (ppi), while the original resolution was 200 ppi., and luminance of each pixel is reduced to 25 % of its original. The normalisation method makes $SQRI_a$ to predict the quality score of a given image regardless the level of adapting spatial frequency. Since the overall dynamic range of $nSQRI_a$ in eq. (16) may be changed due to the normalisation process, it was re-scaled to a 9-category subjective scale (Sun & Fairchild, 2004) using a least-square method for each surround luminance condition. The rescaling process can be written as

$$J' = pJ + q \tag{17}$$

where J' represents a re-scaled 9-category value of J, i.e., $nSQRI_a$ of an image. The scaling factors are denoted as p (slope) and q (offset) and the optimum scaling factors can be determined through the subsequently discussed psychophysical test.

3.4 Subjective experimental setup

In total, five test images were selected for image quality evaluation in this study. They contained sky, grass, water, facial skin (Caucasian, Black, and Oriental) and fruit scenes, as shown in Fig 13. Those images were displayed on a 22.2-inc. Eizo ColorEdge221 LCD. The maximum luminance producible is approximately 140 cd/m^2 in a dark room and the black level elevates up to 1 cd/m^2 due to the inherent leakage light problem of typical LCDs. The display was illuminated by using an EVL Lighting Colourchanger 250 light source in a diagonal direction. More details about the experimental setting are described in the previous section. The surround luminance and the viewing conditions are summarised in Table 1.

Each image was manipulated in terms of the three attributes, blurrness, brightness and noisiness. For adjusting those attributes, resolution, luminance and noise level of the images were controlled. Specifically, the five images were manipulated by changing their resolution from 200 (original) to 80 ppi with steps of 40 ppi (original + 3 resolution degradations), luminance from 100 (original) to 25% with steps of 25% (original + 3 luminance reductions) and adding the Gaussian noise by changing the variance of the Gaussian function from 0 (original) to 0.006 with steps of 0.002 (original + 3 noise additions).

In total, for each test image, 64 images (4 resolution × 4 luminance × 4 noise) were produced by the image rendition when simultaneous variations are included. However, the

(a) (b) (c) (d) (e)

Fig. 13. Test images (a) Skytower, (b) Picnic, (c) Grass, (d) Ladies, and (e) Fruits

combinations between lower levels of the rendition-parameters resulted in considerably low quality images, which can be rarely seen in real world so were excluded. Figure 14 shows the sampled 22 images out of 64 in an image rendering cube. Each axis represents each of the three rendered parameters: resolution, luminance and noise. The coordinates (0, 0, 0) is the original image and larger numbers represent lower levels of each parameter.

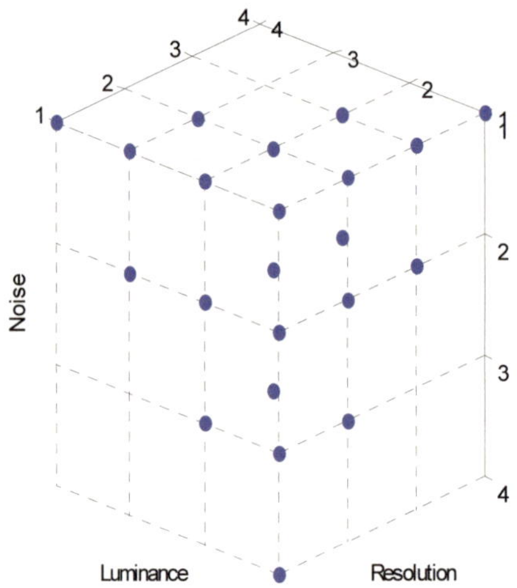

Fig. 14. Sampled images

Among 110 images for 5 distinct test images, only 35 images were randomly selected and used. Those selected images are listed in Table 4, where FR is for 'Fruits', GR for 'Grass', LD for 'Ladies', PC for 'Picnic', SK for 'Skytower'. The four rendition levels for each of the three image parameters (Resolution; R, Luminance; L and Noise; N) are indicated as numbers from 0 to 3, where 0 is the original. The images were processed by the proposed algorithm for the three different surround levels: dark, overcast and bright. A panel of 9 observers with normal colour vision (5 females and 4 males; 26~38 years old) were asked to judge the quality of the rendered images on the mobile LCD from the distance of 25 centimetres (accommodation limit), using a 9-point scale (1 to 9). This subjective image quality judgment procedure was repeated under the three different surround conditions. Therefore, the total number of psychophysical assessments can be 845 (35 images × 9 observers × 3 viewing conditions). The collected subjective data were averaged for each image. This is a ITU-R BT.500-11 method for analysing the category judgment data. (ITU-R, 2002)

FR				GR				LD				PC				SK			
	R	L	N		R	L	N		R	L	N		R	L	N		R	L	N
FR1	0	0	3	GR1	0	0	2	LD1	0	1	0	PC1	0	0	0	SK1	0	0	0
FR2	0	1	0	GR2	0	0	3	LD2	0	1	1	PC2	0	1	1	SK2	0	0	2
FR3	0	1	2	GR3	0	1	1	LD3	0	2	1	PC3	0	2	1	SK3	0	0	3
FR4	0	2	0	GR4	0	3	0	LD4	1	0	1	PC4	1	0	2	SK4	0	3	0
FR5	0	3	0	GR5	1	1	0	LD5	1	1	1	PC5	1	2	0	SK5	1	1	0
FR6	1	0	2	GR6	1	2	0	LD6	2	0	0	PC6	3	0	0	SK6	2	0	0
FR7	1	2	0	GR7	3	0	0									SK7	2	1	0
FR8	2	1	0													SK8	2	1	1

Table 4. The Randomly Selected Test Images

3.5 Results

3.5.1 Observer variation

The mean CV of the all observers participated in this experiment ranged from 20 to 39, and the grand mean CV across the observers and the 5 test stimuli for dark surround condition was 26, which is thought of as acceptable. (Note that CV value of 26 means 26% error of individual from the arithmetic mean.) The mean observer accuracy was found to be 32 for overcast and 30 for bright which are also within the acceptable CV boundary. The results also indicate that there was not much variation in terms of CV values between different experimental phases and image contents. One of the observers showed a relatively higher CV (39) than the other observations, but its impact to the grand mean (29) was not large thus was included for further analysis and modelling procedures.

3.5.2 Prediction accuracy of the proposed algorithm

Figure 15 presents box plots for comparing subjective image quality scores between the 3 surround conditions including dark, overcast and bright. Box is drawn between the lower and upper quartiles and a line across each box represents the median. Whiskers are extended to smallest and largest observations or 1.5 times length of box. In general, the range of subjective data could be decreased as the surround luminance increases. For example, MOS is 5.4 under dark, 4.7 under overcast and 3.5 under bright. It can be seen from the box plots that MOS difference between the viewing conditions is significant.
Scaling factors in eq. (13) optimised for the three viewing conditions are listed in Table 5. Magnitude of them is systematically changed from dark to overcast to bright and could be modelled by an exponential decay fit of surround luminance (see eqs. (18) and (19)). The predicted curves are compared with the computed scaling factors as illustrated in Fig 16.

$$p = 1.16 + 2.36e^{-10^{-4} L_S / 0.35} \quad (18)$$

$$q = 0.35 - 5.38e^{-10^{-4} L_S / 0.29} \quad (19)$$

where L_S is the surround luminance level.

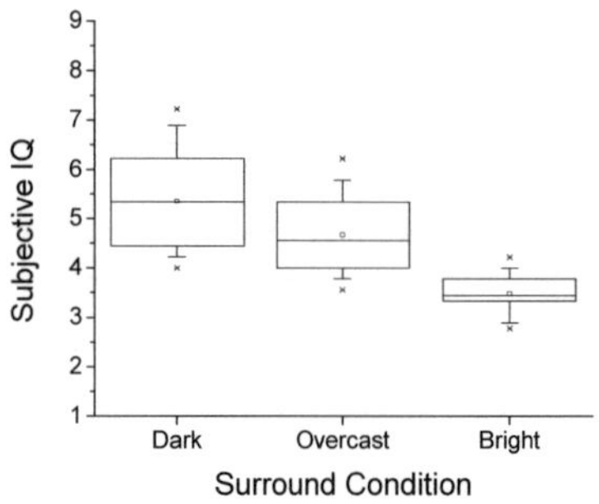

Fig. 15. Box plots for comparing subjective image quality scores between the 3 surround conditions including dark, overcast and bright. Box is drawn between the lower and upper quartiles and a line across each box represents the median. Whiskers are extended to smallest and largest observations or 1.5 times length of box. In general, the range of subjective data could be decreased as the surround luminance increases.

	Dark	Overcast	Bright
Slope	3.93	2.69	1.47
Offset	-6.71	-2.89	-0.11

Table 5. Scaling factors (slope and offset) for the three viewing conditions

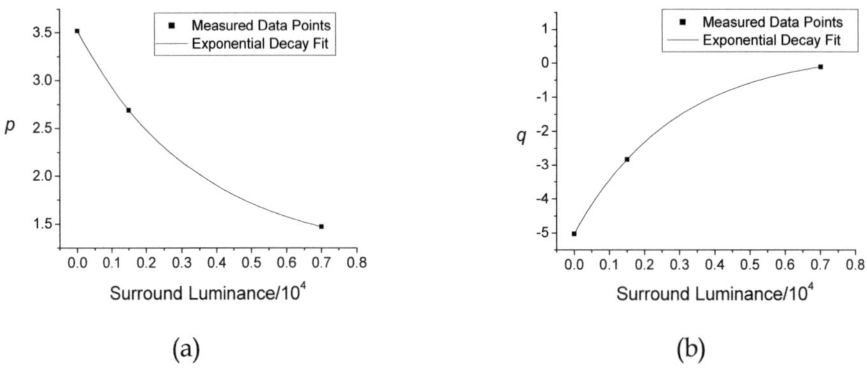

Fig. 16. Scaling factors as a function of surround luminance (a) Slope p (b) Offset q

In Fig 17, the abscissa shows $nSQRI_a$ prediction values, which are re-scaled by the scaling factors listed in Table 5, and the ordinate shows the corresponding MOS. (Note that a 45° line is given for illustrating the data spread.) Different shaped symbols represent different test images. For instance, the filled squares are for "Fruits (FR)", circles for "Grass (GR)", triangles for "Ladies (LD)", crosses for "Picnic (PC)" and diamonds for "Skytower (SK)". The model accuracy for the overall data sets can also be predicted by calculating a CV value between the two axes and it was 15 which is smaller than the mean observer accuracy (29) across the three surround conditions. Specifically, the CV between the two data sets was 18 for dark, 13 for overcast and 9 for bright and all are less than the corresponding mean

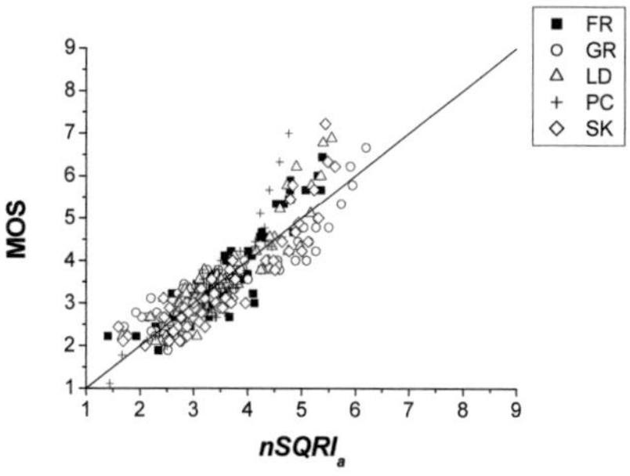

Fig. 17. Comparison between *nSQRIa* and their corresponding MOS across the three surround conditions

observer accuracy. Note that the mean observer accuracy was 26 for dark, 32 for overcast and 30 for bright. Consequently, no significant image dependency of the model prediction was observed due to the spatial frequency normalisation procedure.

3.6 Summary

The current research intends to quantify the surround luminance effects on the shape of spatial luminance CSF and to propose an image quality evaluation method that is adaptive to both surround luminance and spatial frequency of a given stimulus. The proposed image quality method extends to a model called SQRI. (Barten, 1990) The non-linear behaviour of the HVS was taken into account by using CSF. This model can be defined as the square root integration of multiplication between display MTF and CSF. It is assumed that image quality can be determined by considering the MTF of the imaging system and the CSF of human observers. The CSF term in the original SQRI model was replaced by the surround adaptive CSF quantified in this study and it is divided by the Fourier transform of a given stimulus. The former relies upon the surround factor function (φ) shown in eq. (13) and the latter requires a normalization procedure. The model prediction for a certain image is divided by that of its degraded image of which its pixel resolution is manipulated to be 80 ppi and luminance of each pixel is reduced to be 25% of its original. The model accuracy and observer accuracy are comparable in terms of CV. The mean model accuracy is a CV value of 15 and observer accuracy is 29. Consequently, the model accuracy outperformed the observer accuracy and no significant image dependency could be observed for the model performance.

A few limitations of the current work should be addressed and revised in the future study. First, the model parameters should be revised for larger sized images. A 2-inch mobile LCD is used to display images in this study so any image size effect on the model prediction should be verified in the future work. Second, more accurate model predictions may be achievable when the actual display MTF is measured and used instead of the approximation shown in eq. (9). Last but not least, a further improvement to the model prediction accuracy can be made when chromatic contrast loss of the HVS is taken into account.

4. Enhancing image quality

The loss in contrast discrimination ability of the human visual system was estimated under a variety of ambient illumination levels first. Then it was modelled as a non-linear weighting function defined in spatial frequency domain to determine which of parts of the image, whatever their spatial frequency, will appear under a given surround luminance level. The weighting function was adopted as a filter for developing an image enhancement algorithm adaptive to surround luminance. The algorithm aims to improve the image contrast under various surround levels especially for small-sized mobile phone displays through gain control of a 2D contrast sensitivity function.

4.1 Proposed surround luminance adaptive image enhancement

4.1.1 Contrast sensitivity reduction of the HVS

As shown in the earlier section, Fig. 12 illustrates the relation between surround luminance level (cd/m^2) and the surround effect function (φ). The shape of the function is similar to that of the image colour-quality decay function (Kim et al., 2007) that predicts the overall colour-quality of an image based upon measurable image-properties under various outdoor surround conditions. In addition, the change in 'clearness' caused by the illumination increase could also be modelled as an exponential decay function as well. (Kim et al., 2008) CSFs for the three surrounds in total - dark (0 lx), overcast (6100 lx) and bright (32000 lx) - are computed using eqs. 13 and 14 and also plotted in Fig. 18 while other variables such as viewing distance and adapting luminance of a stimulus remain the same. The spatial frequency where the maximum contrast sensitivity occurred was moved toward a lower frequency from dark (4.4 cpd) to bright (3.8 cpd). As a result, the surround luminance increase resulted in approximately 7 and 15% loss in contrast sensitivity of the human visual system for overcast and bright, respectively. (Kim & Kim, 2010)

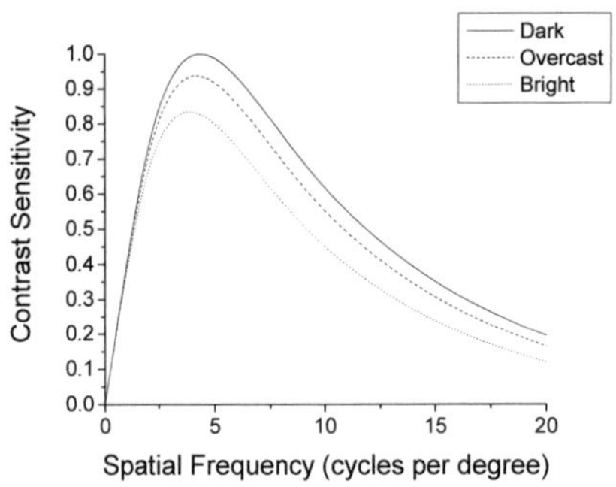

Fig. 18. Comparison of CSFs under dark and ambient illuminations

In order to compensate for the loss in image contrast caused by surround luminance increase and enhance the image quality, an adaptive enhancement gain control algorithm to the surround luminance was developed using an adaptive weighting filter. This filter correlates to the normalised contrast sensitivity difference between the reference (dark) and a target surround luminance level. The contrast sensitivity difference, $D(u,v)$, between the *reference* (dark), $CSF_R(u,v)$, and a given *target* surround, $CSF_T(u,v)$, represents the loss in image contrast caused by increase of the surround luminance which can be expressed as

$$D(u,v)=\left(CSF_R(u,v)-CSF_T(u,v)\right) \tag{20}$$

where u and v are frequency variables.

Since the image enhancement can be achieved, when an enhancement gain greater than 1 is multiplied to the amplitude of a given image, the offset of these weighting filters should be increased up to greater than 1 and a constant value of 1 was added to $D(u,v)$. In addition, the maximum value of $D(u,v)$ is also added to the offset so the adaptive weighting filter can be defined as

$$H(u,v) = D(u,v) + (1 + C) \tag{21}$$

where C = max($D(u,v)$).
The maximum value of $D(u,v)$ implies the change in brightness and the threshold level to be enhanced under a given surround luminance level. Since various spatial frequency levels are mixed in a complex image, the masking phenomenon (Wandell, 1995; Kim et al., 2007) can occur and there might be some contrast loss detectable in unexpected frequencies. The masking commonly occurs in multi-resolution representations and there are cases when two spatial patterns S and S + ΔS cannot be discriminated, while ΔS seen alone, can be visible. Therefore, all frequency regions should be enhanced globally by a certain level of enhancement gain threshold and such significant regions should be enhanced with higher weights. However, the enhancement threshold level was arbitrarily chosen as the maximum value of $D(u,v)$ in this study and more investigations are required in future study.
Figure 19 shows estimates of the adaptive weighting filter for the three surround levels: dark, overcast and bright, when a field size was 5 degrees and the display's adapting (mean) luminance was 89.17 cd/m^2. Since the loss in image contrast becomes larger, as the ambient illumination increases, the weighting filter response for bright surround shows the highest filter response and overcast surround follows. In case of dark surround, the amplitude of original image can be preserved as being multiplied by an enhancement gain of 1 across the all spatial frequencies. The enhancement threshold level is 0 for dark, 0.15 for overcast and 0.31 for bright. Since CSFs are known as smoothly varied band-pass filters, the enhancement gain can also be smoothly changed. The adaptive image enhancement filter can be defined as a weighting function to determine which of parts of the image, whatever their spatial frequency, should have a higher enhancement gain.

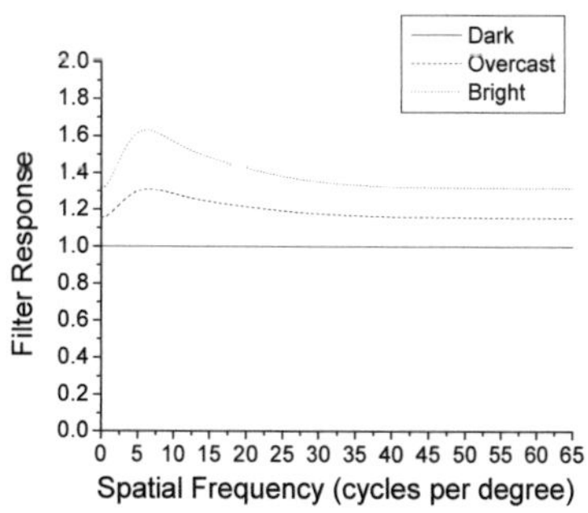

Fig. 19. The adaptive weighting filter estimates

4.2 Results

Figure 20 presents a test image and their enhanced images for the two surround conditions and their histograms of luminance of the composite channel (Luminosity). The input RGB

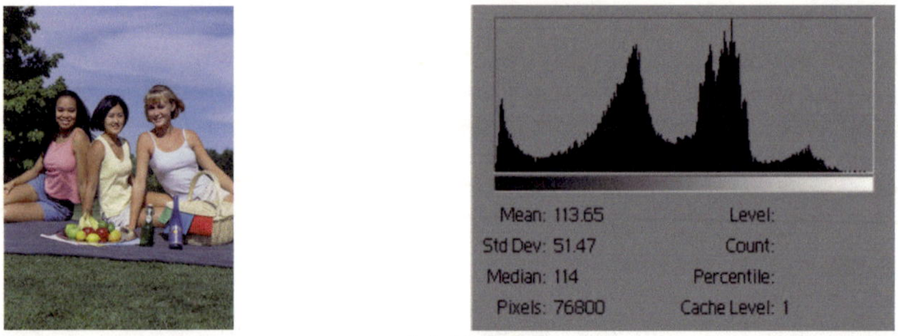

(a) Original

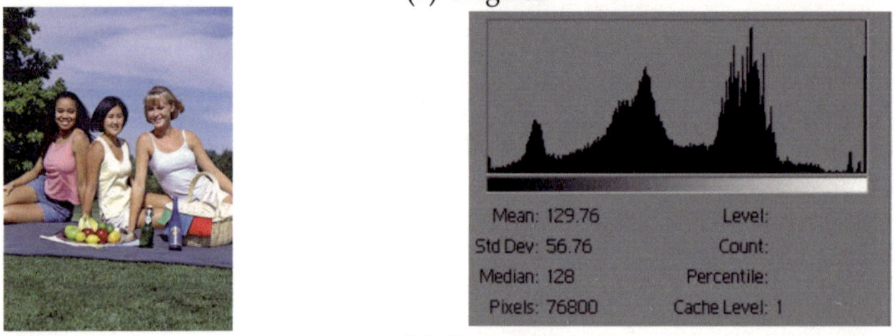

(b) Overcast

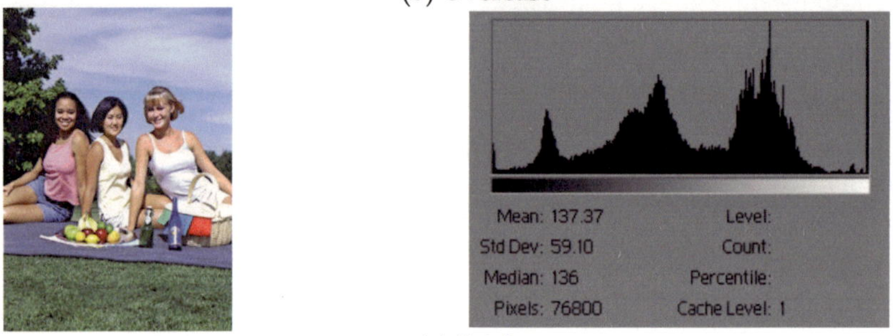

(c) Bright

Fig. 20. Example of enhanced images and their luminosity histogram

values were converted into CIECAM02 (CIE, 2004) perceptual colour attributes such as *Jab* and *J* was then transformed into the recently updated *J'*. (Luo et al., 2006) Only lightness *J'* went into the enhancement procedure while chrominance properties *a* and *b* were preserved. The horizontal axis of each histogram represents the intensity values, or levels, from the darkest (0) at the far left to brightest (255) at the far right; the vertical axis represents the number of pixels with a given value. Moreover, the statistical information about the intensity values of the pixels appears below the histogram: mean, standard deviation (Std Dev), median, the number of pixels in the image and so forth.

As can be seen in Fig. 20, tonal variance in those histograms yields quite spread and both mean and standard deviation were increased as surround luminance increases. The mean was 113.65 for original, 129.76 for overcast and 137.37 for bright. The standard deviation was 51.47 for original, 56.76 for overcast and 59.10 for bright. Consequently, the overcall brightness and contrast of the image were increased. The resultant enhanced images may appear overexposed especially for the enhanced one for bright. However, if those images are seen with the surround luminance levels, they are supposed to show the similar degree of image quality as the original seen under the reference (dark) viewing condition (as if reduced appetite leads to stronger taste of food).

Figures 21 (a) through (b) illustrate the comparison between enhanced and original images in terms of image quality scores judged by the nine observers. The abscissa represents subjective image quality score of the original images under a certain surround condition and the ordinate shows that of their enhanced images. For example, if most of the data points are upper the 45-degree line (red line), the enhanced images were judged as higher image quality. In general, majority of the data points were upper the 45-degree line for all of the surround conditions and it can be said that the enhanced images are rated by higher category values than their original images. When the proposed algorithm was applied for overcast condition data set, 74% (26 out of 35 images) subjective values of the enhanced images were higher than that of the original images (Fig. 21 (a)). In addition, its performance was more or less the same as the original images judged under dark viewing condition. In Fig. 7 (b), the images processed by the proposed algorithm for bright condition were compared with their corresponding original images. As well as overcast, the proposed algorithm produced better quality images than their original images seen under the same condition, 85% (30 out of 35 images). Subjective image quality score of the enhanced images was similar to that of original images judged under overcast surround condition. The 15% reduction caused in image quality could be due to the impairment in chromatic channels. Chromatic contrast should also be decreased under bright surrounds and the chromatic contrast loss effects will be left for the future work.

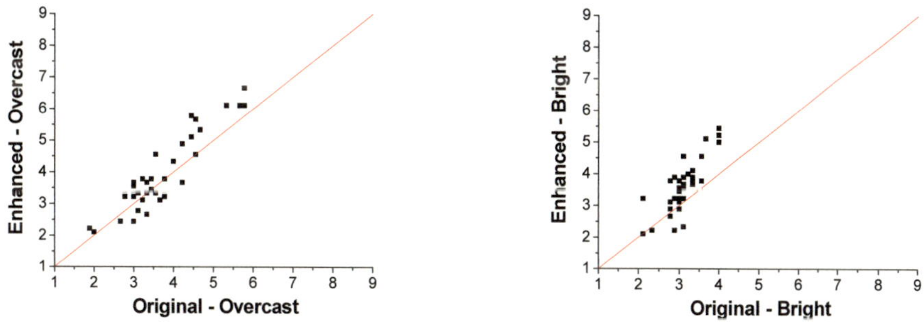

(a) (b)

Fig. 21. Comparison between the original and enhanced image for each surround condition

One of possible artefacts that can be caused by the proposed algorithm is out boundary colours (OBC). Since a gain value larger than one is multiplied to a given image, some colours may lie outside colour gamut of the display. Those colours can be referred to as OBCs and more details can be found in Ref. 22. In this study, OBCs were clipped at the maximum value (255). However, the OBC effect may be overwhelmed by the contrast and brightness compensation so the artefact was not significantly perceptible during the psychophysical evaluations.

4.3 Summary

In this section, an adaptive image enhancement algorithm was proposed and their performance was observed through a set of subjective assessments. The contrast discrimination ability of human observers under ambient illumination was quantified as a weighting function to determine which of parts of the image, whatever their spatial frequency, will appear under a certain surround luminance level. The weighting function was adopted as the image enhancement filter in spatial frequency domain. Most of the enhanced images were rated as higher image quality scores than their original images through a set of subjective validation experiment. The quality of images under bright surround was enhanced up to that of images seen under overcast. Similarly, the quality of images under overcast was reached that of image seen under dark. Further improvement of image contrast can be achieved when chromatic contrast loss is compensated that could be one of the afterthoughts.

5. Acknowledgment

This work is part of the author's PhD thesis at University of Leeds in England. Currently, he is with Samsung Electronics Company in Korea.

6. References

Arend, L. E. and Spehar, B. (1993). Lightness, brightness and brightness contrast: I. Illumination variation. *Perception & Psychophysics*, Vol. 54, No. 4, (February 1933), pp. 446-456, ISSN 0031-5117

Arend, L. E. and Spehar, B. (1993). Lightness, brightness and brightness contrast: II. Reflectance variation. *Perception & Psychophysics*, Vol. 54, No. 4, (October 1993), pp. 457-468, ISSN 0031-5117

Barten , G. J. (1991). Resolution of liquid-crystal displays. *SID Digest*, ISSN 0003-966X

Barten, P. G. (1990). Evaluation of subjective image quality with the square-root integral method. *Journal of the Optical Society of America A*, Vol. 7, No. 10, (October 1990), pp. 2024-2031, ISSN 1084-2529

Barten, P. G. J. (1999). *Contrast Sensitivity of The Human Eye and Its Effects on Image Quality*. SPIE Press, ISBN 978-0819434968,Washington

Bartleson , J. (1984). *Optical Radiation Measurements*, Bartleson, C. J. & Grum, F. (Eds.), Academic Press, ISBN 978-0123049049, NY

Blakeslee, B., Reetz, D. and McCourt, M. E. (2008). Comping to terms with lightness and brightness: Effects of stimulus configuration and instructions on brightness and lightness judgments. *Journal of Vision*, Vol. 8, No. 11, (August 2008), pp. 1-14, ISSN 1534-7362

Braddick, O., Campbell, F. W. and Atkinson , J., (1978). Channels in vision: basic aspects, In : *Handbook of Sensory Physiology*, Held, R., Leibowitz H. W. and Teuber, H. -L. (Eds.), Springer-Verlag, ISBN 0-387-05146-5, New York

Burton, K. B., Owsley, C. and Sloane, M. E. (1993). Aging and neural spatial contrast sensitivity: photopic vision. *Vision Research*, Vol. 33, No. 7, (May 1993), pp. 939-946, ISSN 0042-6989

Campbell, F. W. and Green, D. G. (1965). Optical and retinal factors affecting visual resolution. *Journal of Physiology*, Vol. 181, No. 3 (December 1965), pp. 576-593, ISSN 0022-3751

Campbell, F.W. and Robson, J.G. (1968). Application of Fourier analysis to the visibility of gratings. *Journal of Physiology*, Vol. 197, No. 3, (August 1968), pp. 551-566, ISSN 0022-3751

Choi , S.Y., Luo, M.R. and Pointer, M.R., (2007). The Influence of the relative luminance of the surround on the perceived quality of an image on a large display, *Proceedings of 15th Color Imaging Conference*, ISBN 978-0-89208-294-0, Albuquerque, New Mexico, November 2007

CIE Publication 15.2, Colorimetry, 2nd Ed., Commission Internationale de l'Eclairage, Vienna, 1986.

CIE publication 159-2004, A colour appearance model for colour management systems: CIECAM02, Vienna, 2004.

Cox, M. J., Norma, J. H. and Norman, P. (1999). The effect of surround luminance on measurements of contrast sensitivity. *Ophthalmic and Physiological Optics*, Vol. 19, No. 5, (September 1999), pp. 401-414, ISSN 0275-5408

Dacey, D. M. and Lee, B. B. (1994). The blue-on opponent pathway in the primate retina originates from a distinct bistratified ganglion cell. *Nature*, Vol. 367, No. 6465, (February 1994), pp. 731-735, ISSN 0028-0836

Daly, S. (1993). The Visible Differences Predictor: An Algorithm for the Assessment of Image Fidelity, In: *Digital Images and Human Vision, Watson A. B.* (Ed.), MIT press, ISBN 978-0262231718, Cambridge, Massachusetts

DeValois, R.L., Morgan, H. and Snodderly, D.M. (1974). Psychophysical studies of monkey vision - III. Spatial luminance contrast sensitivity tests of macaque and human observers. *Vision Research*, Vol. 14, No. 1, (January 1974), pp. 75-81, ISSN 0042-6989

Enroth-Cugell, C. and Robson, J. G. (1966). The contrast sensitivity of retinal ganglion cells of the cat. *Journal of Physiology*, Vol. 187, No. 3, (December 1966), pp. 517-552, ISSN 0022-3751

Fairchild, M. D. and Johnson, G. M. (2007). Measurement and modelling of adaptation to noise in image. *Journal of the Society for Information Display*, Vol. 15, No. 9, (September 2007), pp. 639-647, ISSN 1071-0922

Fairchild, M. D. and Reniff, L. (1995). Time-course of chromatic adaptation for color-appearance judgements. *Journal of the Optical Society of America A* Vol. 12, No. 5, (May 1995), pp. 824–833, ISSN 1084-2529

Goldstein , E. B. (2007). *Sensation and Perception, Thomson Wadsworth*, ISBN 978-0-495-27479-7, CA

Graham, N. (1980) Spatial-frequency channels in human vision: detecting edges without edge-detectors, In: *Visual Coding and Adaptability*, Harris, C. S. (Ed.), Erlbaum, ISBN 978098590166, Hillsdale, New Jersey

Heinemann, E. G. (1955). Simultaneous brightness induction as a function of inducing and test-field luminances. *Journal of Experimental Psychology*, Vol. 50, No. 2, (August 1955), pp. 89-96, ISSN 0096-3445

Higgins, K. E., Jaffe, M. J., Caruso, R. C. and deMonasterio, F. (1988). Spatial contrast sensitivity: effects of age, test-retest, and psychophysical method. *Journal of the Optical Society of America A*, Vol. 5, No. 12, (December 1988), pp. 2173-2180, ISSN 1084-2529

ITU-R Rec. BT. 500-10, Methodology for the subjective assessment of the quality of television pictures, Geneva, 2002.

Jameson, D. and Hurvich, L. M. (1959). Perceived color and its dependence on focal, surrounding, and preceding stimulus variables. *Journal of the Optical Society of America*, Vol. 49, No. 9, (September 1959), pp. 890-898, ISSN 108-7529

Jameson, D. and Hurvich, L. M. (1961). Complexities of perceived brightness. *Science*, Vol. 133, No. 1, (January 1961), pp. 174-179, ISSN 1545-1003

Johnston, A. (1987). Spatial scaling of central and peripheral contrast-sensitivity functions. *Journal of the Optical Society of America A*, Vol. 4, No. 8, (August 1987), pp. 1583-1593, ISSN 1084-2529

Katoh, N., Nakabayashi, K., Ito, M., and Ohno, S. (1998). Effect of ambient light on the color appearance of softcopy images: Mixed chromatic adaptation for self-luminous displays. *Journal of Electronic Imaging*, Vol. 7, No. 4, (October 1998), pp. 794-806, ISSN 1017-9909

Kim, Y. J. and Kim, H. (2010). Spatial luminance contrast sensitivity: Effects of surround. *Journal of the Optical Society of Korea*, Vol. 14, No. 2, (April 2010), pp.152-162, ISSN 1226-4776

Kim, Y. J., Bang, Y. and Choh, H. (2010). Gradient approach to quantify the gradation smoothness for output media. *Journal of Electronic Imaging*, Vol. 19, No. 1, (January 2010), pp. 011012, ISSN 1017-9909

Kim, Y. J., Bang, Y. and Choh, H. (2010). Measurement and modelling of vividness perception and observer preference for color laser printer quality. *Journal of Imaging Science and Technology*, Vol. 54, No. 1, (January 2010), pp. 010501, ISSN 1062-3701

Kim, Y. J., Luo, M. R., Choe, W., Kim, H.S., Park, S.O., Baek, Y, Rhodes, P., Lee, S. and Kim, C. (2008). Factors Affecting the Psychophysical Image Quality Evaluation of Mobile Phone Display: the Case of Transmissive LCD. *Journal of the Optical Society of America A*, Vol. 25, No. 9, (September 2008), pp. 2215-2222, ISSN 1084-2529

Kim, Y. J., Luo, M. R., Rhodes, P, Westland, S., Choe, W., Lee, S., Lee, S., Kwak, Y., Park, D. and Kim, C. (2007). Image-Colour Quality Modelling under Various Surround Conditions for a 2-inch Mobile Transmissive LCD. *Journal of the Society for Information Display*, Vol. 15, No. 9, (September 2007), pp. 691-698, ISSN 1071-0922

Kitaguchi, S., MacDonald, L., and Westland, S. (2006). Evaluating contrast sensitivity, *Proceedings of SPIE, Human vision and electronic imaging XI*, ISBN 9780819460974, Westlandm, Stephen, February 2006

Koenderink, J. J., Bouman, M. A., Bueno de Mesquita, A. E. and Slappendale, S. (1979). Perimetry of contrast detection thresholds of moving spatial sine wave patterns, Parts I-IV. *Journal of the Optical Society of America*, Vol. 68, No. 6, (June 1978), pp. 845-865, ISSN 108-7529

Lee, H. J., Choi, D. W., Lee, E., Kim, S. Y., M, Shin., Yang, S. A., Lee, S. B., Lee, H. Y. and Berkeley, B. H. (2009). Image sticking methods for OLED TV applications, *Proceedings of International Meeting on Information Display*, ISBN 1738-7558, Ilsan, Korea, October 2009

Li, Z., Bhomik, A.K., and Bos, P.J. (2008). Introduction to Mobile Displays, In: *Mobile Displays Technology and Applications*, Wiley, ISBN 978-0470723746, Chichester , UK

Liu C. & Fairchild M.D. (2004). Measuring the relationship between perceived image contrast and surround illumination, *Proceedings of IS&T/SID 12th Color Imaging Conference*, ISBN 978-0-89208-294-0, Scottsdale, Arizona, November 2004

Liu, C. & Fairchild, M.D. (2007). Re-measuring and modeling perceived image contrast under different levels of surround illumination, *Proceedings of IS&T/SID 15th Color Imaging Conference*, ISBN 978-0-89208-294-0, Albuquerque, New Mexico, November 2007

Luo, M. R., Clarke, A. A., Rhodes, P., Schappo, A., Scrivener, S. A. R., and Tait, C. J. (1991). Quantifying Colour appearance. Part 1. Lutchi colour appearance data. *Color Research and Application*, Vol. 16, No. 3, pp. 166, ISSN 0361-2317

Luo, M.R., Cui, G. and Rigg, B. (2001). The development of the CIE 2000 colour-difference formula: CIEDE2000. *Color Research and Application*, Vol. 26, No. 5, (August 2001), pp. 340-350, ISSN 0361-2317

Martinez-Uriegas, E. (2006). Spatial and Temporal Problems of Colorimetry, In: *Colorimetry: Understanding the CIE System*, CIE, ISBN 978-0-470-04904-4, Switzerland

Martinez-Uriegas, E., Larimer J.O., Lubin J. and Gille, J. (1995). Evaluation of image compression artefacts with ViDEOS, a CAD system for LCD color display design and testing, Proceeding series

Mornoney, N., Fairchild, M. D., Hunt, R. W. G., Li, C., Luo, M. R. and Newman, T. (2002). The CIECAM02 color appearance model, *Proceedings of 10th Color Imaging Conference*, Scottsdale, Arizona, November 2002

Owsley, C., Sekuler, R. and Siemsen, D. (1983). Contrast sensitivity throughout adulthood. *Vision Research*, Vol. 23, No. 7, (July 1983), pp. 689-699, ISSN 0042-6989

Palmer, S. (1999). *Vision science: Photons to Phenomenology*, MIT Press, ISBN 978-0262161831, Cambridge, Massachusetts

Pardhan, S. (2004). Contrast sensitivity loss with aging: sampling efficiency and equivalent noise at different spatial frequencies. *Journal of the Optical Society of America A*, Vol. 21, No. 2, (February 2004), pp. 169-175, ISSN 1084-2529

Park, Y., Li, C., Luo, M.R., Kwak, Y., Park, D. and Kim, C. (2007). Applying CIECAM02 for mobile display viewing conditions, *Proceedings of 15th Color Imaging Conference*, ISBN, Albuquerque, New Mexico, November 2007

Patel, A.S. (1966). Spatial resolution by the human visual system. *Journal of the Optical Society of America*, Vol. 56, No. 5, (May 1966), pp. 689-694, ISSN 108-7529

Peli, E. (1996). Test of a model of foveal vision by using simulations. *Journal of the Optical Society of America A*, Vol. 13, No. 6, (June 1996), pp. 1131-1138, ISSN 1084-2529

Peli, E. (2001). Contrast sensitivity function and image discrimination. *Journal of the Optical Society of America A*, Vol. 18, No. 2, (February 2001), pp. 283-293, ISSN 1084-2529

Post, D.L. and Calhoun, C.S. (1989). An evaluation of methods for producing desired colors on CRT monitors. *Color Research and Application*, Vol. 14, No. 4, (month and year of the edition), pp. 172-186, ISSN 0361-2317

Rohaly, A. M. and Buchsbaum, G. Global (1989). Spatiochromatic mechanism accounting for luminance variations in contrast sensitivity functions. *Journal of the Optical Society of America A*, Vol. 6, No. 2, (February 1989), pp. 312-317, ISSN 1084-2529

Rohaly, A. M. and Owsley, C. (1993). Modeling the contrast-sensitivity functions of older adults. *Journal of the Optical Society of America A*, Vol. 10, No. 7, (July 1993), pp. 1591-1599, ISSN 1084-2529

Rovamo, J., Virsu, V. and Nasanen, R. (1978). Cortical magnification factor predicts the photopic contrast sensitivity of peripheral vision. *Nature*, Vol. 271, No. 5640, (January 1978), pp. 54-56, ISSN 0028-0836

Rudd, M. E. and Popa, D. (2007). Stevens's brightness law, contrast gain control, and edge integration in achromatic color perception: a unified model. *Journal of the Optical Society of America A*, Vol. 24, No. 9, (September 2007), pp. 2766-2782, ISSN 1084-2529

Schade, O. H. (1956). Optical and photoelectric analog of the eye. *Journal of the Optical Society of America*, Vol. 46, No. 9a, (September 1956), pp. 721-739, ISSN 108-7529

Snodderly, D. M., Weinhaus, R. S. and Choi, J. C. (1992). Neural-vascular relationships in central retina of macaque monkeys (Macaca fascicularis). *Journal of Neuroscience*, Vol. 12, No. 4, (April 1992), pp. 1169-1193, ISSN 1529-2401

Stiehl, W. A., McCann, J. J. and Savoy, R. L. (1983). Influence of intraocular scattered light on lightness-scaling experiments. *Journal of the Optical Society of America*, Vol. 73, No. 9, (September 1983), pp. 1143-1148, ISSN 108-7529

Sun, Q. and Fairchild, M. D. (2004). Image Quality Analysis for Visible Spectral Imaging Systems. *Journal of Imaging Science and Technology*, Vol. 48, No. issue number, (month and year of the edition), pp. 211-221, ISSN 1062-3701

Tulunay-Keesey, U., Ver Hoever, J. N. and Terkla-McGrane, C. (1988). Threshold and suprathreshold spatiotemporal response throughout adulthood. *Journal of the Optical Society of America A*, Vol. 5, No. 12, (December 1988), pp. 2191-2200, ISSN 1084-2529

Van Nes, F. L. and Bouman, M. A. (1967). Spatial modulation transfer in the human eye. *Journal of the Optical Society of America*, Vol. 57, No. 3, (March 1967), pp. 401-406, ISSN 108-7529

Wallach, H. (1948). Brightness constancy and the nature of achromatic colors. *Journal of Experimental Psychology*, Vol. 38, No. 3, (June 1948), pp. 310-324, ISSN 0096-3445

Wandell, B.A. (1995). *Foundations of Vision*, Sinauer Associates, ISBN 0878938532, Sunderland, Massachusetts

Wang, Z. & Bovik , A. C. (2006). *Modern image quality assessment*, Morgan & Claypool Publishers, ISBN 978-1598290226, New Jersey

Watson, A.B. (2000). Visual detection of spatial contrast patterns: Evaluation of five simple models. *Optics Express*, Vol. 6, No. 1, (January 2000), pp. 12-33, ISSN 1094-4087

Westheimer, G. and Liang, J. (1995). Influence of ocular light scatter on the eye's optical performance. *Journal of the Optical Society of America A*, Vol. 12, No. 7, (July 1995), pp. 1417-1424, ISSN 1084-2529

Westland, S., Owens, H., Cheung, V. and Paterson-Stephens, I. (2006). Model of luminance contrast-sensitivity function for application to image assessment. *Color Research and Application*, Vol. 31, No. 4, (June 2006), pp. 315-319, ISSN 0361-2317

Woodworth, R. S. & Schlosberg, H. (1954). *Experimental psychology*, Holt, ISBN 101-179-848, New York

Wright, M. J. and Johnston, A. (1983). Spatiotemporal contrast sensitivity and visual field locus. *Vision Research*, Vol. 23, No. 10, (October 1983), pp. 983-989, ISSN 0042-6989

Yoon, G. and Williams, D. R. (2002). Visual performance after correcting the monochromatic and chromatic aberrations of the eye. *Journal of the Optical Society of America A*, Vol. 19, No. 2, (February 2002), pp. 266-275, ISSN 1084-2529

Zhang, X.M. & B. A,Wandell. (1996). A spatial Extension to CIELAB for digital color image reproduction, *Proceedings of the SID Symposiums*, ISSN 0003-966X